Environmental Justice Analysis

Theories, Methods, and Practice

Feng Liu

LEWIS PUBLISHERS

Boca Raton London New York Washington, D.C.

Library of Congress Cataloging-in-Publication Data

Liu, Feng.
    Environmental justice analysis : theories, methods, and practice / by Feng Liu.
        p.   cm.
    Includes bibliographical references and index.
    ISBN 1-56670-403-0 (alk. paper)
    1. Environmental justice. 2. Environmental policy. 3. Equity. I. Title.

GE170 .L58 2000
363.7′02—dc21

00-041232
CIP

This book contains information obtained from authentic and highly regarded sources. Reprinted material is quoted with permission, and sources are indicated. A wide variety of references are listed. Reasonable efforts have been made to publish reliable data and information, but the author and the publisher cannot assume responsibility for the validity of all materials or for the consequences of their use.

Preface

A classroom debate at the Wharton School turned out to be a preface to this book. The professor, an economist, handed out an internal memo prepared by Lawrence Summers, then chief economist of the World Bank. He wrote:

> Just between you and me, shouldn't the World Bank be encouraging more migration of the dirty industries to the LDCS [less developed countries]? I can think of three reasons:
>
> 1. The measurement of the costs of health-impairing pollution depends on the foregone earnings from increased morbidity and mortality. From this point of view a given amount of health-impairing pollution should be done in the country with the lowest cost, which will be the country with the lowest wages. I think the economic logic behind dumping a load of toxic waste in the lowest-wage country is impeccable and we should face up to that.
> 2. The costs of pollution are likely to be non-linear as the initial increments of pollution probably have very low cost. I've always thought that under-populated countries in Africa are vastly *under*-polluted; their [air pollution] is probably vastly inefficiently low compared to Los Angeles or Mexico City. Only the lamentable facts that so much pollution is generated by non-tradable industries (transport, electrical generation) and that the unit transport costs of solid waste are so high prevent world-welfare-enhancing trade in air pollution and waste.
> 3. The demand for a clean environment for aesthetic and health reasons is likely to have very high income elasticity. The concern over an agent that causes a one-in-a-million change in the odds of prostate cancer is obviously going to be much higher in a country where people survive to get prostate cancer than in a country where under-5 mortality is 200 per thousand. Also, much of the concern over industrial atmospheric discharge is about visibility-impairing particulates. These discharges may have very little direct health impact. Clearly, trade in goods that embody aesthetic pollution concerns could be welfare-enhancing. While production is mobile the consumption of pretty air is non-tradable.

The problems with the arguments against all of these proposals for more pollution in LDCS (intrinsic rights to certain goods, moral reasons, social concerns, lack of adequate markets, etc.) could be turned around and used more or less effectively against every Bank proposal for liberalization.

The professor started the debate by defending these arguments, and we were to come up with counter arguments. Initially, there were some voices against these arguments. A few minutes later, I found myself the lonely voice. Finally, I heard some agreements with the professor. This school produces leaders of national and international business management. The memo and debate have troubled me since.

Years later, on the morning of May 12, 1999, I read David Harvey's *Justice, Nature & the Geography of Difference*, Chapter 13 of which began with a description of the Summers episode. During the same morning, coincidentally, it was reported that Robert Rubin was resigning his post as Treasury secretary, after more than 6 years as a member of the Clinton Administration. During the afternoon, the president named Rubin's deputy secretary, Lawrence Summers, as his replacement. The source speculated that Summers might face some opposition from Republicans on Capitol Hill as he was viewed as more liberal than the market-oriented Rubin.

Years later, I still find myself puzzled and concerned about the debate. While writing this book, I am thinking about the small debate in the context of a large debate on environmental justice. I am thinking about the perspective that I would like to offer to my readers. Is this only an economic issue? No. Is this a social issue? Is this a moral issue? Is this a political issue? Is this a scientific issue? It is all of them. This is what I would like to present to you: How to analyze environmental justice issues using a multi-perspective, a multi-disciplinary and inter-disciplinary approach.

This book is a comprehensive and analytical treatment of theories and methods for analyzing and assessing environmental justice and equity issues. I strived to keep this book well-balanced, carefully and critically examining both sides of the debate and contributing to the debate with first-hand analysis. A lot of attention is focused on the debate on various methodological issues of environmental justice research.

This book provides readers with a holistic framework for conducting rigorous equity analysis, and particularly demonstrates how cutting-edge technologies and methods such as the Internet, Geographic Information Systems, and modeling tools can contribute to better equity analysis and policy evaluations. It covers a wide range of policy areas such as air pollution, solid waste management facilities, hazardous waste management facilities, toxic release facilities, Superfund sites, land use, and transportation and a wide variety of geographic scales. It is a reference resource for professionals, undergraduate and graduate students, academics, activists, and any other individuals who are interested in environmental justice issues.

Author

Feng Liu has been working and conducting research in the environmental and planning areas since the early 1980s. His work has embraced a wide spectrum of environmental and planning issues such as air and water pollution, environmental impact assessment, GIS, environmental modeling, land use and transportation modeling, environmental justice and equity, transportation planning, land use planning, and smart growth. His recent papers have appeared in *Environmental Science & Technology*, *Environmental Management*, and *Journal of the Air & Waste Management Association*.

Dr. Liu has worked in a variety of organizations, including academic research institutions, environmental organizations, regional planning agencies, and state government. He currently works at the Maryland Department of Planning, the lead agency in the implementation of Maryland's nationally renowned Smart Growth policies and programs. Before joining MDP, he worked for the Baltimore Metropolitan Council, Environmental Defense, University of Pennsylvania, and Beijing Normal University. He received a B.S. from Zhongshan University, an M.S. from Beijing Normal University, and an M.A. and a Ph.D. in city and regional planning from the University of Pennsylvania.

Credits

We gratefully acknowledge receipt of permission to use material from the works cited below.

CHAPTER 2

Reprinted by permission of Springer-Verlag GmbH & Co.KG. Liu, F. 1997. Dynamics and Causation of Environmental Equity, Locally Unwanted Land Uses, and Neighborhood Changes. *Environmental Management* 21(5):643–656. Copyright 1997 Springer-Verlag GmbH & Co.KG.

Reprinted by permission of the Air & Waste Management Association. Liu, F. 1996. Urban Ozone Plumes and Population Distribution by Income and Race: a Case Study of New York and Philadelphia. *Journal of the Air & Waste Management Association* 46(3):207–215.

CHAPTER 12

Reprinted by permission of the *Journal of the American Planning Association*, Vol. 50, No. 4, Figure 1 on page 461, 1984.

Reprinted by permission of The Yale Law Journal Company and Fred. B. Rothman & Company from *The Yale Law Journal*, Vol. 103, pages 1383–1422.

CHAPTER 3

Reprinted by permission of the *Journal of Planning Education and Research,* Vol. 6(2):86-92 1987.

CHAPTER 9

Reprinted by permission of the *Journal of the American Planning Association*, Vol. 60, No. 1, Figure 2 on page 22, 1994.

Reprinted by permission of the *Journal of the American Planning Association*, Vol. 60, No. 1, Table 1 on page 23 1994.

CHAPTER 10

Reprinted with permission from *Environ. Sci. Technol.* 1998, 3, 32–39. Copyright 1998 American Chemical Society.

Reprinted by permission of the Air & Waste Management Association. Liu, F. 1996. Urban Ozone Plumes and Population Distribution by Income and Race: a Case Study of New York and Philadelphia. *Journal of the Air & Waste Management Association* 46(3):207–215.

Chapter 12

Acknowledgments

Needless to say, this book would be impossible without people around me, physically or virtually, over the years. I wrote this in memory of my mother, Jianhua Liu, the greatest mother, grandmother, and teacher in the world.

I would like to thank those who spent their time reviewing portions of the manuscript. Professor Robert B. Wenger at the University of Wisconsin at Green Bay provided invaluable insights and advice for the original book proposal and made detailed editing and comments on Chapters 1 to 3. Mark Goldstein, a demographer and my colleague at the Maryland Department of Planning, carefully examined and commented extensively on Chapters 5 and 6. Dr. Collin Wu, Associate Professor of Statistics and Mathematical Sciences at the Johns Hopkins University, carefully scrutinized Chapter 7, provided invaluable comments, and made Chapter 7 more rigorous. Stephanie Fleck, a GIS specialist and my colleague at the Maryland Department of Planning, carefully reviewed and commented on Chapter 8 on GIS. Dr. Jian Zhang of Caliper Corporation made useful suggestions on urban modeling (Chapter 9) from a developer and practitioner perspective. Stephen K. Rapley of the Federal Highway Administration provided useful comments about Chapter 13 on transportation equity.

I would like to thank all those who have helped me in one way or another during my research over the years. In particular, I am very grateful to Professors Stephen H. Putman, Tom Reiner, and Roger K. Raufer of the University of Pennsylvania, and Professor Charles N. Hass of Drexel University. Professor Putman provided me with a lot of encouragement and support during my three and a half years of study at Penn. For this research, he provided not only data and models but also very insightful advice and comments. Professor Reiner was always there when I needed his help. Professor Raufer introduced me to the field of environmental equity or justice. He made very constructive comments on the risk perception perspective. Professor Haas helped me better understand what I was doing in this research. I would also like to thank Professors Arie P. Schinnar and Aileen Rothbard for providing me with a policy-modeling research environment at the Wharton School and the medical school. I thank my colleagues there.

This book draws upon my work published in *Environmental Science & Technology*, *Environmental Management*, and *Journal of the Air and Waste Management Association*. I would like to thank the editors and editorial staff of these journals and the reviewers. Several people commented on the initial drafts of these chapters or helped me with data collection. Dr. Benjamin Davy, then visiting Harvard University, provided constructive notations on my work on urban ozone plume and population distribution. Mr. Jing Zhang, then at the Department of Statistics at Penn, commented on my statistical methods for dynamics analysis. Mr. John R. Kennedy of EPA Region IX provided air quality data for the Los Angeles case study. Ms.

Kathleen Brown and her colleagues at EPA Region III helped me with ozone data for the Houston area. Mr. Bryan Lambeth of the Texas Natural Resource Conservation Commission provided monitoring station information in the Houston area. I thank all of them.

This book grows out of my long interest in the distributional issue of environmental pollution and the use of modeling in scientific inquiry. I am indebted to Professor Y. Tang of Zhongshan University, and to Professors (deceased) Peitong Liu and Huadong Wang of Beijing Normal University, who inspired my interests in environmental research.

I would like to thank the team at CRC/Lewis Publishers who worked on this project. Robert Hauserman, then publisher, brought this book to life. Arline Massey, acquisitions editor, oversaw the book's publication. Mimi Williams and Randi Gonzalez provided timely and patient editorial support.

My daughter, Ivy, reminds me of my family role every day. She always says "baba" or "daddy" when she passes by my study room. Every time, I cannot resist her call, not because she is "spoiled," I just know that she is eager to help me out and speed the whole thing up when she sits on my lap and types wildly on my keyboard. She does not know that she and her mother, Vivien, have already helped me understand a lot of things much better. For example, I learned, first hand, that susceptibility varies with the life-cycle (infant, toddler, pregnant woman, senior) and with race/ethnicity. Indeed, there is no average person, and each person should be treated as an individual. Each individual should be treated with compassion, even though rationality might be compromised.

Feng Liu
Baltimore, MD

Dedication

In Memory of My Mother Jianhua Liu

Contents

1 Environmental Justice, Equity, and Policies

1.1 THE ENVIRONMENTAL JUSTICE MOVEMENT

Aurora Castillo was a soft-spoken woman in her early sixties. A resident of East Los Angeles, she led the efforts during the late 1980s to defeat California's plan to locate the state's first hazardous-waste incinerator near her predominantly Hispanic neighborhood (Russell 1989). Not far away in south-central Los Angeles, Sheila Cannon, a single parent and part-time nurse, spent several hours a day mobilizing her community in an attempt to scrap the city's siting proposal for a garbage incinerator in the predominantly African-American neighborhood. It is people like Aurora Castillo and Sheila Cannon who have defined the concept of environmental justice.

The environmental justice movement is "a national and international movement of all peoples of color to fight the destruction and taking of our lands and communities" (Lee 1992). It represents a diverse, multi-racial, multi-national, and multi-issue coalition and calls for equal protection of all people from environmental harms, regardless of their race, ethnicity, origin, and socioeconomic status. "As with other social movements (i.e., antiwar, civil rights, women's rights, etc.), the environmental justice movement emerged as a response to industry and government practices, policies, and conditions that many people judged to be unjust, unfair, and illegal" (Bullard 1996:493). It has emerged from grassroots activism and organizations and penetrated national and international arenas. It is this grassroots environmental justice movement starting from the early 1980s that pushes the environmental justice and equity issue into the national and international environmental policy agenda. It is this grassroots environmental justice movement that has made a difference in the environmental thinking and policy making in the U.S. and will continue do so worldwide well into the 21st century.

The environmental justice movement originated in the struggles of people of color against toxic waste dumps and waste facility sitings in their communities. A milestone event occurred in a rural, low-income, predominantly black community in Warren County, North Carolina, in 1982. At that time, the State of North Carolina had decided to site a polychlorinated biphenyl (PCB) disposal landfill facility in Warren County. This siting decision sparked strong opposition from local communities. Local residents, grassroots organizations, regional and national civil rights groups, and politicians joined together to protest the decision. The protest to block the PCB-laden trucks resulted in the arrest of more than 500 people, including Walter E. Fauntroy, then Congressman from the District of Columbia; Dr. Benajamin F. Chavis Jr., then Executive Director of the United Church of Christ (UCC) Commission for Racial Justice (UCC 1987).

Many believe that this was a seminal event for the environmental justice movement, and triggered a series of studies investigating the association between environmental risks and population distribution by income and race. The most influential of these studies is a 1987 nationwide study of treatment, storage, and disposal facilities (TSDFs) and uncontrolled toxic waste sites conducted by the UCC Commission for Racial Justice. The report, titled *Toxic Wastes and Race in the United States*, found that minorities and the poor bear a disproportionate burden of these waste sites in their neighborhoods, and "(r)ace proved to be the most significant among variables tested in association with the location of commercial hazardous waste facilities" (UCC 1987:xiii).

When the Warren event and the UCC study gained national prominence, grassroots organizations protesting against environmental hazards in their communities sprang up across the nation. Community protest organizations arose from Los Angeles to New York and from Houston to Chicago. In 1983, the city of Los Angeles proposed the Los Angeles City Energy Recovery Project (LANCER), a network of three 1,600-ton-per-day waste-to-energy incinerators (Russell 1989). The first incinerator, LANCER1, was to be constructed on a site in south-central Los Angeles, a predominantly African-American neighborhood. Residents in the neighborhood formed the Concerned Citizens of South-Central Los Angeles (CCSLA) to fight this siting decision. They mobilized citizens in the community, and forged coalitions with several grassroots groups, public interest law groups, and national environmental organizations. Their efforts led to the final withdrawal in 1987 after a commitment of 5 years and $12 million. While the CCSLA was battling the city of Los Angeles, the Mothers of East Los Angeles (MELA) was formed in 1984 to fight against the state of California and the United States Environmental Protection Agency (EPA). The state's first state-of-the-art hazardous-waste incinerator was sited in Vernon, which was only a mile and upwind from several neighborhoods in East Los Angeles, the city's largest Latino community. Like CCSLA, MELA mobilized its residents, garnered allies, and took to the street during the campaign. They targeted permit-granting authorities, challenged the decision in court, and pressured legislators. In 1991, the project was withdrawn (Bullard 1993).

While residents of Los Angeles were targeting facility siting issues, local communities elsewhere in the country were also mobilizing and organizing to combat various other types of environmental problems in their backyards. In the south side of Chicago, predominantly African-American community members founded People for Community Recovery to deal with a wide range of environmental risks from air and water pollution to toxics. In Harlem, New York, African-Americans formed the West Harlem Environmental Action (WHEACT) to educate and mobilize residents on the issues of air and water pollution, toxics, open space, transportation, and landmark preservation. In Albuquerque, New Mexico, the Southwest Network for Economic and Environmental Justice (SNEEJ) was formed in 1990 to represent over fifty grassroots and indigenous organizations from eight states in the Southwest (Moore and Head 1993). SNEEJ led the regional efforts in educating, researching, organizing, lobbying, campaigning, demonstrating, and petitioning about a wide range of environmental and economic justice issues in the communities of color throughout the Southwest. In southern Louisiana, a bi-racial coalition of community, civil rights, religious, labor, and environmental groups was formed to combat environmental problems in the so-called Cancer Alley (Bryant 1989). Cancer Alley refers to a roughly 80-mile industrial corridor along the Mississippi River between Baton Rouge and New Orleans. In 1988, hundreds of people embarked together on the 11-day, 100-mile "Great Louisiana Toxics March" through the area. In the following year, Louisiana passed its first air quality law (Goldman 1991).

The environmental justice issue intensified in early 1990 and was elevated to the national arena. The environmental justice movement challenged major national establishments, including mainstream environmental groups and the EPA. In 1990, the Gulf Coast Tenants Organization and the Southwest Organizing Project sent letters to the ten largest environmental organizations, "calling for equitable distribution of resources and for representation of people of color on the boards and staffs of the major environmental players" (Moore and Head 1993:119). The letters also challenged the environmental groups for their lack of accountability to minority communities in the United States and Third World countries. In July 1991, the NEEJ submitted an open letter to the EPA "that documented more than a decade's lack of enforcement of environmental regulations in communities of color in the region" (Moore and Head 1993:120). In January 1990, a small group of scholars, researchers, and grassroots activists attended the Michigan Conference on Race and the Incidence of Environmental Hazards. After the conference, the Michigan Coalition was formed and drafted a memorandum to the United States Department of Health and Human Services, EPA, and the Council on Environmental Quality, which called for the agencies' involvement in environmental justice by means of research, risk communication, policy impact assessment, education, and policy development (Bryant and Mohai 1992).

These local and regional efforts led to national organizing and coalition building. In October, 1991, the First National People of Color Environmental Leadership Summit was held in Washington, D.C. "This was the first time that over five hundred participants came together from a variety of traditions and cultural and economic backgrounds to engage each other for the purpose of building unity and effective agenda for environmental justice and action" (Bryant and Mohai 1992:215). The three-day Summit produced the Principles of Environmental Justice (Lee 1992). The conference received widespread media attention. It signified that a national and international movement for environmental justice had taken hold. Indeed, when the first "Directory for People of Color Environmental Groups" was published in 1992, it identified 205 organizations in 35 states, the District of Columbia, and Puerto Rico.

In response to the calls from the environmental justice movement, EPA began dialogues with environmental justice advocates. In July 1990, EPA created its Environmental Equity Workgroup. The Workgroup was charged to (1) "review and evaluate the evidence that racial minority and low-income people bear a disproportionate risk burden," (2) "review current EPA programs to identify factors that might give rise to differential risk reduction, and develop approaches to correct such problems," (3) "review EPA risk assessment and risk communication guidelines with respect to race and income-related risks," (4) "review institutional relationships, including outreach to and consultation with racial minority and low-income organizations, to assure that EPA is fulfilling its mission with respect to these populations" (U.S. EPA 1992a:7-8). In June 1992, the Workgroup released the final, two-volume report titled *Environmental Equity: Reducing Risk for All Communities*. In November 1992, EPA established the Office of Environmental Equity, which was later renamed the Office of Environmental Justice. On November 4, 1993, EPA formally announced creation of a National Environmental Justice Advisory Council to "provide advice, consultations and make recommendations ... directed at solving environmental equity problems." Environmental justice has evolved from the grassroots level and entered the national environmental policy agenda.

As part of government response to the environmental justice challenges, government-sponsored conferences were held to address the environmental justice issue. In December 1990, the National Minority Health Conference: Focus on Environmental Contamination was held and represented the first comprehensive effort to look at environmental justice issues from a scientific perspective (Johnson, Williams, and Harris 1992). EPA, ATSDR (Agency for Toxic Substances and Disease Registry), and the National Institute of Environmental Health Sciences (NIEHS) sponsored a workshop on "Equity in Environmental Health: Research Issues and Needs" in North Carolina in August 1992 (Sexton et al. 1993). The initial manuscripts discussed at the workshop were published as a special issue in *Toxicology and Industrial Health* in 1993. Other academic and non-academic journals also began to devote entire issues to the subject of environmental equity and justice. The 1992 March/April issue of *EPA Journal* was titled "Environmental Protection — Has It Been Fair?" (Heritage 1992). The *National Law Journal* published a special report in September 1992. The report argued that environmental laws were not enforced equitably.

In the U.S. Congress, bills were introduced which addressed environmental justice issues. The "Environmental Justice Act of 1992" was drafted by then Senator Albert Gore and later re-introduced by Senator Max Baucus and Representative John Lewis. The "Environmental Equal Rights Act of 1993" was introduced by Representative Cardiss Collins. Gore's bill ordered EPA to scrutinize human health in the 100 counties containing the highest total weight of toxic chemicals and if adverse health effects were found, to impose a moratorium on future siting that would compound the problem. Collins' bill sought to "amend the Solid Waste Disposal Act to allow petitions to be submitted to prevent certain waste facilities from being constructed in environmentally disadvantaged communities." These bills did not succeed in passing the Congress.

Some state governments also responded to the environmental justice call. In April of 1993, the Arkansas legislature enacted the first state law on environmental justice and equity (Hart 1995). The Siting High Impact Solid Waste Management Facilities Act recognized the disproportionate concentration of disposal facilities in low or minority communities and sought to "prevent communities from becoming involuntary hosts to a proliferation of high impact solid waste management facilities" (Ark. Code Ann. 8-6-1501 (b)). The Act established a rebuttable presumption that prevents permitting the construction or operation of a high impact waste management facility within twelve miles of an existing one. In addition to Arkansas, Florida, Louisiana, Tennessee, and Virginia enacted environmental justice legislation, mandating data collection and analysis of environmental justice issues. Environmental justice bills were also introduced in Georgia, South Carolina, New York, and California.

On February 11, 1994, President Clinton issued Executive Order 12898, "Federal Actions to Address Environmental Justice in Minority Populations and Low-Income Populations." He ordered each federal agency to "make achieving environmental justice part of its mission" and identify and address "disproportionately high and adverse human health or environmental effects of its programs, policies, and activities on minority populations and low-income populations in the United States and its territories and possessions" (President Clinton 1994). In implementing this order, federal agencies subsequently issued their environmental justice strategies. EPA has made environmental justice one of its top priorities (see Table 1.1).

Originating in the facility siting issue, the environmental justice movement has since broadened its agenda. The environmental justice movement has continued to challenge the governments for unequal enforcement of environmental, civil rights, and public health laws and differential exposure to environmental risks. Environmental justice advocates have also attacked current policies and practice in risk assessment, "discriminatory zoning and land use practices," and "exclusionary policies and practices that limit some individuals and groups from participation in decision making" (Bullard 1996:493). Recently, attention has also been given to distributional impacts of transportation systems and urban sprawl.

1.2   ENVIRONMENTAL JUSTICE POLICIES

Executive Order 12898 identified several areas for addressing environmental justice: strategic planning, research, public participation, and information dissemination. It established an Interagency Working Group on Environmental Justice for providing guidance to and coordination among federal agencies, and directed each federal agency to develop an agency-wide environmental justice strategy. In April of 1995, the U.S. Environmental Protection Agency released the document titled "Environmental Justice Strategy: Executive Order 12898." EPA's guiding principles for environmental justice include (U.S. EPA 1995a):

1. Environmental justice begins and ends in our communities. EPA will work with communities through communication, partnership, research, and the public participation processes.
2. EPA will help affected communities have access to information that will enable them to meaningfully participate in activities.
3. EPA will take a leadership and coordination role with other federal agencies as an advocate of environmental justice.

These agency strategies addressed major areas identified in the Executive Order as relevant to their specific domains. The major themes for addressing environmental justice concerns in these strategy plans include research, public participation, information dissemination, development, enforcement, and pollution prevention.

Executive Order 12898 embodied a fundamental federal policy philosophy for environmental justice: addressing environmental justice in the existing framework of laws and regulations. Two laws that are very important for addressing environmental justice are the National Environmental Policy Act (NEPA) of 1969 (42 U.S.C. §4321 et seq.) and Title VI of the 1964 Civil Rights Act (42 U.S.C. §§2000d to 2000d-7). In December 1997, the Council on Environmental Quality (CEQ), which has oversight of the federal government's compliance with NEPA, issued *Environmental Justice Guidance under the National Environmental Policy Act*. In February 1998, EPA issued an *Interim Guidance for Investigating Title VI Administrative Complaints Challenging Permits* (U.S. EPA 1998a).

NEPA's fundamental policy is to "encourage productive and enjoyable harmony between man and his environment" (42 U.S.C. §4321). A primary purpose of NEPA is to ensure that federal agencies consider the environmental consequences of their actions and decisions as they conduct their respective missions (U.S. EPA 1998b). For "major Federal actions significantly affecting the quality of the human environment," the federal agency must prepare a detailed environmental impact statement (EIS) that assesses the full range of potential effects of the proposed action and all reasonable alternatives on human health and the environment. Regulations established by both CEQ (40 CFR Parts 1500-1508) and EPA (40 CFR Part 6) require that socioeconomic impacts associated with significant physical environmental impacts be addressed in the EIS.

In the memorandum that accompanied Executive Order 12898, President Clinton specifically recognized the importance of NEPA for addressing environmental justice concerns. The memorandum identifies four important ways to consider environmental justice under NEPA (Council on Environmental Quality 1997):

- Each federal agency shall analyze the environmental effects, including human health, economic and social effects, of federal actions, including effects on minority communities and low-income communities, when such analysis is required by NEPA.
- Mitigation measures identified as part of an environmental assessment (EA), a finding of no significant impact (FONSI), an environmental impact statement (EIS), or a record of decision (ROD) should, whenever feasible, address significant and adverse environmental effects of proposed federal actions on minority populations, low-income populations, and Indian tribes.
- Each federal agency must provide opportunities for effective communication participation in the NEPA process.
- Review of NEPA compliance must ensure that the lead agency has appropriately analyzed environmental effects on minority populations, low-income populations, or Indian tribes, including human health, social, and economic effects.

The CEQ guidelines offer six principles for considering environmental justice under NEPA (Council on Environmental Quality 1997:8-9):

1. Agencies should determine whether, in the area affected by the proposed action, "there may be disproportionately high and adverse human health

or environmental effects on minority populations, low-income populations, or Indian tribes."
2. Agencies should consider "the potential for multiple or cumulative exposure to human health or environmental hazards in the affected population and historical patterns of exposure to environmental hazards, to the extent such information is reasonably available."
3. "Agencies should recognize the interrelated cultural, social, occupational, historical, or economic factors that may amplify the natural and physical environmental effects of the proposed agency action."
4. "Agencies should develop effective public participation strategies."
5. "Agencies should assure meaningful community representation in the process."
6. "Agencies should seek tribal representation in the process in a manner that is consistent with the government-to-government relationship between the United States and tribal governments, the federal government's trust responsibility to federally-recognized tribes, and any treaty rights."

Notwithstanding these environmental justice considerations in the NEPA process, the power of NEPA to address environmental justice concerns has its limitations. As the CEQ guidelines point out, "[t]he Executive Order does not change the prevailing legal thresholds and statutory interpretations under NEPA and existing case law. For example, for an EIS to be required, there must be a sufficient impact on the physical or natural environment to be 'significant' within the meaning of NEPA. Agency consideration of impacts on low-income populations, minority populations, or Indian tribes may lead to the identification of disproportionately high and adverse human health or environmental effects that are significant and that otherwise would be overlooked" (Council on Environmental Quality 1997:9-10). CEQ regulations require that significance be evaluated in terms of "intensity" or "severity of impact." The narrowed focus because of environmental justice considerations could affect the determination (U.S. EPA 1998b). Again, these considerations might not change the final decision, but they may introduce a very important dimension that would make federal decision-making better and socially beneficial. The CEQ guidelines clearly state the boundary: "Under NEPA, the identification of a disproportionately high and adverse human health or environmental effect on a low-income population, minority population, or Indian tribe does not preclude a proposed agency action from going forward, nor does it necessarily compel a conclusion that a proposed action is environmentally unsatisfactory. Rather, the identification of such an effect should heighten agency attention to alternatives (including alternative sites), mitigation strategies, monitoring needs, and preferences expressed by the affected community or population" (Council on Environmental Quality 1997:10). NEPA's role is also limited for its *future* dimension, only dealing with the *proposed* actions and decisions of federal governments.

Title VI of the 1964 Civil Rights Act is an important avenue for addressing environmental concerns associated with the *past, present, and future* actions and decisions undertaken by any entity that is a recipient of federal financial assistance. This represents a much broader scope than that contained in NEPA. Title VI and other related federal regulations prohibit recipients of federal financial assistance from taking actions that discriminate on the basis of race, sex, color, national origin, religion, age, or disability.

> No person in the United States shall, on the ground of race, color, or national origin, be excluded from participation in, be denied the benefits of, or be subjected to discrimination under any program or activity receiving Federal financial assistance.

Both the Equal Protection Clause of the United States Constitution and Title VI prohibit *intentional* discrimination. The United States Supreme Court has ruled that Title VI authorizes federal agencies to adopt implementing regulations that prohibit discriminatory *effects*. On July 14, 1994, Attorney General Janet Reno issued a memorandum, reiterating that "administrative regulations implementing Title VI apply not only to intentional discrimination but also to policies and practices that have a discriminatory *effect*." The memorandum further states:

> Individuals continue to be denied, on the basis of their race, color, or national origin, the full and equal opportunity to participate in or receive the benefits of programs from policies and practices that are neutral on their face but have the effect of discriminating. Those policies and practices must be eliminated unless they are shown to be necessary to the program's operation and there is no less discriminatory alternative.

The Presidential memorandum accompanying Executive Order 12898 directs Federal agencies to ensure compliance with the nondiscrimination requirements of Title VI for all federally funded programs and activities that affect human health or the environment. EPA's *Interim Guidance for Investigating Title VI Administrative Complaints Challenging Permits* demonstrates EPA's commitment to use Title VI as a means of redress for environmental justice. "Facially-neutral policies or practices that result in discriminatory effects violate the EPA's Title VI regulations unless it is shown that they are justified and that there is no less discriminatory alternative" (U.S. EPA 1998a:1-2)

A closely watched, high-profile case under Title VI is Seif vs. Chester Residents Concerned for Quality Living. In this case, the U.S. Supreme Court agreed to decide whether a group of Chester residents may sue the State of Pennsylvania for permitting waste-treatment facilities in their communities that allegedly resulted in discriminatory *effects*. The legal issue the court is asked to address is whether private citizens can sue over a government agency's regulations that allegedly result in racial discrimination. Did the U.S. Congress intend to create a private course of action in federal court that bypasses a federal agency's review and enforcement process under Section 602 of Title VI of the Civil Rights Act of 1964 simply by alleging a discriminatory effect in the administration of programs and activities of a federally funded state or local agency? The Supreme Court ruled in 1983 that the federal civil rights law allows such private lawsuits when *intentional* discrimination is alleged, but it has never ruled directly on whether they are allowed over discriminatory *effects*.

Chester is a predominantly African-American, low-income community located southwest of Philadelphia. It is home to four hazardous and municipal waste facilities: the nation's fourth largest trash-to-steam incinerator, the nation's largest medical waste autoclave, a sewage treatment plant, and sludge incinerator. EPA has found that blood lead-levels in Chester's children are "unacceptably high," and that "both cancer and non-cancer risks, e.g., kidney and liver disease and respiratory problems, from the pollution sources at locations in the city of Chester exceed levels which EPA believes are acceptable."

Chester Residents Concerned for Quality Living (CRCQL) is the local grassroots community organization that has been fighting for environmental justice since 1994. In May 1996, CRCQL filed a complaint in the Federal District Court for the Eastern District of Pennsylvania, accusing the Pennsylvania Department of Environmental Protection (DEP) of discrimination by permitting waste facilities in this black community. Their suit was dismissed on the ground that there was no intentional discrimination on the part of the DEP. CRCQL appealed his decision to the 3rd Circuit Court. The Chester lawsuit makes no claims of adverse health effects suffered by residents living near the plants, even though the plaintiffs cite statistics showing higher mortality rates in Chester, when compared with all of Delaware County. Instead, the lawsuit argues a narrower issue, that when the Pennsylvania Department of Environmental Protection granted an operating permit to a fifth waste-treatment plant in one Chester black-majority neighborhood, the residents suffered discrimination. In December 1997, the 3rd U.S. Circuit Court of Appeals reversed the judge's ruling and reinstated the suit. "We hold that private plaintiffs may maintain an action under discriminatory-effect regulations promulgated by federal administrative agencies," the appeals court said. The decision has already set precedent for the nation: A community group has the right to seek enforcement of the civil rights statute and more importantly they do not have to prove the intention to discriminate, which is extremely difficult. On August 17, 1998, the Supreme Court dismissed the case as moot at the request of the plaintiffs, who learned that the state agency had recently revoked the permit for the proposed facility at the request of the permittee. Thus, this important legal issue is left to a future case for resolution in the Court.

On May 1, 1997 the Nuclear Regulatory Commission's (NRC) Atomic Safety and Licensing Board rejected, until further review, the application of the Louisiana Energy Services (LES) for a license to build and operate a privately owned uranium enrichment plant in the mostly poor and African-American Forest Grove and Center Springs, Louisiana communities. Citizens Against Nuclear Trash (CANT), a local citizens group, charged the company with environmental racism and filed a lawsuit to block the plant. The judges found that the NRC staff had failed to consider environmental and socioeconomic impacts of the proposed plant on the two nearby African-American communities as required under NEPA and the 1994 Environmental Justice Presidential Executive Order 12898. The decision is significant in that it is the first court ruling in the country whereby a permit was denied on environmental justice grounds.

These developments turn into a demand for rigorous analysis of environmental justice implications of public policies, programs, projects, and plans. This analysis is equally important for governments at federal, state, and local levels, environmental groups, environmental justice advocates, civic and grassroots groups, environmental consulting companies, and industries. Any policies, programs, projects, and plans, without careful analysis of their distributional impacts, could be and have been challenged.

1.3 ENVIRONMENTAL JUSTICE ANALYSIS

The importance of analysis and research has been demonstrated in Executive Order 12898, agency environmental justice strategies, court decisions, and the environmental justice movement. Executive Order 12898 recognizes the importance of research, data collection, and analysis in Section 3-3. It orders environmental human health studies of "segments at high risk from environmental hazards, such as minority populations, low-income populations and workers who may be exposed to substantial environmental hazards," and analysis of "multiple and cumulative exposures" to environmental hazards. It directs each federal agency to "collect, maintain, and analyze information assessing and comparing environmental and human health risks borne by populations identified by race, national origin, or income." The order specifically identified data collection and analysis needs for areas surrounding federal facilities and "areas surrounding facilities or sites expected to have substantial environmental, human health, or economic effect on the surrounding populations, when such facilities or sites become the subject of a substantial Federal environmental administrative or judicial action."

While environmental justice policies were formulated at the national and state levels, the debate also intensified on several grounds. The 1992 EPA report on Environmental Equity acknowledged "clear differences between racial groups in terms of disease and death rates" and higher-than-average exposure of racial minority and low-income populations to "selected air pollutants, hazardous waste facilities, contaminated fish and agricultural pesticides in the workplace" (U.S. EPA 1992a:3). However, the report did not find a clear cause–effect relationship between differential environmental exposure and differential incidences, except for lead poisoning. "In fact, there is a general lack of data on environmental health effects by race and income" (U.S. EPA 1992a:3).

These findings apparently clashed with the belief of environmental justice advocates, who saw their communities loaded with environmental hazards and their residents suffering from a variety of illnesses and even cancers (Grossman 1992; Lavelle 1992). This clash partially reflects two different perspectives on environmental justice research (Sexton, Olden, and Johnson 1993). The community perspective, represented by environmental justice advocates, believes that there is strong and sufficient evidence for environmental inequity and that environmental racism is the cause. Therefore, policy intervention for redressing inequities and injustice is justified, and environmental justice research is to help such intervention. The scientific perspective, as held by some scientists, holds that there is still a paucity of data for proving the cause–effect relationships and that the goals of environmental justice are to identify, assess, compare, manage, and communicate environmental health risks (Sexton, Olden, and Johnson 1993).

The fundamental differences between these two perspectives have created great tension and debate between environmental justice advocates and the scientific community. Environmental justice advocates perceive scientists' pursuit of the cause-and-effect relationship as an excuse for delaying governmental action for addressing their problems. They are frustrated that the scientific community cannot provide clear answers to their questions, and confused when scientists themselves disagree and provide conflicting answers to their questions. These factors lead to "a crisis in scientific belief" (Bryant 1995). This crisis has led to the environmental justice community's loss of confidence in and mistrust of the government and the scientific community. Environmental justice advocates have questioned the EPA's commitment to solve environmental inequities. They believe that the 1992 EPA report failed to respond to a wide range of challenges.

This clash also reflects a fundamental difference between two paradigms in epistemology: positivism and participatory research. As will be discussed in detail in Chapter 3, this debate enters deeply into some fundamental questions about inquiry. How can we best know the real world? Should we rely on our calculation or our communication? What is really environmentally just? (See Chapter 2.)

The participants in the debate also wrestle with a wide range of detailed technical issues. What is the appropriate unit of analysis to define a neighborhood? (See Chapter 6.) What is the appropriate control population as a benchmark? How can we effectively measure environmental impacts? (See Chapter 4.) Who are the disadvantaged groups in a society? How can we quantify their distribution? (See Chapter 5.)

While this debate goes on, federal agencies wrestle with the same kinds of issues when developing implementation regulations and guidelines. In addition, different analysts and stakeholders also disagree on the definitions of various terms in the environmental justice debate.

1.4   THE DEBATE ON TERMINOLOGY

Several terms have been used in the discussion of environmental justice issues, including environmental equity, environmental justice, environmental discrimination, and environmental racism. For the same term, there are different definitions, as illustrated in the following examples.

EPA's Office of Environmental Justice defines environmental justice as

> The fair treatment and meaningful involvement of all people regardless of race, color, national origin, or income with respect to the development, implementation, and enforcement of environmental laws, regulations, and policies. Fair treatment means that no group of people, including racial, ethnic, or socioeconomic group should bear a disproportionate share of the negative environmental consequences resulting from industrial, municipal, and commercial operations or the execution of federal, state, local, and tribal programs and policies.

This official definition embraces both procedural and outcome aspects of justice. It emphasizes both public participation in environmental decision making and distribution of environmental outcomes. This is clearly broader than EPA's earlier conceptualization of the issue. In response to the challenges of the environmental justice movement, EPA initially adopted the term environmental equity, defined as "the distribution of environmental risks across population groups and to our policy responses to these distributions" (U.S. EPA 1992a:2). The focus of the equity concept on distribution or outcome is in line with the popular use of equity in the public policy domain. Distribution or distributional impacts are often a major concern in discussion of government taxation and spending. Economists and policy analysts often discuss the incidence of a proposed tax with regard to different income groups. In the policy debate, politicians are concerned with how benefits, burdens, and costs of public policies are distributed among different social and economic groups.

Some researchers distinguish the procedural and outcome aspects using the two terms. According to Zimmerman (1994:633), environmental equity "typically refers to the distribution of amenities and disadvantages across individuals and groups. Justice, however, focuses more on procedures to ensure fair distribution. Fairness refers to where one group or individual disproportionately bears the burdens of an action." As another example, environmental justice is "broadly defined as the goal of achieving adequate protection from the harmful effects of environmental agents for everyone, regardless of age, culture, ethnicity, gender, race, or socioeconomic status" (Perlin et al. 1995:69). Clearly, for these researchers, environmental equity focuses more on the outcome, while environmental justice more heavily emphasizes the goals, policies, laws, and legal procedures to ensure fair distribution of environmental risks across social groups.

More controversial are definitions of environmental racism and discrimination. Two types of definitions emerge in the debate: "pure-discrimination models" and "institutional models" (Downey 1999). The environmental justice movement, right from the beginning, has used the institutional model to allege racism and discrimination in the environmental domain. This model is more broadly based; it emphasizes intentional and unintentional forms, personal and institutional forms. In the landmark study, the UCC Commission on Racial Justice (UCC 1987:ix-x) defines racism as follows:

> Racism is racial prejudice plus power. Racism is the intentional or unintentional use of power to isolate, separate and exploit others. This use of power is based on a belief in superior racial origin, identity or supposed racial characteristics. Racism confers certain privileges on and defends the dominant group, which in turn sustains and perpetuates racism. Both consciously and unconsciously, racism is enforced and maintained by the legal, cultural, religious, educational, economic, political, environmental and military institutions of societies. Racism is more than just a personal attitude; it is the institutionalized form of that attitude.

Bullard (1996:497) defines environmental racism as referring to "any policy, practice, or directive that differentially affects or disadvantages (whether intended or unintended) individuals, groups, or communities based on race or color." Clearly, this implies an all-embracing spectrum of causes for any inequitable outcome of environmental distribution. These causes may include overt racial intent in environmental decision making, racial discrimination in housing and labor markets, racial bias in education systems, and non-intentionally discriminatory operation of social institutions. According to the environmental justice movement, any current disproportionate distribution of environmental risks is "evidence, in and of itself, of environmental racism" (Downey 1999:770). There is no need to prove the racial intent.

The pure-discrimination model, on the other hand, defines environmental racism and discrimination on the basis of racist intent (Been 1994). This narrow definition arises from the legal doctrine. To legally challenge a decision on the ground of environmental racism or discrimination, plaintiffs can make claims under the Equal Protection Clause of the United States Constitution and the equal protection guarantees in state constitutions. In this case, they have to prove the *intention* to discriminate, which is often extremely difficult. They can also make claims under Title VI of the Civil Rights Act of 1964, showing unjustifiably disproportionate adverse effects. This pure-discrimination model relies heavily on the strict doctrine of the equal protection clause.

These terms can also be differentiated in the following aspects. Environmental racism is a narrow term with race as its core. In contrast, environmental equity and justice can cover a much broader range, including age, culture, ethnicity, gender, socioeconomic status (income), as well as race, although these terms are more often used in the context of race and income. Similarly, environmental discrimination can be broadly used but is often used in terms of race. Environmental racism and discrimination are negative in nature, while environmental equity tends to be neutral. For the negative aspect, the terms environmental inequity or injustice can be used. Environmental racism implies both outcome and causes; that is, people of color bear a disproportionate burden of environmental risks and racism is the cause. Environmental equity, in contrast, assumes no specific outcome and causes, and leaves it for an analyst to determine the relationship between environmental risk distribution and population distribution. To test for environmental inequity is to prove that environmental risks are born unfairly by the disadvantaged population under study.

Different terms reflect different political imperatives and symbolize various icons for mobilizing mass support for public policy objectives (Foreman 1998). Environmental equity is "relatively technical and unprovocative." Environmental racism is "provocative and evocative—an excellent media tool" for mobilizing the attention of people of color (Foreman 1998:10). It also serves as a link between environmentalism and civil rights. Environmental justice is "all-embracing, virtually a bumper sticker attachable to all manner of procedural and policy vehicles and to all community claims for redress" (Foreman 1998:11). It is "positive" and "aspirational," while environmental racism is "incendiary" and "adversarial." Environmental justice has become a favored term during the Clinton presidency.

Different conceptions of environmental justice also affect the research methodologies and interpretation of results. Some studies focus on urban areas or census-defined metropolitan areas. These areas tend to have heavy concentrations of poor and minority populations. Critics argue that the narrow focus fails to compare such populations with the population at large and fails to address the fundamental questions of why poor and minority populations are concentrated in urban or metropolitan areas. They argue that the heavy concentration of minorities and the poor and of industries in urban and metropolitan areas is itself evidence of inequity. Environmental justice advocates insist on a broader comparison or control population. They argue that impacted communities should be compared with every community elsewhere. More detailed debate on methodological issues is presented in Chapter 3.

1.5   OVERVIEW OF THIS BOOK

This book focuses on methods for analyzing equity impacts of environmental issues, and is not about the origin, organization, history, and development of the environmental justice movement. As indicated in the subtitle, the book has three major parts: theories, methods, and practice. Chapter 2 covers relevant theories, Chapters 3 through 9 cover methods, and Chapters 10 through 14 cover practice and case studies.

As argued in Chapter 2, theories of environmental justice and equity are in the preparadigm period. Currently, there is no consensus on a single paradigm, but instead competing theories and hypotheses offer differing explanations for environmental justice issues. These theories come largely from traditional social and scientific theories such as theories of justice, theories of risk, economic theory, location theory, health theories, and theories of neighborhood change. In this chapter, we review and synthesize major theories, and deduce from them some hypotheses about the statics and dynamics of environmental justice and equity.

In Chapters 3 through 9, major methodological issues in environmental justice analysis are carefully and critically examined, and a systematic, analytical framework and methodology for analyzing the equity issue is offered. Chapter 3 starts with a discussion of two paradigms in epistemology: positivism and the phenomenological or interpretive perspective. Then, the issues of analytical validity and fallacy in research are reviewed. In particular, major fallacies and controversies in environmental justice research are critically examined. Finally, an integrated framework for environmental justice analysis is explored that links theories in Chapter 2 with technical methods in Chapters 4 through 9.

In Chapter 4, major types of environmental impacts and the methods and models for measuring these impacts are reviewed. The strengths and weaknesses of these methods and their implications for equity analysis are discussed. Finally, critiques and responses of a risk-based approach to environmental justice analysis are presented. Chapter 5 examines how population in general and disadvantaged populations in particular are measured for environmental justice analysis. We discuss the censuses that are the predominant source for data about current and past population. Methods and models that are used for estimating current population, and projecting and forecasting future population are also reviewed. Chapter 6 illustrates the hierarchical structure of census geography, carefully examines both sides of the debate on geographic units of analysis, and presents alternative approaches for defining a unit of analysis in environmental justice analysis. It emphasizes the pitfalls of census geography and the limitations of our ability to use census geography to define environmental impact boundaries.

Chapters 7 through 9 demonstrate how cutting-edge technologies and methods such as Geographic Information Systems (GIS) and modeling tools can contribute to better equity analysis and policy evaluations. In Chapter 7, major statistical methods for environmental justice analysis are summarized, and major issues and pitfalls using statistical methods in environmental justice analysis are discussed. In Chapter 8, basic GIS concepts and methods are reviewed, and actual and potential applications of GIS techniques to environmental justice analysis are examined. Chapter 9 reviews integrated urban models that simulate multiple urban systems and elaborates on how they can contribute to environmental equity analysis. The potential for linking GIS and urban and environmental models is discussed in Chapters 8 and 9.

Chapters 10 through 13 cover a wide range of policy areas such as air pollution, solid wastes, hazardous and toxic wastes, land use, and transportation. A systematic view of recent evidence of environmental justice in a variety of policy areas is presented. These chapters also cover a wide variety of policy scales such as those at the national level, at the city and regional level, and at the site level. This diverse scope is essential for meeting the needs of a diverse audience concerned with environmental justice issues at different levels.

As shown in Table 1.1, equity concerns really started with issues concerning distributional impacts of air pollution. At the beginning of Chapter 10, relevant theories as presented in Chapter 2 are discussed and used to examine the relationship between air quality and population distribution. Research methods and evidence obtained from the air pollution domain are reviewed. In particular, we present a spatial interaction modeling approach to testing the statics of environmental equity. This approach takes into account the interactions among environmental risks, residential location, employment location, and land use and transportation, by means of a multivariate nonlinear spatial interaction model and GIS. Two case studies involving the Houston and Los Angeles metropolitan areas are used to illustrate this modeling approach. Finally, the distributional impacts of national ambient air quality standards are analyzed.

Hazardous waste facilities are the most controversial among environmental justice concerns. Chapter 11 deals with three types of waste facilities: hazardous waste facilities, Superfund sites, and toxics release facilities. For each one, we briefly discuss basic concepts about these wastes and waste facilities. Then we review major environmental justice studies associated with each type of facility, with particular attention given to the debate in the literature. Finally, we discuss some methodological issues and the potential for developing improved methodological approaches.

These facilities are part of Locally Unwanted Land Uses (LULUs). In Chapter 12, we first conduct a methodological critique of existing dynamics studies. Then we develop a conceptual framework for equity dynamics analysis that incorporates technical methods presented in previous chapters and theories and hypotheses of neighborhood changes presented in Chapter 2. Using this framework, we re-visit the Houston case, testing the market dynamics hypothesis and examining alternative hypotheses.

In Chapter 13, equity issues in transportation planning are discussed. In this chapter, we look first at environmental impacts of transportation systems. We focus on analytical tools that can be used in analyzing distributional impacts of transportation projects, programs, and policies. We devote much attention to equity analysis of mobility and accessibility, discussing basic concepts, measurement methods, their uses and limitations, empirical evidence about mobility and accessibility disparity, and spatial mismatch. Then we discuss the use of the hedonic price method for evaluating distributional impacts on property values, outline an integrated GIS and modeling approach for assessing differential environmental impacts, and examine a few analytical issues. Finally, we review the

TABLE 1.1
Milestones in Environmental Justice and Equity

Year	Event	Sources
1971	CEQ's annual report found environmental inequity by income	CEQ (1971)
1972	Inequitable distribution of air pollution by race and income was found in three urban areas and racial disparity was more striking	Freeman (1972)
1982	Residents of Warren County, NC, protested a proposed PCB disposal landfill and sparked media and congressional attention	UCC (1987)
1983	The GAO report found inequitable distribution of hazardous waste landfills by race and income in eight southeastern states	GAO (1983)
1983	Inequitable distribution of solid waste facilities by race was found in Houston	Bullard (1983)
1987	UCC released a nationwide study finding inequitable distribution of hazardous waste facilities by race and income	UCC (1987)
1987	Los Angeles residents successfully blocked the construction of a garbage incinerator (LANCER) in a black community	Lee (1992)
1990	Conference on Race and the Incidence of Environmental Hazards was held in Ann Arbor, MI, and the Michigan Coalition was formed for environmental justice issue	Mohai & Bryant (1992)
1991	The First National People of Color Environmental Leadership Summit convened in Washington, D.C.	Lee (1992)
1992	*EPA Journal* published an issue "Environmental Protection — Has It Been Fair?"	Heritage (1992)
1992	Senator Albert Gore introduced "Environmental Justice Act of 1992," which was reintroduced in 1993	Sexton et al. (1993)
1992	EPA issued a report "Environmental Equity: Reducing Risks for All Communities" and established the Office of Environmental Equity	U.S. EPA (1992a)
1993	National Environmental Justice Advisory Council was created	
1993	Environmental justice became one of EPA's top priorities	
1994	President Clinton issued Executive Order 12898 "Federal Actions to Address Environmental Justice in Minority Populations and Low-income Populations"	President Clinton (1994)
1994	A study of commercial hazardous waste facilities reached a conclusion contradicting the UCC study, leading to an intensified debate on methodological issues of environmental justice analysis	Anderton et al. (1994)
1995	Federal agencies issued environmental justice strategies	U.S. EPA (1995a)
1997	As the nation's first case to deny a permit on environmental justice grounds, the Nuclear Regulatory Commission rejected the application of the Louisiana Energy Services (LES) for a license to build and operate an uranium enrichment plant in the mostly poor and African-American Louisiana communities	U.S. NRC (1997)
1998	The Supreme Court heard the case Seif et al. vs. Chester Residents Concerned for Quality Living, deciding whether a private group could sue a state or local government for allegedly discriminatory effects of its policies. The case was dismissed as moot	

equity implications of some transportation policies such as congestion pricing, and discuss the Los Angeles Metropolitan Transit Authority (MTA) case and some analytical issues involved.

In Chapter 14, the use of Internet technology in environmental justice analysis is discussed. Recent trends in environmental justice analysis are examined, and the book concludes with a summary.

2 Theories and Hypotheses

Kuhn's work on the development of scientific thought raises many insights and questions in both the physical and non-physical sciences. In his influential book, *The Structure of Scientific Revolution*, Kuhn (1970) describes a scientific development process as one which covers a life cycle of paradigms full of random discoveries of anomalies, crises, and scientific revolutions. Kuhn defines a paradigm in two senses: first, as "the entire constellation of beliefs, values, techniques, and so on shared by the members of a given community," and second, as "one sort of element in that constellation, the concrete puzzle-solutions which, employed as models or examples, can replace explicit rules as a basis for the solution of the remaining puzzles of normal science" (Kuhn 1970:175). The scientific development process can be summarized as periods of preparadigm, paradigm development, paradigm articulation, paradigm extension, and paradigm crisis (Galloway and Mahayni 1977).

Applying the Kuhn structure, theories of environmental justice and equity are apparently in the preparadigm period. Currently, there is no consensus on a single paradigm, but instead competing theories and hypotheses offer different explanations for environmental justice issues. These theories are derived largely from traditional social and scientific work and include theories of justice, theories of risk, economic theory, health theories, location theory, theories of urban development, and theories of neighborhood change. In this chapter, we will review relevant bodies of literature concerning these theories and deduce from them some hypotheses about environmental justice.

The theories reviewed here are in themselves subjects of many years' research, which has generated rich bodies of literature. The focus here is a selective examination of the literature that is relevant to our central area of interest: environmental justice and equity analysis. Omitted from this review is the literature on the sociological and political perspectives on the environmental justice movement, which the reader can easily find in the current environmental justice literature.

2.1 THEORIES OF JUSTICE AND EQUITY

Theories of justice and equity provide principles and guidelines for deciding what makes acts equitable or inequitable. Over the past several hundred years, philosophers and social scientists have been debating various perspectives on equity and justice, but have been unable to reach a unified theory of equity. There are different ways to classify these perspectives.

Young (1994) distinguishes three basic principles of substantive fairness: equal distribution of resources among all constituents (egalitarianism); distribution of resources according to each person's merits or input (proportionality to contribution);

distribution of resources according to some priority principle, such as each person's needs (distribution rule).

Howe (1990) made a classification of normative ethical theories, which is helpful in current discussions of environmental justice. She classifies these theories along two dimensions: teleological vs. deontological, to act vs. to rule. The two major perspectives in the first dimension have been very influential in the debate on equity or ethics. A major distinction between the two perspectives is the goodness and rightness of action. The basic principle of the teleological perspective, commonly referred to as consequentialism, is that the goodness of the consequences of action decides what is the right action. In contrast, the deontological perspective focuses on the rightness of action itself — what we ought to do no matter what.

The other dimension distinguishes "the level of application of a theory's principles" (Howe 1990:127). In the "rule" case, broad "principles are used to generate rules that, in turn, serve as standards for guiding the behavior of individuals" (Howe 1990:127). In the "act" case, principles are used directly to make particular decisions.

Wenz (1988) provides an extensive treatment of theories of justice in the environmental context. He examines property and virtue, the libertarian theory, the efficiency theory, human rights, animal rights, the utilitarian theory, and John Rawls' theory of justice. In the following sections, we look at four major theories: utilitarianism, contractarianism, egalitarianism, and libertarianism.

2.1.1 UTILITARIANISM

Utilitarianism is the most commonly used form in the teleological perspective. Classical utilitarianism states that goods and services should be produced and distributed so as to maximize the total welfare or aggregate social utility. The goal is to achieve the greatest possible balance of good over bad for society as a whole. According to Davy (1996), "Justice is what is beneficial to the most, or: Maximize happiness!"

Classical utilitarianism, as often credited to Jeremy Bentham ([1789] 1948) and John Stuart Mill ([1863] 1985), is "universalistic," concerned with aggregate consequences to everyone. In their view, "the good" is pleasure or happiness, and "the bad" is pain. For utilitarians, it is a matter of calculating the good and bad and identifying the greatest net benefit for all people in the long run. In neoclassical economics, "the good" is measured in utiles, and for policy analysis, is translated into benefits and costs.

Classical utilitarianism has both strengths and weaknesses. It is intuitively appealing, quantitative, and "attractively egalitarian." Its quantitative emphasis provides a common umpire to deal with conflicts of obligations and is particularly attractive for policy analysis.

However, its strengths are also its weaknesses. Its quantification techniques are far from being simple, straightforward, and objective. Indeed, they are often too complicated to be practical. They are also too flexible and subject to manipulation. They are impersonal and lack compassion. More importantly, they fail to deal with the issue of equity and distributive justice.

Seemingly, you cannot get fairer than this. In calculating benefits and costs, each person is counted as one and only one. In other words, people are treated equally.

For Mill, "justice arises from the principle of utility" (Howe 1990:130). Utilitarianism is concerned only with the aggregate effect, no matter how the aggregate is distributed. For almost all policies, there is an uneven distribution of benefits and costs. Some people win, while others lose. The Pareto optimality world is almost nonexistent. A policy's outcome is Pareto optimal if nobody loses and at least one person gains.

The Kaldor-Hicks optimality relaxes the Pareto criteria and would justify a policy if the gainers could potentially compensate the losers (Kaldor 1939; Hicks 1939). In theory, this seems to be fair and attractive. The emphasis on **potential,** but not required compensation, is where various problems start. Does the market or the government provide any mechanism for such compensation? Not necessarily. If there is a compensation mechanism, what is a fair compensation? More fundamentally, is a loss compensable? Some economists would argue that any loss can be compensated because compensation transactions happen every day, from compensation for damage in a car accident to compensation for a life lost in an airplane crash. If this is so, then by simply resorting to those compensation schemes proposed by economists, we would not have such an impasse on waste facility siting issues as we have today. There is clear evidence that the true cost of hazardous waste facilities is uncompensable (Bacow and Milkey 1982; Weisskopf 1992).

Summers' proposal, as discussed in the preface, is utilitarianistic. Concerning this proposal, *The Economist* (1992b:18) stated: "He supposes that the value of a life, or of years of life expectancy, can be measured by an objective observer in terms of incomes per head — in other words, that an Englishman's life is worth more than the lives of a hundred Indians. This is naive utilitarianism reduced to an absurdity. It is so outlandish that even a distinguished economist should see that it provides no basis for World Bank policy."

Other than this false assumption, however, *The Economist* stated that Summers' economics was hard to answer. The economic premise underlying Summers' three arguments is "environmental policy involves trade-offs, and should seek a balance between costs and benefits" (*The Economist* 1992b:18). This balance should determine the "right" non-zero level of pollution, which varies with local circumstances. *The Economist* believes that for the environmental policy arena, "economic method — the weighing of costs and benefits — is indispensable." *The Economist* further argues that because poverty is the greatest cause of mortality in the third world and cleaner growth sometimes means slower growth, which means slower elimination or mitigation of poverty, "most poor countries will rightly want to tolerate more pollution than rich countries do, in return for more growth. So the migration of industries, including 'dirty' industries, to the third world is indeed desirable."

If this is a win–win strategy for both developed and developing countries, why are there as many as ninety countries that prohibit transboundary import of hazardous wastes? If this migration-of-pollution-to-LDCs strategy is indeed desirable in the international environmental policy arena, why do developed countries insist on a meaningful participation of developing countries in implementing an international treaty for curbing global warming? Could developed countries solve the global warming problem by simply dumping their CO_2-generating industries into developing countries? Similarly, could developed countries solve other transboundary environmental problems such as acid deposition through dumping into their neighbors?

If this economic logic behind dumping on the poor is "impeccable" in the international policy arena, it should work similarly in the domestic area. The domestic counterpart of this logic is to dump pollution into the poorest areas in the country. More specifically, the U.S. EPA should, according to this logic, encourage migration of pollution industries from California and the Northeast to the Appalachia region, the Deep South, and urban areas with large slums, where America's poorest are concentrated. Why does the EPA bother with regulations about regional transport of ozone in the eastern United States?

Besides these ridiculous policy implications in the United States and in the world, the logic underlying Summers' proposal represents "cultural imperialism," the capitalist mode of production and consumption, and "a particular kind of political-economic power and its discriminatory practices" (Harvey 1996:368). Except for its beautiful guise of economic logic, the proposal is nothing new to those familiar with the history. The capitalistic powerhouses in Europe practiced material and cultural imperialism against countries in Africa, America, and Asia for years. They did it by raising the banner of trade and welfare enhancement. They did it through guns and powder. Of course, they had their logic for exporting opium to Canton (Guangzhou) in China through force. Now, we see a new logic. This time, it is economic logic and globalization. This time, the end is the same, but the means is not through guns and powder. Instead, it is political-economic power.

This example illustrates clearly the danger of using the utilitarian perspective as the only means for policy analysis. Fundamentally, the utilitarian disregards the distributive justice issue altogether and espouses the current mode of production and consumption and the political-economic structure, without any attention to the inequity and inequality in the current system. Even worse and more subtly, it delivers the philosophy of "it exists, therefore it's good." However, "just because it sells, doesn't mean we have to worship it" (Peirce 1991).

2.1.2 Contractarianism and Egalitarianism

In *A Theory of Justice*, Rawls develops a contractarian theory of justice. In a hypothetical society, individuals reach a consensus on a social contract that includes basic institutions and guiding principles for the society to distribute resources. Individuals make their choices "behind a veil of ignorance" that obscures them from knowing their abilities, history, and socioeconomic position. This veil of ignorance provides incentives for individuals to create a society that is fair to all. Under these conditions, two principles of justice would emerge. The first principle is that "each person is to have an equal right to the most extensive basic liberty compatible with a similar liberty for others." The second is the well-known maxim or difference principle: "Social and economic inequalities [of primary goods such as liberty and opportunity, income and wealth] are to be arranged so that they are both (a) to the greatest benefit of the least advantaged... (b) attached to offices and positions open to all under conditions of fair equality of opportunity" (Rawls 1971:302). In other words, the least well-off group in society should be made as well off as possible. In the words of Davy (1996), "Justice is what is beneficial to the poor, or: Minimize pain!"

Contractarianism provides strong moral rules pertaining to the dignity and autonomy of human beings. These rules are particularly concerned with collective goods and are useful for policy analysts and planners. However, they do not provide a single method akin to cost–benefit analysis for evaluating policy alternatives. Although emphasizing the rightness of an action, contractarianism pays inadequate attention to the consequences of action or policy outcomes. The principles and rules are "impractically extreme and rigid" (Howe 1990:140).

People often confuse contractarianism with egalitarianism. Both recognize existing inequity, but they differ in their ultimate end goal. The egalitarian has the goal of eliminating the existing inequality altogether, while the contractarian does not have such a goal. The contractarian would choose an alternative, among many options, that benefits both the poor and the rich, and, as a result, still does not help reduce any existing inequality. Even worse, an action that could exacerbate the existing inequality would still be acceptable to a contractarian as long as it is to the greatest benefit of the poor.

The egalitarian emphasizes the existing inequality and evaluates any action based on the degree to which such action can reduce the level of inequality. According to Sager (1990), Bedau (1967) summarizes egalitarianism as follows:

- All social inequalities are unnecessary and unjustifiable, and ought to be eliminated
- All men are equal — now and forever — in intrinsic value, inherent worth, and essential nature
- The concept of justice involves that of equality
- Social equalities need no special justification, whereas social inequalities always do
- All persons are to be treated alike, except where circumstances require different treatment

2.1.3 LIBERTARIANISM

Libertarianism emphasizes freedom of individuals. People should be able to do whatever they want in the absence of force or fraud, as long as they respect the equal right of others to do the same (Wenz 1988). Justice results from the free market, where individuals make their choices freely. In the words of Davy (1996), "Justice is what is beneficial to the strong: or, Maximize liberty!" Libertarian justice is illustrated by Friedrich von Hayek's "catallaxy" (a spontaneous order produced by the market) and Robert Nozick's theory of a minimal state: All the state can do is to prevent aggression and fraud; otherwise, it would violate individual's rights.

Contractarianism and libertarianism can be traced to the same root in Kant (Howe 1990). Kant, in his *Groundwork of the Metaphysic of Morals* ([1785] 1964), developed two purely rational criteria for the validity of moral rules. The first criterion is the well-known categorical imperative: "Act only on that maxim through which you can at the same time will that it should become a universal law" (Kant [1785] 1964:88). The second criterion is the humanity principle: "Act in such a way that you always treat humanity, whether in your own person or in the person of any other, never simply as a means, but always at the same time as an end" (Kant [1785] 1964:96). Therefore, Kantian theories emphasize dignity and freedom. Rawls's contractarian theories draw on both dignity and freedom, while Nozick's minimal state theory relies only on freedom.

Libertarianism stresses the importance of free market, private property right, voluntary transaction, and free choices. This theory provides an underlying rationale for settling conflicts between individuals such as nuisance and trespass (Wenz 1988). However, when the conflicts involve a large number of people, the libertarian theory often fails to provide any just remedy. Environmental pollution often involves many people, and cannot be solved through voluntary bargaining among individuals. Environmental resources such as air and water are usually collective goods, which one person consumes without taking into account costs or benefits for other people that his/her consumption creates. In these cases, market failures often occur. A more fundamental problem with the libertarian theory is the justification for the initial assignment of property rights. One underlying assumption for the libertarian theory is that property rights are well and rightfully defined. In addition, original acquisition must take place without force and fraud. This is often not the case; "there are few, if any parts of the earth whose environmental resources have reached their current owners without force or fraud" (Wenz 1988:75). The libertarian theory "fails to provide adequate underlying justification for contemporary property rights and for the view that all issues of justice should be decided solely by reference to such rights" (Wenz 1988:77).

2.1.4 Which theory?

There is no doubt that policy-makers and planners holding different perspectives of justice will make different decisions. Figure 2.1 illustrates four hypothetical proposals with different distribution of net benefits (Beatley 1984). The utilitarian will choose Proposal I based on the maximum net benefits to the community as a whole. However, this proposal exacerbates the existing inequality. Under this proposal, the rich get richer, and the poor get poorer. Those who emphasize equal share of benefits and burdens would choose Proposal II. The aggregate net benefit to society is smaller in Proposal II than that in Proposal I, but different income groups share the net benefit more evenly. Proposal II would also deepen existing inequalities, although to a lesser degree. Both proposals are unacceptable to the egalitarian, who seeks to minimize inequalities. The egalitarian would choose Proposal III, because it would reduce the relative level of inequality to the greatest extent among four proposals. Please note that this proposal has the smallest aggregate net benefit and does not contribute the greatest benefits among four proposals to low and lower-middle income groups. However, the low and lower-income groups benefit from Proposal III more than high and upper-middle income groups. To maximize the benefit to the least advantaged group of a society, the Rawlsian would favor Proposal IV, which generates the greatest net benefit among four proposals to the low income group. In Proposal IV, however, the high income group would benefit the most, and thus the level of inequality would be increased.

Each of the theories discussed above has its strengths and weaknesses. Each one is powerful and "impeccable" in the eyes of its beholders. On the other hand, each

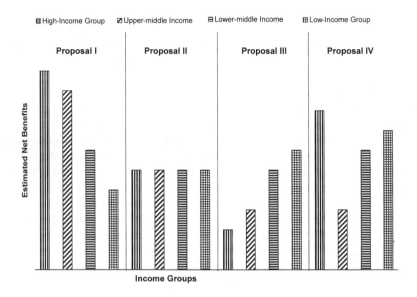

FIGURE 2.1 Different theories of justice select different policies. (Adapted from Beatley, T., Figure 2, 22. *J. Am. Planning Assoc.* 50, 4, 22, 1984. With permission.)

is subject to strong criticism from other camps. In fact, Wenz (1988: 310) concludes that "each theory failed when taken by itself." Noting the failure in searching for a singular unified theory to deal with a diversity of environmental issues, Wenz (1988) suggests that different theories should be used in different situations. Indeed, in a controversial situation, we often see application of different theories by different participants. In the case of waste facility siting (Davy 1996) or lead-paint poisoning (Harvey 1996), different stakeholders use different notions of justice to support their interests. Landlords and developers use the libertarian view of justice and seek maximum individual freedoms with respect to their rights invested in their properties under minimal government intervention. Residents in a facility-hosting community or living in houses with lead paint take the contractarian view of justice, laying claim as (potential) victims of the corporate greed for profit and asking for protection from exposure to environmental risks.

Some government officials attempting to act on behalf of the public interest assume a utilitarian view of justice. If one takes lead paint as a public health issue or facility siting as a public policy issue, such officials attempt to balance the benefits and costs of siting the facility in a community or removing lead paint, thereby making decisions based on the aggregate net gains to the society as a whole.

Some environmental justice advocates take an egalitarian perspective. They reject any inequality, whether social, economic, environmental, or political. They reject siting efforts not by raising the banner of "not-in-my-backyard" but by the principle of "not-in-anybody's-backyard." According to Bullard (1994:206), a pioneer in the environmental justice movement, "the solution to unequal environmental protection is seen to lie in the struggle for justice for all Americans. No community, rich or poor, black or white, should be allowed to become an ecological 'sacrifice zone.'"

It is no wonder that different stakeholders speak different languages in such conflicts. Numerous failures in the waste facility siting efforts lead some observers to conclude that the conflict due to these different perspectives of justice is irrecon-cilable. Davy (1996) argues that in this pluralist world of justice concepts, adoption of any justice concept will be seen as unjust or unfair to somebody with a different justice concept. Any siting will be unjust or unfair to somebody. He labels this phenomenon as "essential injustice." From a philosophical point of view, the conflict of different perspectives of justice is also irreconcilable, with some qualifications (Howe 1990). Which principle of justice prevails in such an irreconcilable conflict? Harvey (1996) argues that it is a "class struggle." He quotes Marx as saying "Between equal rights, force decides."

What should a policy analyst or any individual do in the pluralist world of justice concepts? An analyst can choose a perspective according to a specific situation (Wenz 1988) or a role he or she holds (Held 1984). This situation-specific strategy is not without difficulty. A lack of available criteria for choosing one perspective over the other creates a flexible situation where any choice may be subject to "acts of power" (Harvey 1996).

Some argue that a policy analyst should apply multiple perspectives to a policy issue and analyze the conflicts involved from different angles (Anderson 1979; Moore 1981). This enables the analyst to offer public officials a multi-faceted analysis of the strengths and weaknesses of each perspective. This information is particularly useful to search for "policies that move most surely in the direction of commonweal" (Moore 1981:14). In the case of facility siting, Davy (1996) argues for "a fair balance between different justices," and for eliminating and avoiding injustice to the extent possible.

Justice to whom? In the policy or planning domain, groups of people that share common interests and concerns are often the subject of inquiry for justice. In philosophy, however, the focus is on the individual. In measuring equity concerning groups, disproportionality is often used and equated with inequity or injustice. Perhac (1999), a philosopher, argues that three major theories of justice — utilitarianism, natural rights theory, and (Rawlsian) contractarianism — do not call for "eliminating disproportionality along racial or socioeconomic lines" as a matter of justice. Pro-portionality of risk distribution along racial or socioeconomic lines does not ensure justice to individuals. He suggests a shift of focus from racial or socioeconomic groups to individuals at greatest risk irrespective of their group association. However, in the current political environment, it is not individuals but rather groups that play major roles.

2.2 ECONOMIC AND LOCATION THEORIES

Efficiency is what economists worship. When they talk about efficiency, what they usually mean is Pareto efficiency. It is a state in which no one can be made better off without making someone else worse off. An economy is Pareto efficient if there are "no unexploited gains to trade, no unexploited way of increasing output with the same level of inputs," and no mix of products that do not reflect the preferences of consumers (Stiglitz 1993). These very strong conditions for the Pareto efficiency of the market economy are referred to, respectively, as exchange efficiency, production efficiency, and product-mix efficiency. A competitive market is Pareto efficient, and every Pareto-efficient allocation can be obtained through the market mechanism. A competive market also has strong assumptions: rational participants, perfect information, utility- or profit-maximization, and non-existence of market failure conditions. In the real world, these conditions are seldom realistic. Many environmental issues have defied all these conditions.

2.2.1 EXTERNALITY AND PUBLIC GOODS

An externality arises when one individual's activity imposes costs or benefits on other unrelated parties, but this individual does not take into account these external effects when making his or her decision. As a result, there is a divergence between social costs and private costs, which indicates economic inefficiency. In order to achieve efficiency, such government policies as taxation and subsidy are justified (Pigou 1926). On the other hand, Coase (1960) argued that such government intervention was unnecessary. The well-known Coase theorem states that if property rights are well defined and freely transferable in a world of zero transaction cost, a voluntary bargaining and compensation solution among different users of the environment will result in a Pareto-optimal allocation, and properties will go to their most socially valued uses. Coase argued further that victims of environmental pollution should not only be not compensated but also be taxed, for victims' decisions to locate near the polluter impose an external cost or an increase in damages incurred by the polluter's activity. The Coase theorem has been criticized on many grounds, including the impracticality of bargaining in real world situations involving a large number of parties, relevance of transaction costs, and equity considerations (Baumol and Oates, 1988). In particular, environmental quality is a public good. Public goods are goods for which one person's consumption does not affect another's (nonrivalrous), and it is extremely costly to exclude anyone from enjoying (nonexcludable). In the case of public goods, an individual can take the position of a freerider and never pay the cost for using them.

From economic theories, Hamilton (1995) derives three hypotheses for explaining why exposure to environmental risks might vary by race: pure discrimination, the Coase theorem (differences in willingness to pay), and the theory of collective action (differences in propensity to engage in collective action). In the pure discrimination model, owners of environmentally risky facilities increase their utility by siting their facilities in the communities with the racial groups against which they have prejudice. Interpreting the Coase theorem in the context of siting a hazardous waste facility, he indicates that this type of facility will locate where compensation due to damages is the least. Furthermore, compensation is associated with the characteristics of the host community, such as population affected, income, property values, and residents' willingness to pay for an environmental amenity. In turn, these variables may be associated with race. As discussed above, voluntary negotiations for compensation may be impossible in real world situations. More often, residents in a target community voice their concern and compensation demands through the political process. To the extent that different racial and income groups have different propensities for political participation, the location of the environmentally risky facilities may be associated with the race and income characteristics of the host community. In examining capacity expansion decisions of commercial hazardous waste facilities during the period from 1987 to1992 by using population data at the ZIP code level, Hamilton (1995) found that collective action, measured in terms of actual voter turnout in a county, offered the best explanation for which neighborhoods were selected for capacity expansion. After controlling for other variables, the percentage of nonwhite population was not statistically significant.

2.2.2   WELFARE ECONOMICS

Economic effects of any environmental quality changes such as air and water quality can be measured in terms of public welfare changes. Welfare changes occur as "changes in the prices they pay for goods bought in markets; changes in the prices they receive for their factors of production; changes in the quantities or qualities of nonmarket goods (for example, public goods such as air quality); and changes in the risks individuals face" (Freeman 1993:39).

Ordinary consumer's surplus, also called Marshallian consumer's surplus, is a welfare measure and is simply defined as the difference between the price which the consumer is willing to pay and the actual price he or she pays. It measures the surplus of satisfaction that is derived from the consumer's paying a lower price than his or her maximum willingness to pay. This interpretation is very intuitive. However, the Marshallian consumer's surplus measure does not offer much help for quantifying welfare changes associated with public policies. It does not measure either utility changes, except for some special cases, or gains or losses of potential compensation. Following Freeman (1993), let us look at four other welfare measures: compensating variation, equivalent variation, compensating surplus, and equivalent surplus.

Compensating variation measures the compensating payment that is required to make a consumer indifferent to his or her original consumption bundle at the old price and new consumption bundle at the new price. The consumer's initial utility is the reference point; that is, the consumer should be as happy as he or she is now in the face of proposed changes in price. This measure implies that the consumer has the right to the status quo. If a proposed policy increases the price for the consumer and thus decreases his or her utility, the consumer has to be compensated with a certain income to maintain his or her original utility level. For a price increase, compensating variation is the minimum amount that the consumer is willing to accept (WAC) to prevent a utility decrease. On the other hand, if a proposed policy decreases the price for the consumer and thus would increase his or her utility, the consumer would be willing to pay a certain amount to purchase at the new price level. The maximum amount that the consumer would be willing to pay is where he or she still maintains his or her original utility level and thus exhausts any potential welfare gain. For a price decrease, compensating variation is the consumer's maximum willingness to pay (WTP) for consuming at the new price. Essentially, compensating variation is the income change required to maintain the status quo when prices change.

Equivalent variation measures the income change that would cause the consumer's utility to change at the same amount as the change in the price. The consumer's final utility is the reference point; that is, we should make the consumer as happy as he or she will be when price changes occur. This implies that the consumer has the right to the change. If a proposed policy decreases price and thus increases his or her utility, the consumer would be better off and the consumer would have to be paid to forgo the opportunity for potential welfare gain. For a price decrease, equivalent variation is the minimum payment that the consumer would be willing to accept (WAC) to forgo the opportunity to consume at the new low price. For a price increase, it is the maximum amount that the consumer would be willing to pay to avoid the price increase. Basically, equivalent variation is the income change required to reach or avoid the final utility in the future when prices change.

When prices change, consumers will adjust their consumption behaviors, including the quantity they will consume. These changes are reflected in compensating and equivalent variation measures. In contrast, the compensating and equivalent surplus measures for price changes hold the consumption at the specified level. Compensating surplus measures the compensating payment that is required to keep the consumer at the original utility level and maintain the consumption quantity at the new level that would be purchased at the new price. For a price decrease, compensating surplus is the consumer's WTP to consume the new quantity at the new price. The way this differs from compensating variation is that the consumer is not allowed to consume other than the fixed quantity at the new price.

Equivalent surplus measures the income change that would be required to hold the consumption quantity at the original level, but change the consumer's utility to the new level associated with a price change. For a price decrease, equivalent surplus is the minimum payment that the consumer would be willing to accept (WAC) to forgo the opportunity to consume at the new low price. The difference between equivalent variation and equivalent surplus is that equivalent surplus has a fixed quantity of consumption at the original level.

Earlier in the section on utilitarianism, we discussed Pareto optimality and Kaldor-Hicks optimality. These are two social welfare criteria that are used to evaluate alternative public policies. Because a policy that makes no one worse off and at least one person better off is non-existent, the Pareto criterion is too restrictive to be useful in any policy evaluation. The Kaldor-Hicks criteria are concerned with the potential Pareto improvement at the aggregate level. The criteria are twofold. First, if all of the winners can potentially compensate all of the losers from a proposed policy, then the Kaldor version of the criterion is satisfied (Kaldor 1939). In the framework of the welfare measures discussed above, this is the condition whereby the sum of compensating variation or compensating surplus measures for all individuals is positive. Second, all losers from a proposed policy could potentially compensate the winners so that the proposed policy can be forgone. This Hicks test is satisfied if the sum of all winners' equivalent variations is greater than the sum of all losers' equivalent variations.

While these two criteria are efficiency criteria of welfare economics, they do not address the equity issue. Subsequent modifications of the Kaldor-Hicks criteria introduce the equity dimension and address the question of whether the proposed policy improves the distribution of income. One such approach utilizes the social welfare weighting function, which introduces different weights to different individuals for their welfare changes. The rich may be given a low weight, while the poor may be given a high weight. The key issue is how we can appropriately determine the weights to different individuals or groups. One may design the weighting function in such a manner that policy proposals would redistribute income toward the poor so as to reduce the existing inequality. This explicit consideration of equity will justify policy alternatives that would result in negative aggregate welfare changes, which are not possible under the Pareto or Kaldor-Hicks criteria (Freeman 1993).

2.2.3   RESIDENTIAL LOCATION THEORY

Studies that focus on how human activities consume space as a limited resource fall under a variety of labels, including spatial economics, regional science, urban and regional studies, urban economic theory, location theory, and economic geography. Spatial economics addresses two principal questions: "how economic agents of various types choose their locations in a spatially extensive economy, and how the market areas of these agents are determined" (Norman 1993). Agents include, among others, residents who choose a residence, industries that select a site for manufacturing, commercial sectors that decide where to provide their services to consumers, and government agents who determine the locations for public services.

The seminal and a monumental contribution to location theory is attributed to von Thünen (1826), who developed a theory of agricultural land use. Land within an urban area is allocated according to the rents that the competing users are able and willing to bid. In von Thünen's world of a homogeneous, featureless plain containing a single town at its center in which only manufactured goods are produced, an urban form emerges which has a concentric ring pattern of land use with a declining gradient of bid rents from the center.

After more than a century's neglect, von Thünen's theory has been revived since the end of World War II. Alonso (1964) generalized von Thünen's central concept of bid rent curves to an urban context. Locational choice of the household was initially seen as a problem of *trade-off between accessibility and space*. Under assumptions similar to von Thünen's, a monocentric residential distribution pattern arises in an urban space where households with higher income locate further away from the central business district (CBD) than those with lower income, *ceteris paribus*. Later development along this line introduced such locational factors as the time cost of commuting, family structure, and housing consumption, and produced more realistic spatial patterns of residential location. According to Fujita (1989:42), "Wage-poor and wage-rich households with few dependents (such as singles and working couples with few children) will tend to reside close to the city center. Beyond them and out toward the suburbs, middle-income households with large families and few commuters will be found. Farther away, asset-rich households with larger families and few commuters will locate." Furthermore, the competitive land market is proven to lead to a unique equilibrium of residential land use, which is socially efficient.

More recent developments have incorporated *externality* into residential location decisions; namely, in making a residential choice, households weigh three basic factors: *accessibility, space, and environmental amenities*, under budget and time constraints (Fujita 1989). Four types of externalities are introduced: externalities from producers to households, externalities among households, externalities among producers, and externalities associated with urban transportation (Kanemoto 1987). It has been shown in a spatial model of industrial pollution that location-dependent Pigouvian taxation and land use control may be necessary to achieve an efficient allocation when industrial production imposes externalities on city residents. Land rent has jumps at the boundary between the industrial and residential zones: "The industrial rent is higher (lower) than the residential rent at the boundary if the social benefit of increased dispersion in the industrial zone due to shifting the boundary farther from the firms is larger (smaller) than the social cost of increased residential pollution caused by moving the boundary closer to residents" (Kanemoto 1987:53). In a small open city with externalities from producers to households, the existence of a spatial equilibrium may be proved under reasonable assumptions, but it may be unstable for certain rates of adjustment in the labor and land markets, unless the externality is sufficiently small. If the industrial zone expands beyond the equilibrium boundary, pollution in the residential zone intensifies and thus lowers the residential rent. If this impact is strong enough, the industrial rent exceeds the residential rent at the new boundary and a further expansion of the industrial zone occurs. As a result, the equilibrium boundary is unstable. One implication is that polluting industries could drive away residents, whether white, black, brown, or red.

Externalities among households occur when one social group has prejudice against another social group and the group with prejudice feels it suffers from the presence of the other group in their neighborhoods. Three types of models have been developed to examine the properties of equilibrium spatial structures resulting from such externalities: border models, local externality models, and global models (Kanemoto 1987; Fujita 1989). The border models assume a completely segregated residential pattern with black (white) households occupying the inner (outer) ring and whites preferring to live away from the white-black border. The local externality models assume that households are concerned with the racial composition of their own location but not that of other locations. While the assumptions are restrictive in these two types of models, the global models assume that the total amount of externalities received by a white is the weighted sum of blacks living in the same city where weights are given by a decreasing function of distance between a white and a black. If whites (and blacks) have racial prejudice, the models show that the spatial pattern in equilibrium is complete segregation with blacks in the central location. If blacks prefer racial mixing, it is possible that this pattern is not an equilibrium. If the preference is not too strong, the segregated pattern can still be an equilibrium. It is also shown theoretically that there is a possibility of dynamic instability in racially mixed cities; that is, a small increase in the number of blacks (or the poor) in the neighborhood may drive away whites (or the rich), causing a sudden change in the racial (social) composition of the neighborhood. This phenomenon is called "neighborhood tipping." This theory offers us a model for explaining how and why neighborhoods change over time. Other models are discussed in Chapter 2.4.

The role of public goods in a residential location choice is addressed in the well-known *Tiebout model*. A residence locator chooses a residence community "which best satisfies his preference pattern for public goods" (Tiebout 1956:418). Assume that individual locators are fully mobile among a large number of communities and have perfect knowledge of these communities. Assume that there are no restrictions on employment opportunities and no externalities among these communities due to the public goods they provide. Communities compete with each other by providing their own package of public goods. Individuals will "vote with their feet" for the community with the bundle of advantages and disadvantages that best fits their preferences. Consequently, provision of public goods will be efficient and a community will reach its optimal size.

Empirical tests have provided supporting evidence for the Tiebout hypothesis. Oates (1969: 960) found that, *ceteris paribus*, "property values would be higher in a community the more attractive its package of public goods." Jud and Bennet (1986) demonstrated that school quality has been a significant factor in shaping interurban locations, even when holding racial composition constant.

Location theory can provide us with some hypotheses about the relationship between environmental risk distribution and population distribution. In a Tiebout type of world, where levels of environmental quality are continuously distributed in space, individuals choose residential locations that offer them desired environmental quality in accordance with their demands. Assuming environmental quality to be a normal good, we would expect that the more affluent would "buy" more of it. Empirical evidence appears to suggest that the more affluent place a higher value on environmental quality. Under the condition of perfect information, upper-income households would tend to locate *ex ante* where there is little environmental risk or pollution, *ceteris paribus*. The poor would choose their residences in the communities with the lowest environmental quality and the most serious environmental risks. This association between population distribution and environmental risk distribution can also arise *ex post*. After an environmentally risky facility is sited and operating in a community, environmental quality will decline and environmental risk will increase to the extent that some residents placing a higher value on environmental quality will be discontented. These dissatisfied residents would shop for the communities that can satisfy their demands on environmental quality and "vote with their feet." Given the normality of environmental quality, these residents tend to be the rich in the host communities and have the resources and many choices for relocation. They could relocate to other sections of a metropolitan area or migrate out of the metropolitan area. Those remaining in the host communities tend to be those with low incomes. In either the *ex ante* or *ex post* case, we would expect that the poor disproportionately populate the areas with the poorest environmental quality and the most serious environmental risk. The rich would most likely live outside these areas. Dynamically, we would now expect a larger proportion of the poor and a smaller proportion of the rich in the areas with serious environmental pollution or risks than ten years ago. In another view, the median household income in an area with serious environmental pollution or risks would have a slower growth rate than those without; and may even decrease over the decade. Therefore, we have the following hypotheses:

Hypothesis 2.2.1: The more serious environmental risks are (perceived to be) in a community, the smaller the proportion of the rich and the greater the proportion of the poor who live there.

Hypothesis 2.2.2: The proportion of the rich decreases and the proportion of the poor increases over a period of time in a community with serious environmental risks.

It should be pointed out that stringent assumptions are required for the two hypotheses. For example, people should have perfect information, which really never exists. Furthermore, people should behave rationally, or respond to or perceive environmental risks rationally. In fact, the public's perception of risks, as will be discussed in detail in Section 2.3, differs from "objective" assessment of the actual risks. Furthermore, it is very difficult, and most often impossible, to truly assess the "objective" risks. Therefore, the hypotheses have to be relaxed in terms of perceived environmental risks.

2.2.4 INDUSTRIAL LOCATION THEORY

Neoclassical theory of industrial location has evolved from the work of Weber (1909), and has sometimes been labeled as minimum (or least) cost theory. A firm makes a location decision by making trade-offs between labor costs and transport cost. Assuming production and consumption take place at concentrated points, Weber (1909) developed a theory of industrial location, where transportation, labor, and agglomeration were key factors for industrial location. According to his reasoning, production which reduces the weight of raw materials will be located near raw material sites or ports, while production that results in an increased weight of the products will be located near consumption sites. Industries also choose their locations by taking into account the agglomeration effect and sources of cheap labor, land, and other localized materials. Weber's theory has a number of modern implications. First, it indicates that hazardous waste management facilities will tend to be located where hazardous waste generators are located so as to reduce waste transport costs and decrease the risks associated with waste transport, or where there is cheap land and labor. From this, we may hypothesize that industrial waste management facilities may be closely associated with an industrial complex in an urban area.

Hotelling (1929), in his seminal work on spatial competition, demonstrated that spatial agglomeration can also result from spatial oligopolistic competition. In Hotelling's bounded linear market with evenly distributed buyers and with two sellers of a homogeneous product, the duopolists will choose a location close to the market center. Hotelling's original analysis has been extended extensively, including applications to political science. In political campaigns, politicians fight for the middle ground.

Weber's theory was subsequently extended to include the costs of other production factors (e.g., land price) and scale economies (Lever 1985), taking into account the fact that land prices or rents vary sharply over relatively short distances. Assuming a concentric city with its Central Business District (CBD) and customers at its center, and a land price structure which varies inversely with distance from the center, Alonso (1967) showed that profit-maximizing firms had their equilibrium locations away from the center, where their revenues were low, operating costs were higher, but land costs were low enough to more than offset them. Another development is the introduction of substitution of major production factors (land, labor, and capital). This was used to explain the suburbanization of manufacturing, whereby firms used more capital for which the price was often spatially invariant, more land that was cheapest at a city's edge, and less labor that had become increasingly expensive since the World War II.

A firm decides to locate at a given location based on comparative advantages with other locations. Factor endowments vary from one location to another, and thus locations have different factor prices. Firms maximize their profits by locating where they have an optimal factor mix to generate their optimal revenues. In a competitive market, firms bid for their optimal location. Distinct firms require alternative factor mixes, and thus various industrial activities will choose locations at different points within the city. Therefore, we may have the following hypothesis.

Hypothesis 2.2.3: The hazardous waste management facilities in a community are positively related to the volume of hazardous wastes in the area encompassing that community. To the extent that the middle class and the poor are associated with the hazardous waste production operation, they are also associated with the hazardous waste management facilities in their communities.

Neoclassical location theory is built upon some strict assumptions, such as perfect information and profit maximization. In reality, a firm's decision-makers do not have perfect information upon which to base their location choice. The location choice might be regarded as utility maximizing in which profit is only one element. Social or environmental attributes might be taken into account in location decision-making processes. Furthermore, firm managers may have multiple goals including growth, security, minimization of risk, or entrepreneurial satisfaction. This is the behavioral location theory.

The structuralist theory emphasizes the conflict between capital and labor in a capitalist society, the role of large corporations in using their power to achieve authority over their workforces, the role of organized labor in responding to this control, and the overall patterns of change in the world. These conflicts and changes are reflected in the location patterns of industries and their changes over time.

2.3 THEORIES OF RISK

Theories of risk are built upon various traditional disciplines, particularly physical and health sciences, economics, geography, psychology, sociology, anthropology, and political science. In the following, four paradigms are reviewed: psychometric, expected utility, cultural, and sociological theories of risks.

2.3.1 PSYCHOMETRIC THEORY

It is well documented that the public perception of risks differs from the experts' "objective" evaluation of actual risks. The intuitive judgments that the public uses to evaluate environmental risks are called risk perceptions (Slovic 1987). The psychometric paradigm of risk perception research uses an experimental approach and quantitative techniques to produce the public's "cognitive maps" of risk perception. Consider two major factors of hazard: "dread risk" and "unknown risk" (Slovic, 1987; Slovic, Fischhoff, and Lichtenstein, 1985). The high degree of dread risk is associated with catastrophic potential, fatal consequences, perceived lack of control, and perceived unequal distribution of risks and benefits, while the high degree of unknown risk is associated with those that are new, unobservable, and long latent. Lay persons' risk perceptions are closely related to the position of a hazard within this two-factor space, particularly in the dread risk dimension.

Earlier studies in this paradigm were limited to small, homogeneous, convenient samples, such as college students or certain social groups. Recent studies have begun to investigate the issue of how psychometric perception of risks varies across demographic, social, and economic groups and across countries. It has been found in most studies that women tend to perceive greater technological risks than men (Slovic 1992; Pilisuk and Acredolo 1988; Savage 1993). Using a random sample of 450 people in California, Pilisuk and Acredolo (1988) found greater concern for technological risk among ethnic minorities, less educated, and poorer people. A national survey of Canada revealed that perception of risk and benefit was strongly related to region of residence, age, sex, education, and degree of political activism (Slovic 1992). Using a random sample of about 800 residents in Chicago, Savage (1993) found that women, people with lower levels of schooling and income, younger people, and blacks had more dread of hazards. It was hypothesized that people with greater perceived exposure to a hazard were more fearful. In a study conducted in Connecticut and Arizona, however, education, political leaning, and gender had little affect on attitudes toward risks (Gould et al. 1988).

In short, the public perception of environmental risks deviates from any "objective" evaluation of risks, and the perception of and response to environmental risks may vary with the socioeconomic characteristics of individuals and communities. Environmental quality is valued differently by various population groups, who may have different degrees of access to imperfect information. Therefore, Hypotheses 2.2.1 and 2.2.2 should be relaxed in terms of perceived risks. Furthermore, we may have the following hypothesis:

Hypothesis 2.3.1: To the extent that risk perception is related to residential location choices, environmental risks with a larger degree of dread and unfamiliarity tend to be associated with the poor. Environmental risks with a small degree of dread and unfamiliarity are not necessarily associated with any particular social group.

2.3.2 EXPECTED UTILITY THEORY

A major principle of expected utility theory is "the transformation of physical harm or other undesired effects into subjective utilities" (Renn 1992:61). This transformation allows measurements of all consequences and comparisons between risks and benefits across different actions. In spite of these obvious advantages and others, this theory has been criticized for its assumption of rational behavior and utilitarian ethics. Specifically, application of utilitarian ethics will lead to inequitable distribution of environmental risks across social groups. Through an auction or compensation mechanism suggested by this theory, a hazardous waste facility is more likely to end up in a poor community than in a rich community. Other ethicists, however, believe that "a fair distribution of risks and benefits is a value in itself and should not be subject to bargaining" (Renn 1992:63).

2.3.3 CULTURAL THEORY

According to cultural theory, "risks are defined, perceived, and managed according to principles that inhere in particular forms of social organization" (Rayner 1992:84). Cultural theory has made contributions to our understanding of risks, particularly as to why people in various cultures select different risks to worry about, ignore others, and present alternative prescriptions in response to these risks. Cultural theorists believe that the risk selectivity in various social organizations helps to reinforce the social solidarity of these organizations. Cultural theorists have developed a grid/group analysis, which is used to characterize cultural patterns (Rayner 1992). The group variable describes the degree of social incorporation of individuals in a social unit, while the grid variable represents the degree of interactions among individuals. Given the high–low dichotomy of each variable, we have four prototypes of social organizations: competitive individualists, egalitarian groups, hierarchical groups, and stratified individuals. Entrepreneurs in the competitive individualists prototype regard risk taking as an opportunity and they are less concerned about equity. Egalitarians care very much about equity and cooperation rather than competition and freedom. The authoritarians in the hierarchical prototype rely on rules and procedures to cope with risks. Stratified individuals tend to be the most vulnerable to risks. Each cultural pattern emphasizes those risks "that reinforce the moral, political, or religious order that holds the group together" (Rayner 1992).

In their work, cultural theorists take into account the relationship between equity and risks. Douglas (1990) argues that it is "inherently difficult" to take care of the risks imposed on socially and economically disadvantaged groups "in a society organized under the principles of competitive individualism." These groups are either ignored or stigmatized as "human derelicts." The more inequitable the distribution of wealth within a society, the more danger and hazards the poor suffer. Rayner (1992) proposes the fairness hypothesis and defines risk as probability × magnitudes + TLC (trust, liability, and consent). He points out the different concerns for various social groups in risk decision-making processes. While public/private decision-makers focus their concerns on economic risks and technological aspects of environmental risks, the public is particularly concerned about equity.

Although cultural theory of risks has received widespread attention in the scholarly community, it has seen little application in risk management and decision-making processes. There has been insufficient empirical evidence to verify its validity. Some critics state that cultural theory will lead to "cultural determinism" or "cultural relativism" (Rayner 1992; Renn 1992).

2.3.4 SOCIOLOGICAL THEORY

Sociologists have enhanced our understanding of risk from several perspectives, promulgating the rational actor concept, social mobilization theory, organizational theory, systems theory, neo-Marxist and critical theories, and social constructionist concepts (Renn 1992). In spite of their differences, all of these approaches emphasize "social processing of risk" via social influences, such as family, friends, co-workers, and some public officials. Sociologists have helped place the public perception of risk in the context of the social fabric (Cutter 1993; Short 1984; Renn 1992). Social influences have been shown to strongly affect human responses to hazards (Short 1984). Thus, an individual's response to environmental risks may be only partially related to his/her own assessment of the risks. Human response to environmental risks is also influenced by the salience of risk in comparison with a host of competing social values. Thus, people may be concerned with some environmental risks but give them lower priority than other social concerns (Cutter 1993). This is verified by some studies on air quality perception that are reviewed in Chapter 10.

These approaches also have a common interest in the issue of social justice from the positive and normative perspectives (Renn 1992). From a positive perspective, social inequity of risk is regarded as a reflection of the class structure of society and the inequity of the distribution of power and social influence (neo-Marxist and critical theories), or potential violations of group interests (rational actor concept), or a social construct which can be used to demand corrective actions (social constructionist concepts). From a normative perspective, these sociological approaches demand that risk policies be based on "the experience of inequities, unfairness, and — to a lesser degree — perceived social incompetence" (Renn 1992:72).

2.4 THEORIES OF NEIGHBORHOOD CHANGE

The theories reviewed above are mostly static in nature. Theories of neighborhood changes can offer us some insights as to how the relationship between population distribution and environmental risks changes over time. Theories reviewed here include the classical invasion-succession model, the neighborhood life-cycle model, the push-pull model, and institutional theory. In the following, LULU refers to locally unwanted land use and environmentally risky facilities such as hazardous and solid waste management facilities, uncontrolled toxics and solid waste sites, etc., which impose negative externalities on the host communities.

2.4.1 CLASSICAL INVASION–SUCCESSION MODEL

The Chicago school of human ecology developed the classical invasion-succession model, drawing the concept from the field of biological ecology. In ecology, an invasion-succession is a process whereby a new type of plant invades a new habitat, gradually adapts to the habitat, and eventually succeeds previous species in dominating the habitat. A similar process happens in human society. A pioneer in a minority group may, for some reason, enter a new neighborhood. Initially, the neighborhood conditions are not compatible with this socially and racially different outsider. Inevitably strong resistance and hostility occur. If individuals can survive the initial hard time, they may "cultivate the land" and make it more attractive to themselves. Then, others with similar attributes will enter this new world. At the same time, the original inhabitants perceive the neighborhood as less and less attractive. They respond by fighting more strongly against invasion by others. If they should fail, they would withdraw. Their departure would encourage entry of more members from the new group. As a result, succession of the neighborhood from one social group to another would eventually occur.

The classical invasion-succession model has been extensively applied as a tool for characterizing neighborhood racial and social-status transitions in American cities. Duncan and Duncan (1957) identified four stages in this process: penetration, invasion, consolidation, and piling up. This model implies that racial changes in a neighborhood may have nothing to do with LULUs. Rather, racial transition in a neighborhood is viewed as an ecological process in which competition, conflict, and accommodation characterize the relationships among different social groups.

This kind of racial relationship was more evident in the past than at present. Historically, African-Americans and other minorities had the lowest socioeconomic status in American society. There was substantial racial prejudice against African-Americans and other minorities. This is substantiated in some early public opinion surveys. In 1964, in a poll of white respondents, 60% agreed with the statement: "White people have a right to keep blacks out of their neighborhoods if they want to, and blacks should respect that right" (Farley and Frey 1994). In 1958, a Gallop poll found that 44% of the respondents in a national sample of whites would leave if a black person moved next door. In addition to these attitudes toward those of a different race, there were widespread fears that racial transitions would lower property values dramatically and increase crime significantly. Racial prejudice and fear of racial transitions had a self-fulfilling effect. Because whites feared that the racial mix in their neighborhoods was changing, they were willing to sell their houses for low prices and, in the process, neighborhood property values were lowered (Taub, Taylor, and Dunham 1987).

As can be expected, whites at the edge of a black neighborhood were most insecure in their residential status (Schwirian 1983). The increase in urban black population due to rural to urban migration and natural growth early this century and following WW II has been well documented. The black enclaves were natural places for them to look for residences. As the black population increased, the black enclaves could initially accommodate the growth with increased density, but, eventually, it would have to expand into contiguous white areas. Therefore, racial transition in a neighborhood could be a spatial diffusion process (O'Neill 1981).

This suggests that the black population could expand into neighborhoods with or without LULUs. The possibility of expansion and fears of whites could drive down property values in those neighborhoods adjacent to the black enclaves. LULUs may happen to have been located in these adjacent neighborhoods. In other words, the post-siting decline in the neighborhoods hosting LULUs might be a result of spatial expansion of black enclaves, which had started before siting.

In addition to widespread applications to racial transitions in neighborhoods, the invasion-succession model was also used to explain changes in land-use activities in neighborhoods over time. The encroachment of commercial and industrial activities upon adjacent residential areas can contribute to conversion from residential land use to commercial and/or industrial land uses (Burgess 1925; Smith and McCann 1981). Therefore, we would expect that a LULU would drive away residents in the hosting communities, regardless of their racial composition. Moreover, land use would become more and more dominated by industrial use.

2.4.2 Neighborhood Life-Cycle Model

The life-cycle model, formally devised by Hoover and Vernon (1959), takes the view that a neighborhood has a natural life-cycle, just like any natural and life process. Through this life cycle, a neighborhood ages and declines, and its inhabitants become members of successively lower socioeconomic groups. Hoover and Vernon identified five stages in the life cycle: residential development, transition, downgrading, thinning out, and renewal. As a neighborhood moves through this cycle, land-use density and population density increase through the downgrading stage and then decrease thereafter. Also, cycling through the stages are the socioeconomic status of inhabitants and housing conditions. The age profile of residents may grow older in concert with housing age (Myers 1992). Birch (1971) reformulated this model using six stages: rural, first wave of development, fully developed high-quality residential, packing, thinning, and recapture. His test using various indices provided support for his and the original formulation.

This theory supports the view that age and density are the fundamental factors which determine neighborhood structure and characteristics and their changes over time. As the neighborhood ages *without any renewal and renovation*, it is natural that its physical structure deteriorates gradually and its property values go down over time. The benefits that the neighborhood can offer to its inhabitants also decline with the aging process. Accordingly, the neighborhood becomes undesirable to upper-income households and more attractive and affordable to lower-income households. Therefore, we would find that the neighborhood becomes poor and the poorest people live in the oldest areas. Intrinsically, this declining process has nothing to do with the presence of LULUs. In other words, the presence of LULUs is not a necessary condition for this declining process. Rather, the aging itself leads naturally to deterioration, and aging is inevitable. In fact, this view of neighborhood change as an evolutionary process pays little attention to the role of external or triggering events. Rather, it is the life process itself that matters.

Whether the presence of LULUs can accelerate this aging process is unclear. If there is an effect, we would expect that the host neighborhood declines faster after the siting of LULUs than would be expected without a LULU. Assuming the aging process is gradual and linear, we would expect that the host neighborhood declines faster post-siting than pre-siting if the LULU has some effects on the host neighborhood.

Furthermore, rather than viewing the neighborhood change as a negative process, this theory perceives a net social benefit associated with the cycle. As Birch (1971) demonstrated with an illustrative example of family movement, a family could always gain residential satisfaction by moving to another neighborhood in an upper stage of neighborhood development, even though the city as a whole was declining. Evidently, neighborhood change opens up more housing opportunities for lower-income households, which are not affordable to them otherwise.

2.4.3 PUSH–PULL MODEL

There are two groups of forces — pull and push — that affect the behavior of individual households in their residential mobility preferences and cause neighborhood change (Kolodny 1983). The push forces are an existing or anticipated set of neighborhood conditions that make a neighborhood a less desirable place in which to stay. These may include fear of and/or hostility toward minority and lower-income groups, the aging, obsolescence, and deterioration of the standing housing stock, community facilities and infrastructures, the encroachment of nonresidential land uses, and reduction in employment opportunities.

The pull forces are an existing or anticipated set of attractive opportunities outside the current neighborhood. These may include a preference for residential homogeneity based on race and income, the attractiveness of single-use, lower-density residential areas with more open space and other amenities, and shifts in employment locations.

Been's hypothesis (1994) emphasizes the push effect of a LULU for those people who could afford to move and the pull effect of a LULU for the poor. The forces that a LULU can exert on a host community are twofold: making people who can afford to leave and making it desirable for the poor.

> First, an undesirable land use may cause those who can afford to move to become dissatisfied and leave the neighborhood. Second, by making the neighborhood less desirable, the LULU may decrease the value of the neighborhood's property, making the housing more available to lower income households and less attractive to higher income households. The end result of both influences is likely to be that the neighborhood becomes poorer than it was before the siting of the LULU (Been 1994:1389).

To the extent that income is associated with race, the host community may become more populated by people of color. In addition, to the extent that racial discrimination in the housing market "relegates people of color (especially African-Americans) to the least desirable neighborhoods" (Been 1994:1389), the neighborhood hosting a LULU could also become more populated by people of color.

This theory predicts a long-term inevitable outcome of a disproportionate burden upon the poor and people of color. Been (1994) clearly presents this inevitability.

> As long as the market allows the existing distribution of wealth to allocate goods and services, it would be surprising indeed if, over the long run, LULU did not impose a disproportionate burden upon the poor. And as long as the market discriminates on the basis of race, it would be remarkable if LULUs did not eventually impose a dispro-portionate burden upon people of color (Been 1994:1390).

Just like most other push factors, the LULUs are concrete and easily identifi-able. However, whether the LULUs will have a significant impact on the socio-economic structure of the host neighborhoods depends on how the residents in the neighborhoods as a whole respond to the LULU or how they perceive the risks imposed by the LULUs upon the neighborhoods. The factors influencing the neighborhood's attitudes toward environmental risk and pollution may include the socioeconomic characteristics of the residents, the psychological makeup of the residents, and the nature and magnitude of environmental pollution or risk (Bullard 1987). Furthermore, this perception of LULUs is put into a context of comparison with the amenities and other disadvantages in the neighborhood. Whether a LULU can actually depreciate property values in the host communities is affected by these factors and the neighborhood context. See Chapter 4 for property value studies and empirical evidence about the impacts of a LULU on property values in the host neighborhood.

2.4.4 INSTITUTIONAL THEORY OF NEIGHBORHOOD CHANGE

The institutional theory indicates that institutions such as banks, insurance compa-nies, or universities play a very important role in the neighborhood change (Taub, Taylor, and Dunham 1987).

- As individual actors, these institutions' decisions to stay or relocate have much more profound impacts on the local communities than individual residents
- As major players in the local economic structure, these institutions con-tribute to the local economic foundations, which affect the potential for future spatial agglomeration and the direction in which the surrounding neighborhoods will develop
- As a resource, these institutions determine their own investments or influence directly the investment decisions of other institutions and individuals in the neighborhoods. An example is redlining by financial institutions, which affects availability of capital to investors and indi-vidual buyers
- As a resource mobilizer, these institutions have substantial influences over local governments and community organizations. They lobby city halls to provide and improve services and the infrastructure of the neighbor-hoods. They influence local development policies and other regulations

These institutional actors, along with ecological variables and individual residents, determine the ups and downs of a neighborhood. In a departure from the traditional ecological approach to neighborhood changes, this institutional theory downplays the role of ecological variables. Ecological variables such as the age of the housing stock in the life-cycle model, location, transportation, amenities, and liabilities do not unilaterally determine neighborhood change as predicted by the previous models. Rather, these ecological variables act as a genetic background, which shapes the direction of and reveals the potential for neighborhood changes. What matters for ecological variables are individuals' perceptions of them rather than the actual situations, as is demonstrated in the literature on crime and environmental risk.

> If ecological facts are overwhelming, it is because of the effect of these facts on the perceptions and actions of individual and corporate actors. In a neighborhood that goes up or down, it is ultimately the actions of these residents that make the outcomes real (Taub, Taylor, and Dunham 1987:186).

Institutional decisions significantly affect the perceptions of individual residents. Institutions can contribute to neighborhood decline by withdrawing from the investment market in a neighborhood. Their withdrawal will likely induce further reduction in investment by small investors and individual residents. On the other hand, they can provide necessary investments in the neighborhood so that inhabitants in the neighborhood feel secure about the housing market. This helps alleviate the fear and the sense of hopelessness that individual inhabitants have concerning potential neighborhood decline. Moreover, institutional investments can lower the threshold level at which individual residents begin their own investments in their homes (Taub, Taylor, and Dunham 1987).

> Corporate decisions take into account the ecological facts; individual decisions take into account the ecological facts as well as corporate decisions; community organizations depend on corporate and individual contributions for survival ... (Taub, Taylor, and Dunham: 1987).

This theory implies that the presence of LULUs does not necessarily lead to neighborhood decline. The siting of a LULU in a neighborhood contributes negatively to the ecological foundations of the neighborhood. This, however, does not impose an insurmountable obstacle for a neighborhood to remain stable. As long as corporate actors and residents see the LULU as an opportunity and act to seek remedies of possible externalities produced by the LULU, the fear and concern of the LULU could be reduced. As long as the major institutions show commitment and support to the neighborhood, individual residents will be less likely to feel a sense of being abandoned. The confidence of institutional actors and individual residents in their neighborhood is crucial for stabilizing the neighborhood.

The role that a LULU plays in neighborhood development might be similar to that for crime. Taub, Taylor, and Dunham (1987) have demonstrated that crime is not a deterrent to neighborhood development. Some communities have gentrification and dramatic economic development in spite of high crime rates and the perception of high risk among their residents. Neighborhood safety is only one attribute that one seeks in a neighborhood. This factor may be overshadowed by other attributes that a neighborhood can offer to its residents. Similarly, a LULU may produce some nuisance to its surrounding neighborhoods, but this may be trivial, compared with other attractions of the neighborhood.

In a positive respect, this theory may offer an alternative explanation to the decline of neighborhoods hosting LULUs. Before siting, institutional actors such as banking and insurance companies might, for some legitimate reasons, lose confidence in the housing market in the host neighborhood and decide to withdraw from the market. This might increase the insecurity of individual residents and induce them to leave. Without necessary investments from both individual residents and corporate actors, the neighborhood would decline. This decline would continue after siting of LULUs. The LULU may contribute little to the neighborhood decline.

2.5 SUMMARY

In this chapter, we have discussed several theories of justice and equity: the utilitarian notion of equity, Rawls' theory of contractarian justice, the egalitarian notion of equality, and the libertarian notion of freedom. None of them can adequately serve as a unified framework to deal with environmental justice issues. In practice, we have seen that different groups espouse different perspectives of justice and as a result, injustice always happens to somebody. Within the pluralist world of equity concepts, we examine what substantive theories can offer for a better understanding of static and dynamic relationships between population distribution and environmental risk distribution. This relationship is viewed as an economic relationship from the perspectives of economic and location theories. As economic agents, households maximize their utility, and firms maximize their profits. Their location choices and associated relationship with environmental risk distribution reflect their calculations under various assumptions. These assumptions may not hold true in reality. We always have imperfect information.

Actual and perceived risks often diverge, and this divergence occurs partly because people have different socioeconomic characteristics, cultural backgrounds, and social relationships. This divergence also happens because different risks have different characteristics. How households perceive environmental risks and treat their social relationship affects where they want to move and whether they want to stay. Neighborhoods change over time. Racial relationships, neighborhood life cycle, institutional forces, and ecological factors may contribute to neighborhood changes. The role of a LULU in neighborhood changes must be cast in the larger context.

Collectively, these theories show that social, economic, political, psychological, and environmental factors shape the distributional relationships between subpopulations and environmental risks. They offer us competing hypotheses that are useful for empirical research. None of these theories is all-inclusive and perfect. Each operates under certain assumptions, which may break down in reality.

3 Methodology and Analytical Framework for Environmental Justice and Equity Analysis

As indicated in Chapter 1, the debate concerning research methodology is among the most controversial in the environmental justice debate. This debate touches on some of the fundamental questions regarding scientific inquiry. In this chapter, we first look at two paradigms for inquiry: positivism and the phenomenological perspective. Then, we examine validity and fallacies in scientific research and their manifestations in environmental justice analysis, and we discuss the concept of causality. In Chapter 3.2, we briefly summarize major methodological issues in environmental justice research but leave the detailed discussion for subsequent chapters. Finally, we examine an integrated analytical framework for environmental justice analysis. This framework unifies various perspectives that will be presented in subsequent chapters and provides a bird's eye view of environmental justice analysis.

3.1 INQUIRY AND ENVIRONMENTAL JUSTICE ANALYSIS

3.1.1 POSITIVISM AND PARTICIPATORY RESEARCH

Positivism is a belief in the scientific conception of knowledge. The positivist model or scientific model of knowledge has such underlying principles as "(1) knowledge consists of measurable facts which have an independent reality; (2) these facts may be discovered by a disinterested observer applying explicit, reliable methodology; (3) knowledge also consists of general laws and principles relating variables to one another; (4) such laws can be identified through logical deduction from assumptions and other laws and through testing hypotheses empirically under controlled external conditions; and (5) the true scientist will give up his most cherished ideas in the face of convincing evidence" (de Neufville 1987:87).

Positivism has been the dominant belief underlying scientific research for knowledge creation for more than a century. Recently, the positivist approach has been challenged on several grounds. The emphasis on the universals embodied in the positivist model offers little help for dealing with particular problems in particular places and times. The positivist model stresses the importance of identifying cause–effect relationships, the objectivity and neutrality of researchers, and statistical significance. The knowledge thus produced provides guidance for how people should deal with an average situation. However, it might not be helpful to a particular situation at a particular place. The knowledge produced from the positivist model "offers a poor fit to the world of the decision maker," and "the research function is separate from the world of action and emotion" (de Neufville 1987:87). Moreover, the pursuit of objectivity and value-neutrality has been elusive.

Recognizing these limitations, researchers have called for alternative approaches to knowledge creation such as the phenomenological or interpretive approach and participatory research. The interpretative epistemology emphasizes "the understanding of particular phenomena in their own terms and contexts" and the development of a knowledge creation process that is relevant to practice (de Neufville 1987:88). Rather than relying on quantitative measurement, hypothesis testing, and generalizations, researchers use qualitative and exploratory methods to interpret stories. Rather than being a dispassionate observer, researchers use their subjective capacities, "putting themselves in the place of others to interpret and explain actions" (de Neufville 1987:88). The researchers select concepts, measures, and methods that are shared and agreed upon by the community of study. That is, the community is part of the research team, as well as a user of the knowledge thus generated. Therefore, phenomenological knowledge is more relevant and acceptable to residents in the affected community and to policy makers and planners. It is more realistic and more effectively motivates policy makers and the public.

Similarly, participatory research "involves the affected community in the planning, implementation, evaluation, and dissemination of results" (Institute of Medicine 1999:37). Participatory research methodology is "based upon critical thinking and reflection; it can be the foundation of rigor, giving meaning to truth. Truth is evaluated and confirmed through a practical discourse of planning, acting on the plan, and then observing and reflecting on results" (Bryant 1995:600). Participatory research involves a repetitive cycle of planning, action, observation, and reflection "in a nonlinear process — it is a dialectical progression moving from the specific to the general and back again" (Bryant 1995:600–601).

However, the phenomenological perspective or participatory research has its own limitations and difficulties. It does not provide a way to resolve conflicts in ideology and value (de Neufville 1987). Cultural differences between researchers and minority communities may be a barrier to effective participatory research (Institute of Medicine 1999). The affected communities may distrust and misunderstand the researchers. The time required to undertake participatory research may be longer.

Given the various strengths and weaknesses for the two epistemologies, analysts are better off using a dialectic approach. First, analysts should employ the methods at the scale that is best for them. At the national and other large geographic levels, it is more appropriate and realistic to use the positivist perspective. At the local level, participatory research will shine but needs to be combined with the positivist approach. To mesh these two perspectives, participatory researchers, positivists, and members of the affected community should be "equal partners in setting both research goals and the means by which to obtain those goals" (Bryant 1995:600). These participants should be open and receptive to new ideas and alternatives and recognize the values of contributions from each other. They must be "both teachers and learners in the process of discovery and action" (Bryant 1995:600).

Analysts should also use different perspectives where appropriate. The interpretive approach shows its strength in defining problems, reframing debate, formulating goals and objectives, describing processes, generating alternatives, and negotiating. The positivist perspective contributes most to alternative analysis, outcome prediction, and outcome evaluation.

3.1.2 SCIENTIFIC REASONING

Philosophers of science have identified different types of reasoning in scientific research. Two of the most commonly discussed are induction and deduction. Beveridge (1950:113) described them as follows: "In induction one starts from observed data and develops a generalization which explains the relationships between the objects observed. On the other hand, in deductive reasoning one starts from some general law and applies it to a particular instance." In the traditional model of science (Babbie 1992), scientists begin with already established theories relevant to the phenomenon or problem of interest, then construct some hypotheses about that phenomenon or problem. They develop a procedure for identifying and measuring the variables to be observed, conduct actual observations and measurements, and finally, test the hypotheses. This is a typical deduction reasoning process. In some cases, researchers have no relevant theories to rely on at the beginning. They begin by observing the real world, seeking to discover patterns that best represent the observations, and finally trying to establish universal principles or theories. In a scientific inquiry, deduction and induction are used in an iterative process.

Kuhn (1970) described scientific development as consisting of cumulative and non-cumulative processes. In the steady, cumulative process, scientists determine significant facts, match facts with theory, and articulate theory in the framework of the existing paradigm. By solving puzzles during the process, scientists advance normal science and may discover anomalies that contradict the existing paradigm. This discovery leads to a crisis in the existing paradigm, and ultimately novel theories emerge, leading to the shift to a new paradigm. Kuhn called this non-cumulative process a scientific revolution.

3.1.3 VALIDITY

As noted earlier, research methodology is a focus of the environmental justice debate. Recent studies have challenged the validity of previous research that shaped environmental justice policies. Validity of a measurement is defined as "the extent to which a specific measurement provides data that relate to commonly accepted meanings of a particular concept" (Babbie 1992:135). Similarly, validity of a model or a study can be defined as the extent to which the model or study adequately reflects the real world phenomenon under investigation. It is obvious, by definition, that validity is a subjective term subject to a judgmental call. Although validity can never be proven, we have some criteria to evaluate it. These criteria include face validity, criterion-related validity, content validity, internal validity, and external validity (Babbie 1992). We can first look at how valid the measurements in the study are and then consider how valid the model is, if it possesses any validity at all. Finally, we need to examine the overall internal and external validity of conclusions drawn from the study.

Carmines and Zeller (1979) identified three distinct types of validity: criterion-related validity, construct validity, and content validity. Bowen (1999) distinguished three forms of validity for models in order of successive stringency: face (content) validity, criterion-related validity, and construct validity. Content validity "refers to the degree to which a measure covers the range of meanings included within the concept" (Babbie 1992:133). Construct validity relies on an evaluation of the manner in which a variable relates to other variables based on relevant theories and how such relationships are reflected in a study. Face validity concerns whether "the model relates meaningfully to the situation it represents" (Bowen 1999:125). Or simply, does the model make sense? Evaluation of the face validity of a model relies solely on judgment. Criterion-related validity consists of two types: concurrent validity and predictive validity. The evaluation of concurrent validity is based on "correlating a prediction from the model and a criterion measured at about the same time as the data are collected for the model (Bowen 1999: 125)." A question often asked of an analyst is: How good is the model fit? To answer this question, the analyst relies on the comparison between the predicted values from a model and the corresponding observed values. A typical measure for evaluating the goodness-of-fit of a least-square regression model is the R-squared value. Predictive validity is concerned with how well the estimated model predicts the future. It is usually evaluated with a different data set than the one used for model estimation.

Both internal validity/invalidity and external validity/invalidity draw from experimental design research and can be applied to other research designs as well. Internal validity refers to the extent to which the conclusions from a research protocol accurately reflect what has gone on in the study itself. Does what the researcher measures accurately reflect what he or she is trying to measure? Can the researcher attribute the effects to the cause? "Did in fact the experimental treatments make a difference in this specific experimental instance?" (Campbell and Stanley 1966:5). External validity refers to the extent to which the conclusions from a research project may be generalized to the real world. Are the subjects representative of the real world? "To what populations, settings, treatment variables, and measurement variables can this effect be generalized?" (Campbell and Stanley 1966:5). Is there any interaction between the testing situation and the stimulus? Most psychometric studies of risk are conducted at a university or college. Not surprisingly, the ability to generalize such studies in the real world has been questioned.

Campbell and Stanley (1966) identify eight sources of internal invalidity in experimental design. Extending this list, Babbie (1992) summarizes the following twelve sources: history, maturation, testing, instrumentation, statistical regression, selection biases, experimental mortality, causal time-order, diffusion or imitation of treatments, compensation, compensatory rivalry, and demoralization. Take statistical regression as an example. When we conduct a regression, we actually regress to the average or mean. If we conduct an experiment on the extreme values, we will run the risk of mistakenly attributing the changes to the experimental stimulus. If you are the world's No. 1 tennis player, you can go no higher and will certainly go down in the ranking someday. When we investigate if a locally unwanted land use (LULU) leads to socioeconomic changes in the host neighborhood, we take the LULU as an experimental stimulus and the host neighborhood as a subject. The host neighborhood with a very low proportion of minorities can only stay low or go up in minority composition. Conversely, a predominantly minority neighborhood does not have much room to increase its minority proportion.

The classical experimental design in conjunction with proper selection and assignments of subjects can successfully handle these internal invalidity problems. A basic experiment consists of experimental and control groups. For both groups, subjects are measured prior to the introduction of a stimulus (pretest) and re-measured after a stimulus has been applied to the experimental group (posttest). Comparisons between pretest and posttest and between experimental and control groups tell us whether there is any effect caused by the stimulus. As will be shown in Chapter 12, this type of design can be used for dynamics analysis of environmental justice issues.

Campbell and Stanley (1966) list four types of factors that may jeopardize external validity: the reactive or interactive effect of testing, the interaction effects of selection biases and the experimental variable, and reactive effects of experimental arrangements. For example, the presence of a pretest may affect a respondent's sensitivity to the experimental stimulus and, as a result, make the respondent unrepresentative of the unpretested universe.

In addition to the above concepts of validity, researchers also use the term analytical validity to refer to the question of whether the most appropriate method of analysis has been applied in the study, producing results that are conceptually or statistically sound and truly representing the data (Kitchin and Fotheringham 1997). Ecological validity concerns whether the inferences can be made from the results of a study.

Fallacies often arise when the analyst fails to make appropriate generalizations across circumstances, times, and areas. Ecological fallacy occurs when the analyst infers characteristics of individuals from aggregate data referring to a population. Often, researchers resort to aggregate data to identify patterns in various groups. The danger is that the analyst makes unwarranted assumptions about the cause of those patterns. When we find some patterns about a group, we are often tempted to draw conclusions about individuals in that group based solely on the patterns of that group. For example, let us assume that we have two data sets at the county level: census data about the county-level minority and low-income population and the number of hazardous waste facilities in each county. Our analysis might show that those counties that have low proportions of minority and low-income population tend to have more hazardous waste facilities than those counties that have high proportions of minority and low-income population. We might be tempted to conclude that a high-income and non-minority population is potentially at a higher risk of exposure to hazardous wastes, and that minority and low-income populations do not bear an inequitable burden of hazardous waste facility. This conclusion would be problematic because it may well be that in those rich counties with a majority population of whites, it is the minority and low-income populations who live near hazardous waste facilities and, therefore, are more likely to be at a higher risk of exposure. The cause for this erroneous conclusion is that we take counties as our units of analysis but draw conclusions about population groups in relation to their potential exposure to risks from site-based hazardous waste facilities. Moreover, we make implied assumptions that both population and hazardous waste facilities are uniformly distributed at such a large geographic level. This assumption seldom holds true.

For more refined units of analysis, there is still a danger of creating ecological fallacies. Anderton et al. (1994) warned of such a danger for studies using ZIP code as a unit of analysis. They noted that using too large a geographic unit of analysis may lead to aggregation errors and ecological fallacies; namely, "reaching conclusions from a larger unit of analysis that do not hold true in analyses of smaller, more refined units" (Anderton et al. 1994:232). To avoid such a danger, they suggested using a unit of analysis as small as practical and meaningful and chose the census tract in their studies. Their work triggered the debate on choice of units of analysis in general and census tract vs. ZIP code units in particular. The issue of units of analysis will be discussed in detail in Chapter 6. While the danger of using too large a geographic unit is warranted, some critics argue that there are also dangers in using units that are too small (Mohai 1995). In addition, avoiding ecological fallacies is not an excuse to devise an individualistic fallacy.

Individualistic fallacy arises when the analyst incorrectly draws "conclusions about a larger group based on individual examples that may be exceptions to general patterns found for the larger group" (Mohai 1995:627). For example, one may notice that some of the richest people in the world do not hold a college degree; this fact cannot lead one to believe that there is no value in higher education. On the contrary, college education has one of the highest returns on investment, and higher education correlates positively with higher income. In other words, Bill Gates is an exception rather than the rule.

As indicated in Chapter 1, the toxic wastes and race study commissioned by the United Church of Christ (UCC 1987) was influential in shaping the environmental justice movement. This landmark study was later challenged by a study conducted by a group of researchers at the University of Massachusetts (UMass), which contradicted conclusions of the UCC study (Anderton et al. 1994). This UMass study led to an intensified debate on environmental justice research. One of criticisms of the UMass studies is their justification for choosing comparison or control populations (Mohai 1995). The UMass studies excluded any Standard Metropolitan Statistical Area (SMSA) that had no treatment, storage, and disposal facilities (TSDFs) and non-SMSA rural areas (Anderton et al. 1994; Anderson, Anderton, and Oakes 1994). The authors of these studies justified this exclusion based on data availability and the siting-feasibility argument.

Before the 1990 census, census tracts were defined only for Standard Metropolitan Statistical Areas (SMSAs), which consisted of cities with 50,000 or more people and their surrounding counties or urbanized areas. Therefore, before 1990, tract-level data were unavailable for many rural areas, small cities, and small towns. About 15% of TSDFs were located outside SMSAs (Anderton et al. 1994). The UMass studies assumed that those SMSAs with at least one TSDF constituted market areas for feasible TSDF sitings while those SMSAs without any TSDF were not feasible for siting TSDFs. "This strategy, for example, might tend to exclude national parks, rural areas without any transportation facilities, cities without an industrial economy that would require local TSDF services, etc." (Anderson, Anderton, and Oakes 1994). However, as Mohai argues, the UMass studies provided no data about the suitability or unsuitability of the areas excluded. Instead, they only cited examples such as national parks and Yankton, SD. Drawing conclusions about the excluded areas based solely on these examples, the UMass studies run the risk of creating an individualistic fallacy (Mohai 1995). In Chapter 11, we elaborate on methodological issues involved in these studies and specifically address the question of whether this danger occurs.

Some other fallacies are also likely to arise in environmental justice studies, such as universal (external) fallacies (use of a non-random sample for generalization), selective fallacies (use of selected cases to prove a general point), cross-sectional fallacies (application of one-time findings to another time), and cross-level fallacies. A cross-level fallacy "assumes that a pattern observed for one aggregation of the data will hold for all aggregation" and fails to recognize the so-called modifiable areal unit problem (MAUP), which means that different zonation systems can potentially affect the research findings (Kitchin and Fotheringham 1997:278).

3.1.4 CAUSALITY

Cause-and-effect relationships are a very important, but also hotly debated topic in the environmental justice literature, as well as in other social science literature. While we are interested in discovering and describing what the phenomena are, we also very much want to explain why the phenomena are the way they are. If we know X causes Y, we can change Y by manipulating X. Therefore, knowledge of cause-and-effect relationships is especially critical for successful public policy intervention.

Explanatory scientific research operates on the basis of a cause-and-effect, deterministic model (Babbie 1992). What causes apples to fall down to earth? Answer: gravity. The answer is simple and deterministic. In the social sciences, we are less likely to find such simple answers. Social scientists also strive to find out why people are the way they are and do the things they do. Because of the complexity of social phenomena, in most cases, social scientists use a probabilistic (rather than deterministic) causal model — X likely causes Y. There are many variables like X that may cause Y. Most of the time, social scientists seek to identify relatively few variables that provide the best, rather than a full, explanation of a phenomenon.

Kish (1959) distinguished four types of variables that are capable of changing a second variable Y:

Type 1. Independent variables X that we are most concerned with and attempting to manipulate.

Type 2. Control variables X that are potential causes of Y, but do not vary in the experimental situations because they are under control or do not happen to vary.

Type 3. Unknown or unmeasured variables X that have effects on Y which are unrelated to X.

Type 4. Unknown or unmeasured variables X that have effects on Y that are related to X.

The last two types of variables give rise to measurement errors. Type 4 variables are especially troubling, because they confound the effects of X. If Type 4 variables exist, what appear to be the effects of X could actually be attributed to covariation between X and Type 4 variables. Statistical significance tests can identify the effects of Type 3 variables, as compared with the effects of X. However, they cannot rule out the effects of Type 4 variables. In experimental design, we can use random sampling to reduce the confounding effects of Type 4 variables. Therefore, the use of control variables, randomization, and significant tests is all important to identifing casual relationships between X and Y.

Lazarsfeld (1959) suggested three criteria for causality, as discussed in Babbie (1992:72):

- The cause precedes the effect in time
- The two variables are empirically correlated with one another
- The observed empirical correlation between two variables cannot be explained away as being due to the influence of some third variable that causes both of them

As will be discussed in detail in later chapters, most environmental justice studies are really concerned with the empirical correlation between environmental risks and race and/or income. This correlation alone is far from proving causality. Recently, researchers have begun to explore the temporal dimension and other confounding variables, which will bring about a better understanding of causality.

3.2   METHODOLOGICAL ISSUES IN ENVIRONMENTAL JUSTICE RESEARCH

In recent years, the debate on environmental justice has become broader in scope. It has not simply been restricted to a discussion of the appropriate geographic unit of analysis. It touches on a wide range of issues that also confront researchers in other fields. It is a philosophical debate as well as a technical debate. It reminds us of two fundamental questions that humankind has faced for a long time: How can we best know the world? How can we best link our knowledge with action? The debate also raises questions about the relationship between politics and science.

Not surprisingly, the scientific community recommends more positivist research to build a strong scientific basis for action, while the environmental justice advocates emphasize immediate action and promote action research. As noted in Chapter 1, the clash between the scientific and community perspectives also reflects fundamental differences between two paradigms in epistemology: positivism and participatory research. These differences are reflected in these two camps' recommendations for research agendas. Bullard and Wright (1993:837-838) recommended participatory

research activities in order to "(e)nsure citizen (public participation) in the development, planning, dissemination, and communication of research and epidemiologic
projects" and to "(i)ndividualize our methods of studying communities."

Bryant (1995) called for adoption of participatory research as an alternative
problem-solving method for addressing environmental justice. He challenged positivism, which has dominated scientific inquiry for more than a century. Acknowledging that the union of positivism and capitalism has brought us modern prosperity
and a high quality of life, he also attributed to this alliance poverty, war, environmental destruction, and a close call with nuclear destruction. For environmental
justice analysis, "positivism or traditional research is adversarial and contradictory;
it often leaves laypeople confused about the certainty and solutions regarding exposure to environmental toxins" (Bryant 1998).

On the other hand, Sexton, Olden, and Johnson (1993) stressed the central role
of environmental health research in addressing environmental justice concerns. They
conclude that "(r)esearch to investigate equity-based issues should be structured to
generate, test, and modify hypotheses about causal relationships between class and
race and environmental health risks" (Sexton, Olden, and Johnson 1993:722).

This clash is not irreconcilable. Indeed, as the two sides interact in the public
policy arena, there are signs that increased mutual understanding is occurring.
Recognizing the complementary roles of the two paradigms, the Institute of Medicine
Committee on Environmental Justice (1999) recommended three principles for public health research related to environmental justice: an improved science base,
involvement of the affected population, and communication of the findings to all
stakeholders. Specifically, it recommends developing and using "effective models
of community participation in the design and implementation of research," giving
"high priority to participatory research," and involving "the affected community in
designing the protocol, collecting data, and disseminating the results of research on
environmental justice" (Institute of Medicine 1999:42).

The debate among those using the positivist paradigm focuses on several specific
methodological issues. In addition to the issues discussed above, such as validity,
ecological fallacy vs. individualistic fallacy, comparison (control) populations, and
causality, contested issues also include units of analysis, independent variables, and
statistical analyses. Researchers have challenged the analytical validity of studies
that have influenced environmental justice policies such as the landmark United
Church of Christ (UCC) study. They concluded that "no consistent national level
association exists between the location of commercial hazardous waste TSDFs and
the percentage of either minority or disadvantaged populations" (Anderton et al.
1994:232). This study set off a debate on a series of methodological and epistemo-
logical issues in environmental justice analysis.

As a result of their study, Anderton et al. (1994), as well as authors of other
studies, concluded that a different choice of research methodology may lead to
dramatically different research results. Greenberg (1993) pointed out five issues in
testing environmental inequity: (1) choice of study populations (e.g., minorities, the
poor, young and old, or future generations), (2) choice of LULUs to be studied (e.g.,
types, magnitude, and age of LULUs), (3) the burden to be studied (e.g., health
effects, property devaluation, and social and political stresses), (4) choice of study

and comparison areas (i.e., burden and benefit areas), and (5) choice of statistical methods (parametric vs. nonparametric methods). Through a case study of population characteristics of the WTEF (Waste-to-Energy Facilities) host communities at the town and ZIP code levels, he demonstrated that the results depended on (1) combinations of WTEF capacity and the population size of the host communities, (2) choice of comparison areas (i.e., service areas or the U.S.), (3) choice of evaluation statistics (i.e., proportions, arithmetic means, or population-weighted values), (4) choice of study populations (i.e., minorities, young and old), and (5) choice of analysis units (i.e., towns vs. ZIP codes). The importance of analysis units was also demonstrated in the site-based equity studies of Superfund sites (Zimmerman 1994), air toxic releases (Glickman, Golding, and Hersh 1995), and hazardous waste facilities (Anderton et al. 1994).

Critics charged that some of these challenges have corporate interests behind them (Goldman 1996). Furthermore, critics of the UMass research and other studies believed that the conflicting results could be attributed to the employment of faulty research methodology. They provided counter arguments to the UMass research on several methodological grounds, including control population, unit of analysis, and variables (Mohai 1995). In subsequent chapters, we will discuss these issues in detail.

Different conceptions of environmental justice affect the research methodologies and interpretation of results. Some studies focus on urban areas or census-defined metropolitan areas. These areas tend to have heavy concentrations of the poor and large minority populations. Critics argue that this narrow focus fails to compare the population at large and fails to address the fundamental questions of why poor and minority populations are concentrated in urban or metropolitan areas. They argue that the heavy concentration of minorities, the poor, and industries in urban and metropolitan areas is itself evidence of inequity. They advocate use of a broader comparison or control population and argue that impacted communities should be compared with every other type of community.

The diverse perspectives on justice have also shaped the debate on the research methodology of environmental justice and equity analysis. Not surprisingly, environmental justice advocates have strong objections to the use of cost-benefit analysis and risk assessment. Both methods serve the utilitarian notion of justice. Any strengths and limitations of utilitarianism apply to them. There is a substantial body of literature on critiques of cost-benefit analysis. In theory, cost-benefit analysis and risk assessment fail to deal with rights and duties, ignore problems of inequality and injustice, shut out disadvantaged groups from the political process, and reinforce the administrative state. The rejection of cost-benefit analysis by the environmental justice movement "reflects an intuitive or experiential understanding of how it is that seemingly fair market exchange always leads to the least privileged falling under the disciplinary sway of the more privileged and that *costs* are always visited on those who have to bow to money discipline while benefits always go to those who enjoy the personal authority conferred by wealth" (Harvey 1996:388).

One response to such a critique of distributive justice is to subject cost-benefit analysis to moral rules and equity consideration (Vaccaro 1981; Braybrooke and Schotch 1981). In evaluating a project or policy, equity or moral issues are analyzed first and, if they are met, then move on to cost-benefit analysis.

The debate once again testifies that the objectivity and value-neutrality of scientific research are elusive. Behind the conflicting evidence are investigators who have various ideologies and motivations. Some investigators are scientific researchers with a commitment to knowledge creation through positivism, while others are social activists and policy entrepreneurs, who push the policy agenda for social change and for empowering the disadvantaged with their own knowledge. Some investigators are scientific realists, who acknowledge that science is and will be full of uncertainty and that policy responses should proceed with the help of and despite limited knowledge. Others are scientific idealists, who stick to the rigor and integrity of scientific research and reject any policy response before substantial evidence has been established through rigorous research that meets standards of scientific validity. Some observers characterize society as essentially harboring injustice and laden with struggles between different classes, thereby concluding that the fundamental production–consumption system must be completely changed to achieve equality. Others believe in the capitalistic system and insist that any injustice can be fixed through fine-tuning the existing system.

What will an environmental justice analyst do in such contested situations? In the following, we outline an integrated approach to these diverse perspectives and present a holistic framework for environmental justice analysis.

3.3 INTEGRATED ANALYTICAL FRAMEWORK

An integrated analytical framework takes into account various debated issues in a balanced way. This framework takes advantage of the strengths of both the participatory research and the positivist perspectives (Figure 3.1).

The interpretive/participatory research approach should be used to identify environmental justice problems and research issues and to formulate goals and objectives for research and public policies. Based on the findings of these types of studies, macro-analysis identifies specific issues and spatial patterns at national or other large geographic levels. Macro-analysis relies heavily on the positivist research approach, but it starts with issues, goals, and objectives defined through participatory research. An example of macro-analysis is shown in Figure 3.2. The details of this framework will be discussed later. Of course, the specifics of a research framework can be modified after consultation with various stakeholders. These studies are generally quantitative, large scale, explanatory, or confirmatory. Such studies help policy makers at the national or state level to formulate policies and strategies that deal with environmental justice issues on large geographic scales. Failure to find inequities from these studies does not mean that inequities do not exist at the local level. In fact, macro-analysis can provide guidance on specific issues and areas that require detailed analysis.

For cases at the micro-level, participatory research should play a pivotal role. For micro-analysis, the analyst shall take on a qualitative and historical research approach, in combination with a quantitative approach. The analyst should take advantage of local knowledge in the affected community. This type of research helps uncover the intimate details of the relationships among race, income, social and economic structure and their transformation, distribution of environmental hazards,

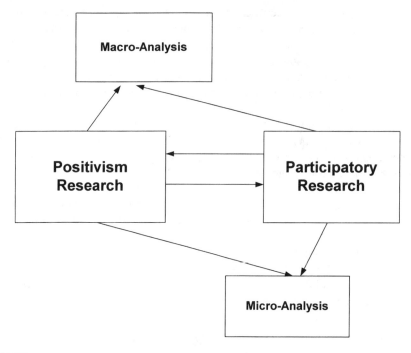

FIGURE 3.1 Integrated analytical framework for environmental justice analysis.

the roles of public policies and planning, community development and evolution, and the historical development of social relationships. This type of research has implications for large-scale policies, but its primary purpose should be to serve the locale in terms of helping to formulate solutions to environmental justice issues.

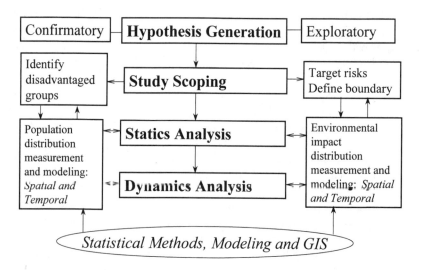

FIGURE 3.2 Macro-analysis framework.

With a couple of exceptions such as the work by Pulido, Sidawi, and Vos (1996), the current literature lacks rigorous studies that take into account these factors.

The macro-analysis framework shown in Figure 3.2 has four major interrelated parts: hypotheses generation, study scoping, statics analysis, and dynamics analysis. A purpose of an environmental justice analysis is either to test hypotheses that are deduced from established theories or to generate some hypotheses based on actual data through exploratory data analysis. Some studies may contain a mixture of both. Research methods depend on the purpose of environmental justice and equity analysis.

Analysts at both the macro- and micro-levels should confront major methodological issues up front. First, several questions need researchers' attention. What risks are the public concerned about in the study area? What risks are the experts concerned about? What risks are the policy makers concerned about? These questions help researchers target major potential environmental risks. The risks to be addressed also influence the choice of units of analysis and the way in which an impact boundary is delineated. If researchers try to respond primarily to the public's concerns, then a public perception-based approach to boundary delineation and choice of unit of analysis would be appropriate. A question we may ask is, To what extent is the risk perception of the general public consistent with the perception of the subpopulations that are socially and economically disadvantaged? If researchers adopt concerns of experts, then an exposure-based approach would be a natural choice. However, it is difficult, if not impossible, to choose a single unit of analysis that satisfies many sources in an urban area, many risky chemicals, and various exposure pathways. If data, resources, and current knowledge do not permit implementation of an exposure-based approach, a proximity-based approach can be a substitute. This inevitably introduces additional uncertainties. In all these cases, there are some uncertainties for the impact boundary and unit of analysis. It is essential to conduct sensitivity analyses for alternative impact boundaries and units of analysis. As will be discussed in Chapter 8, GIS technology can provide much help in this area. In Chapter 6, we address the issues of choosing an appropriate geographic unit of analysis.

As will be demonstrated in Chapters 4 and 5, population distribution measurement/modeling and environmental risk measurement/modeling are two distinct and independent areas of research. As shown in Chapter 9, urban models have been a subject of academic research and practical application for at least three decades. Work has also been done in the area of environmental risk modeling, although there is much to be desired, particularly in dose-response relationships. What has not been done is to integrate the two areas for analyzing the relationship between environmental risk distribution and population distribution. As discussed in Chapter 4, environmental risk assessment, as currently practiced, emphasizes the aggregate effect of risk and fails to address the spatial or distributional effects. Environmental concerns have not been addressed in urban models until recently. Similarly, equity issues have not been adequately addressed in urban models.

Despite the many caveats and limitations, the two independent paradigms — risk assessment and location analysis — can be integrated into a framework that provides more rigorous equity analysis. They can be connected through several links. By incorporating population distribution modeling into a risk assessment, we will be able to improve measurement/modeling of environmental risks particularly from area and line sources. By incorporating a spatial dimension into the risk assessment paradigm, we will be able to evaluate the distribution of environmental risks. By incorporating environmental risk distribution into location models, we will be able to test environmental equity hypotheses by taking into account interactions among key location factors, such as employment and residential location, land uses, transportation, and environmental factors. In Chapter 10, we will present the first known attempt to incorporate an air quality index into a location model, which is used to test equity hypotheses. Although criticized by environmental justice advocates, the methods of risk assessment can be improved to serve the purpose of conducting a more rigorous environmental justice analysis.

We must be aware of the existing constraints to a successful integration of the two paradigms. The constraints come from the limitations in both paradigms. As will be discussed in Chapters 4 and 5, there is much to be desired in both paradigms. Currently no study has gone far enough in measuring/modeling environmental risks for a whole urban area. There is also a lack of empirical studies that investigate the role of environmental risks in urban models. Several authors have pointed out the constraints facing a risk-based approach to environmental equity analysis (Sexton et al. 1992; Perlin et al. 1995). These constraints include databases, resources available for conducting an analysis, and incomplete knowledge.

While testing environmental equity is not easy, examining the causation or dynamics is even more difficult. It requires historical data, which might be difficult, even impossible to obtain. However, the causation and dynamics issue is crucial for making sound environmental policies for remedying any environmental inequity. Only when we know why there is a problem can we decide how to effectively deal with it. For a site-based analysis, for example, we want to know whether LULU-induced market dynamics cause or contribute to the current inequitable distribution of environmental risks. For this purpose, we must know how the host communities change their socioeconomic characteristics over time and how these changes are related to the LULUs. To examine this issue, we suggest a two-step process. First, we should test whether there is any significant difference between the pre-siting changes and the post-siting changes in terms of the socioeconomic characteristics of the host communities, and whether there is any significant difference between the LULU and control communities in terms of the changes in the socioeconomic characteristics of the communities. These tests will serve to answer whether there is a "treatment" effect associated with the LULUs. To pinpoint the real reasons behind the dynamics of the host communities, we need to examine many competing hypotheses. As discussed in Chapter 2, theories of neighborhood changes provide us with some hypotheses, which should be considered in a causation/dynamics analysis. In Chapter 11, an integrated framework for dynamics analysis and a case study using this framework are presented.

It is necessary to emphasize the importance of two research tools in an environmental equity analysis: statistical methods/models and geographic information systems (GIS). Statistical methods/models are an integral component of every part of the four-part framework. The validity of an environmental equity analysis depends heavily upon the choice of statistical methods/models and the correct interpretation of the results. We should pay particular attention to the underlying assumptions of different statistical methods/models and examine the degree to which the research issue meets these assumptions. The nature of the space dimension complicates the use of ordinary statistical methods/models. The space dimension is an inherent feature of the relationship between population distribution and environmental risk distribution. Ironically, previous environmental equity studies mostly ignored space interaction and the spatial relationship. GIS provides a powerful tool to deal with spatial data. It is particularly useful for delineating the impact boundary, choosing appropriate units of analysis, preparing data for population and risk modeling, testing hypotheses, and presenting the modeling results.

In summary, environmental justice analysis should consider a wide range of variables. The analytical framework presented above simplifies the analytical process and streamlines the relationships among various variables. Table 3.1 lists factors and variables to consider for a comprehensive environmental justice analysis.

TABLE 3.1
Factors to Consider in Environmental Justice Analysis

Category	Factors	Variables
Potential exposure and risk	Demographic	Race, ethnicity, income, age, gender, disability
		Susceptible and highly exposed populations
		Population density
		Population literacy
		Population/economic growth
	Geographic	Climate
		Land use/land cover
		Topographic and geomorphic features
		Hydrologic features
	Economic	*Individual Economic Conditions*
		Income level/health care access
		Infrastructure conditions such as water and sewage
		Life-support resources such as subsistence living situations
		Distribution of costs to pay for environmental projects by user fees for necessary goods and services
		Community Economic Base
		Industrial
		Brownfields
		Natural resources
	Human health and risk	Proximity to environmentally risky facilities
		Public perception of risks
		Toxics, pollutants, and pesticides
		Emission sources, amount and distribution
		Ambient concentrations and their distribution
		Exposures: locations, multiple, cumulative, synergistic
		Health status and effects
		Research gaps (e.g., subsistence consumption, dietary effects)
		Data collection/analysis reliability and validity

continued

TABLE 3.1 (CONTINUED)
Factors to Consider in Environmental Justice Analysis

Category	Factors	Variables
Cultural and ethnic differences and communications concerns		Public access to the decision-making process
		Cultural expectations and understanding of the decision-making process
		Meaningful information about risk assessment and management
		Job security
		Literacy rate for consideration in choosing the right communication materials
		Translations for non-English speaking audience
		Community representation
		Community identification
		Indigenous populations
Historical and policy issues		Industrial concentration
		Inconsistent standards in enforcement and site selection
		Research gaps
		Program gaps
		Non-inclusive processes
		Past practices
		Cultural diversity
		Obligations

Source: U.S. EPA, *Final Guidance for Incorporating Environmental Justice Concerns in EPA's NEPA Compliance Analyses,* Washington, D.C. 1998b. Available at http://es.epa.gov/oeca/ofa/ejepa.html, October 15, 1998.

4 Measuring Environmental and Human Impacts

Executive Order 12898 orders each Federal agency to identify and address, as appropriate, disproportionately high and adverse human health or environmental effects of its programs, policies, and activities on minority populations and low-income populations. What are human health or environmental effects?

The concept of environmental impacts has been broadened considerably over the past century. The initial focus is human health. From time immemorial, people recognized that certain plants are toxic to human health. There are also natural hazards that are detrimental to human health and well-being. The modern industrial revolution not only led to prosperity and enhanced human capability to fight hazards but also generated a harmful by-product, environmental pollution. People realized that pollution could be deadly from the tragic episodes of air pollution in Donora, Pennsylvania in 1949 and in London, England in 1952. Carson's *Silent Spring* raised the public's awareness of environmental and ecological disasters caused by modern industrial and other human activities. Now, we know that environmental impacts can occur with respect to both the physical and psychological health of human beings, public welfare such as property and other economic damage, and ecological health of natural systems.

In this chapter, we will examine how environmental impacts are measured, modeled, and assessed, and explore the possibility and difficulties of using a risk-based approach in environmental equity studies. First, we will review major types of environmental impacts, which include human health, psychological health, property and economic damage, and ecological health. Then we discuss approaches to measure, model, and simulate these impacts. We will discuss the strengths and weaknesses of these methods and their implications for equity analysis. Finally, we examine the critiques and responses of a risk-based approach to environmental justice analysis.

4.1 ENVIRONMENTAL AND HUMAN IMPACTS: CONCEPTS AND PROCESSES

Environmental impacts occur through interaction between environmental hazards and human and ecological systems. Environmental hazard is "a chemical, biological, physical or radiological agent, situation or source that has the potential for deleterious effects to the environment and/or human health" (Council on Environmental Quality 1997:30).

An environmental impact process is often characterized as a chain, including

- Sources and generation of environmental hazards
- Movement of environmental hazards in environmental media

- Environmental exposure
- Dose
- Effects on human health and/or the environment

Environmental hazards come from both natural systems and human activities. For example, toxics come from stationary sources such as fuel combustion and industrial processes, mobile sources such as car and trucks, and natural systems. Emission level is only one factor for determining eventual environmental impacts. Other factors include the location of emission, time and temporal patterns of emission, the type of environmental media into which pollutants are discharged, and environmental conditions.

After being emitted into the environment, pollutants move in the environment and undergo various forms of transformation and changes. The fate and transport of pollutants are affected by both the natural processes such as atmospheric dispersion and diffusion and the nature and characteristics of pollutants. Some pollutants or stressors decay rapidly, while others are persistent and long-lived. Some environmental conditions are amenable to formation of pollution episodes, such as inversion layers in the Los Angeles Valley and high temperatures in the summer, which facilitate formation of smogs. When undergoing these fate and transport processes, pollutants reach ambient concentrations in environmental media, which may or may not be harmful to humans or the ecosystem. Research has investigated the level of ambient concentrations that impose adverse impacts on the environment and/or human health. These studies provide a scientific basis for governments to establish ambient standards for protecting humans and the environment.

Ambient environmental concentrations of pollutants, no matter how high, will not impose any adverse impacts until they have contact with humans or other species in the ecosystem. Whether or where such contact with humans occurs depends on the location of human activities; it could happen indoors or outdoors. Indoor concentrations could differ dramatically from outdoor concentrations.

Environmental exposure is a "contact with a chemical (e.g., asbestos, radon), biological (e.g., Legionella), physical (e.g., noise), or radiological agent" (Council on Environmental Quality 1997:30). The Committee on Advances in Assessing Human Exposure to Airborne Pollutants of the National Research Council (1991:41) defines exposure as

> contact at a boundary between a human and the environment at a specific contaminant concentration for a specific interval of time; it is measured in units of concentration(s) multiplied by time (or time interval).

In the real world, exposure happens daily and there are generally more than one agent and source. This is called multiple environmental exposure, which "means exposure to any combination of two or more chemical, biological, physical or radiological agents (or two or more agents from two or more of these categories) from single or multiple sources that have the potential for deleterious effects to the environment and/or human health" (Council on Environmental Quality 1997:30). Furthermore, environmental exposure occurs through various environmental media and accumulates over time. Cumulative environmental exposure "means exposure to one or more chemical, biological, physical, or radiological agents across environmental media (e.g., air, water, soil) from single or multiple sources, over time in one or more locations, that have the potential for deleterious effects to the environment and/or human health" (Council on Environmental Quality 1997:30).

Human exposure to environmental hazards can come from many contaminants (for example, heavy metals, volatile organic compounds, etc.) generated from many sources (such as industrial processes, mobile sources, and natural systems), from various environmental media (air, water, soil, and biota), and from many pathways (inhalation, ingestion, and dermal absorption).

As a result of exposure to pollutants, humans receive a certain level of dose for those pollutants. "Dose is the amount of a contaminant that is absorbed or deposited in the body of an exposed organism for an increment of time" (National Research Council 1991:20). Dose can be detected from analysis of biological samples such as urine or blood samples.

Human response may or may not occur with respect to a certain dose level. Different toxics have different dose-response relationships. The response to an exposure includes one of the following (Louvar and Louvar 1998):

- No observable effect, which corresponds to a dose called no observable effect level (NOEL)
- No observed adverse effect at a dose called NOAEL
- Temporary and reversible effects at effective dose (ED), for example, eye irritation
- Permanent injuries at toxic dose (TD)
- Chronic functional impairment
- Death at lethal dose

Human health effects are often classified as cancer and non-cancer, with corresponding agents called carcinogens and non-carcinogens. Cancer endpoints include lung, colon, breast, pancreas, prostate, stomach, leukemia, and others. Non-cancer effects can be cardiovascular (e.g., increased rate of heart attacks), developmental (e.g., low birth weight), hematopoietic (e.g., decreased heme production), immunological (e.g., increased infections), kidney (e.g., dysfunction), liver (e.g., hepatitis A), mutagenic (e.g., hereditary disorders), neurotoxic/behavioral (e.g., retardation), reproductive (e.g., increased spontaneous abortions), respiratory (e.g., bronchitis), and others (U.S. EPA 1987).

Based on the weight of evidence, the EPA's Guidelines for Carcinogenic Risk Assessment (U.S. EPA 1986) classified chemicals as Group A (known), B (probable), and C (possible) human carcinogens, Group D (not classified), and Group E (no evidence of carcinogenicity for humans). Known carcinogens have been demonstrated to cause cancer in humans; for example, benzene has been shown to cause leukemia in workers exposed over several years to certain amounts in their workplace air. Arsenic has been associated with lung cancer in workers at metal smelters. Probable and possible human carcinogens include chemicals for which laboratory animal testing indicates carcinogenic effects but little evidence exists that they cause cancer in people. The Proposed Guidelines for Carcinogenic Risk Assessment (U.S. EPA 1996a) simplified this classification into three categories: "known/likely," "cannot be determined," and "not likely." Subdescriptors are used to further differentiate an agent's carcinogenic potential. The narrative explains the nature of contributing information (animal, human, other), route of exposure (inhalation, oral digestion, dermal absorption), relative overall weight of evidence, and mode of action underlying a recommended approach to dose response assessment. Weighing evidence of hazard emphasizes analysis of all biological information, including both tumor and non-tumor findings.

Estimates of mortality and morbidity as a result of environmental exposure vary with studies. An early epidemiological study attributed about 2% of total cancer mortality in the U.S. to environmental pollution, 3% to geophysical factors such as natural radiation, 4% to occupational exposure, and less than 1% to consumer products (Doll and Peto 1981). Half of total pollution-associated cancer mortality was attributed to air pollution (4,000 deaths annually in 1981). U.S. EPA (1987) used risk assessment to estimate cancer incidences caused by most of 31 environmental problems. Transformation of cancer incidence into cancer mortality, using a 5-year cancer survival rate of 48% and an annual death toll of 485,000 from cancer, shows that EPA's estimates are similar to Doll and Peto's estimates (Gough 1989). EPA's estimates translate to 1–3% of total cancer deaths that can be attributed to pollution and 3–6% to geographical factors. Recent studies show that occupational and environmental exposures account for 60,000 deaths per year (McGinnis and Foege 1993) and particulate air pollution alone could account for up to 60,000 deaths per year (Shprentz et al. 1996).

The environment and ecosystem may respond differently to various chemical, physical, biological, or radiological agents or stressors. Some agents or stressors may pose risks to both humans and the environment, while others affect just one of them. For example, radon is a serious risk for human health but does not pose any ecological risk. Conversely, filling wetland may degrade terrestrial and aquatic habitats but does not have direct human health effects. Two commonly cited ecological effects are extinction of a species and destruction of a species' habitat. Although impacts on humans often focus on the chemical agents or stressors, both physical and chemical stressors often have significantly adverse impacts on the ecosystem. For example, highway construction may cause habitat fragmentation and migration path blockage. Ecological impacts can be assessed according to criteria such as areas, severity, and reversibility of impact (U.S. EPA 1993a).

In addition to health, impacts of environmental hazards on humans also include those on social and economic (sometimes referred to as quality of life) issues. Examples are impacts on aesthetics, sense of community, psychology, and economic well-being. Economic damages have been widely documented and typically include damages to materials, commercial harvest losses (such as agricultural, forest, and fishing and shellfishing), health care costs, recreational resources losses, aesthetic and visibility damages, property value losses, and remediation costs (U.S. EPA 1993a).

Economic impacts, particularly those to property value, have been a major concern as a result of environmental pollution, risks, environmentally risky or noxious facilities. Property value studies widely document property value damages associated with air pollution or economic benefits associated with improving air quality. A meta-analysis of 167 hedonic property value models estimated in 37 studies conducted between 1967 and 1988 generated 86 estimates for the marginal willingness to pay (MWTP) for reducing total suspended particulates (TSP) (Smith and Huang 1995). The interquartile range for estimated MWTP values is between 0 and $98.52 (in 1982 to 1984 dollars) for a 1-unit reduction in TSP (in micrograms per cubic meter). The mean reported MWTP from these studies is $109.90, and the median is $22.40. Local market conditions and estimation methodology account for the wide variations. Studies also report negative impacts of noxious facilities on nearby property values, as will be discussed in detail later in the chapter. Social impacts have received increasing attention. Research has shown some psychological impacts associated with exposure to environmental hazards such as coping behaviors.

Different environmental problems have adverse impacts on humans and the environment on different spatial scales. Some environmental hazards have adverse impacts in microenvironments such as homes, offices, cars, or transit vehicles. Examples include radon, lead paint, and indoor air pollution. Other environmental problems have global impacts such as global warming and stratospheric ozone depletion. Table 4.1 shows some examples of environmental problems and their spatial scales of impacts. It should be noted that some environmental problems can occur at different spatial scales.

4.2 MODELING AND SIMULATING ENVIRONMENTAL RISKS

Environmental risks were often addressed on the basis of human health effects imposed by a single chemical, a single plant, or a single industry in a single environmental medium. Assessing the spatial distribution of environmental risks is

---

TABLE 4.1
Spatial Scales for Various Environmental Problems

Spatial Scale	Home	Community	Metropolitan Area	Region	Continent/ Global
Examples of environmental hazards	Indoor air pollution Radon Lead paint Domestic consumer products	Noise Trash dumping Some locally unwanted land uses Hazardous and toxic waste sites	Traffic congestion Ambient air pollution such as nitrogen oxides, VOCs, Ground-level ozone	Tropospheric ozone Water pollution Watershed degradation Loss of wetlands, aquatic, and terrestrial habitats	Acid rain Global warming Stratospheric ozone depletion

Source: U.S. EPA (1993a).

a rare event. There is in particular a lack of research on the spatial distribution of various environmental risks at the urban or regional level. This gap is partly due to the complexity of urban risk sources and the limitations of ambient monitoring and risk modeling. The few studies that touched on the spatial distribution of environmental risks arose from the early concern for managing total risks to all media in a cost-effective way (Haemisegger, Jones, and Reinhardt 1985). EPA's Integrated Environmental Management Division (IEMD) studies attempted to define the range of exposures to toxic substances across media (i.e., air, surface water, and ground water) in a community, to assess the relative significance, and to develop cost-effective control strategies for risk reduction. These studies did not explicitly explore the spatial distribution of environmental risks in the city, but its results had some spatial dimensions. EPA's Region V conducted a comprehensive study of cancer risks due to exposure to urban air pollutants from point and area sources in the southeast Chicago area (Summerhays 1991). This study explicitly pursued the spatial distribution of environmental risks in the study area.

More recently, EPA initiated various projects studying cumulative impacts. EPA's Cumulative Exposure Project was designed to assess a national distribution of cumulative exposures to environmental toxics and provide comparisons of exposures across communities, exposure pathways, and demographic groups (U.S. EPA 1996b). The first phase of the project studied three separate pathways: inhalation, food ingestion, and drinking water independently, while the second phase was designed to evaluate exposures to indoor sources of air pollution and to develop estimates of multi-pathway cumulative exposure.

Assessing environmental risks generally follows the NRC/NAS paradigm on risk assessment. The National Research Council (NRC) under the National Academy of Sciences (NAS) developed a definition of risk assessment (1983) that is most widely cited. It defines risk assessment to mean "the characterization of the potential adverse health effects of human exposures to environmental hazards. Risk assessments include several elements: description of the potential adverse health effects based on an evaluation of results of epidemiological, clinical, toxicological, and environmental research; extrapolation from those results to predict the type and estimate the extent of health effects in humans under given conditions of exposure; judgments as to the number and characteristics of persons exposed at various intensities and durations; and summary judgments on the existence and overall magnitude of the public-health problem. Risk assessment also includes characterization of the uncertainties inherent in the process of inferring risk" (National Research Council 1983:18).

Risk assessment has four steps: hazard identification, dose-response assessment, exposure assessment, and risk characterization. Models have been used mainly in the two intermediate steps of the risk assessment process: exposure assessment and dose-response assessment. In the following, we review the status of modeling and applications in these two processes.

4.2.1 Modeling Exposure

Exposure assessment describes the magnitude, duration, schedule, and route of exposure, the size, nature, and classes of the human populations exposed, and the uncertainties in all estimates (National Research Council 1983). Human exposure to environmental contaminants can be assessed in different ways (National Research Council 1991):

- Direct Measure Methods: personal monitoring, biological markers
- Indirect Measure Methods: environmental monitoring, models, question- naires, and diaries

Some of these methods can be combined in actual applications. For example, in the IEMD study (Haemisegger, Jones, and Reinhardt 1985), environmental monitor- ing was used to measure the concentrations of pollutants at the influent and effluent points of the sewage treatment and drinking water treatment plants, and to measure ambient air concentrations of pollutants across the city and in the industrial areas. A dispersion model was later used for comparison with the actual monitoring data.

Models for assessing environmental risks have been developed in literature and computer packages and widely used in practice. Modeling human exposure to environmental contaminants generally involves estimation of pollutants' emissions, pollutant concentration in various environmental media, and time-activity patterns of humans. They are discussed in detail in the following.

4.2.1.1   Emission Models

Emission estimation is the first step in the risk quantification process. Although emission of toxics can be measured directly from the emission points, emission models provide an inexpensive alternative. Furthermore, it is extremely difficult, if not impossible, to monitor millions of small area sources. There are generally three types of models to estimate the emissions from point, area, volume, or line sources: species fraction model, emission factor models, and material and energy balance models.

In the species fraction model, the species emissions are estimated via multiplying the estimated total organic emissions or total particulate matter emissions for each emission point by the species fraction appropriate for that type of emission point. EPA has issued compilations of compositions of organic and particulate matter emissions (U.S. EPA 1992b).

Essential to emission factor models are, of course, emission factors. As defined here, the emission factor is the statistical average of the mass of pollutants emitted from each source per unit activity. For point sources, unit activity can be unit quantity of material handled, processed, or burned. For area sources, unit activity can be one employee for a sector of industry, or a resident for a residential unit. For mobile sources, unit activity may be unit length of road.

The basic assumption of the emission factor models is that the emission factor is constant over the specified range of a target (if any). Therefore, they are also referred to as the "constant emission rate" approach. Of course, an emission factor can be a function of various variables. For mobile sources, an emission factor is a constant rate of emission over the length of a road, calculated mainly as a function of traffic flow and speed. In addition, other variables include year of analysis, percentage of cold starts, ambient temperature, vehicle mix, and inspection and maintenance of vehicle engines. This is how EPA's MOBILE series models compute the emission rates for mobile sources (U.S. EPA 1994c), and it is the most common approach in practice. Certainly, emission factors can be further segmented.

EPA has published extensive emission factor data and models for quantification of emissions from various sources, such as EPA's Compilation of Air Pollutant Emission Factors and Mobile Source Emission Factors. Emission factors have also been developed by some industrial organizations, such as the Chemical Manufacturers' Association and the American Petroleum Institute. Most of these emission factors are related to fugitive emissions, and emissions from nonpoint sources, such as pits, ponds, and lagoons, are more difficult to obtain.

The strengths of the emission factor models include the following, among others:

- The methodology is very straightforward and easy to use
- There are a lot of empirical data available for application
- For mobile sources, it is particularly good for uninterrupted flow conditions, and for transportation planning in a large network

Their main weaknesses include, among others:

- An emission factor may change over time, which is hard to predict in the long run
- An emission factor developed for a specific activity in one area may introduce some biases if used in another area without validation
- For mobile sources, it is inadequate for interrupted flow conditions, such as those caused by traffic signalization

The material and energy balance models are based on engineering design procedures and parameters, the properties of the chemicals, and knowledge of reaction kinetics if necessary (National Research Council 1991).

The species fraction and emission factor methods were used to estimate the emissions of 30 quantifiable carcinogenic air pollutants in the Chicago study (Summerhays 1991). The sources include area sources and non-conventional sources such as wastewater treatment plants, hazardous waste treatment, storage and disposal facilities (TSDFs), and landfills for municipal wastes, as well as traditional industrial point sources. For industrial point sources, emission estimates were generally based on questionnaires or derived using the species fraction method. For the area sources, both the species fraction method and the emission factor method were used. Emissions of each area source category were distributed to the receptor regions "according to the distribution of a relevant 'surrogate parameter' such as population, housing, roadway traffic volumes, or manufacturing employment" (Summerhays 1991:845). In the IEMD study, the species fraction method was used to estimate various organic compounds from total volatile organic compound emissions for dry cleaners, degreasers, and other industrial sources. Measured data and pollution inventory provided by facilities and local environmental agencies were used to estimate emission from other area sources. The air toxics component of the EPA's Cumulative

Exposure Project obtains hazardous air pollutants (HAPs) through EPA's Toxics Release Inventory (TRI) and EPA's VOCs and PM emission inventories (Rosenbaum, Axelrad, and Cohen 1999). TRI provides self-reported emissions for large manufacturing sources (see Chapter 11). For non-TRI sources such as small point sources, mobile sources, and area sources, the speciation method was used to derive HAP emission estimates from VOC and PM emission inventories. For area and mobile sources, the county level emissions were allocated to census tracts using a variety of surrogates for different emission source categories such as population, roadway and railway miles, and land use.

4.2.1.2   Dispersion Models

There are four fundamental approaches to dispersion modeling: Eulerian, Lagrangian, statistical, and physical simulation. The *Lagrangian* approach uses a probabilistic description of the behavior of *representative* pollutant *particles* in the atmosphere to derive expressions of pollutant concentrations (Seinfeld 1975, 1986). This approach is the foundation of the Gaussian models, currently the most popular models for modeling the dispersion processes of inert pollutants. The *Eulerian* approach, by contrast, attempts to formulate the concentration statistics in terms of the statistical properties of the Eulerian fluid velocities, i.e., the velocities measured at fixed points in the fluid. The Eulerian formulation is very useful to reactive pollution processes. The *statistical* approach tries to establish the relationships between pollutant emissions and ambient concentrations from the empirical observations of changes in concentrations that occur when emissions and meteorological conditions change. The models are generally limited in their applications to the area studied. The *physical* simulation approach is intended to simulate the atmospheric pollution processes by means of a small-scale representation of the actual air pollution situation. This approach is very useful for isolating certain elements of atmospheric behavior and invaluable for studying certain critical details. However, any physical model, however refined, cannot replicate the great variety of meteorological and source emission conditions over an urban area.

EPA categorizes air quality models into four classes: Gaussian, numerical, statistical or empirical, and physical (U.S. EPA 1993b). Within each of these classes, there are a lot of "computational algorithms," which are often referred to as models. When adequate data or scientific understanding of pollution processes do not exist, statistical or empirical models are the frequent choice. Although less commonly used and much more expensive than the other three classes of models, physical modeling is very useful, and sometimes the only way, to classify complex fluid situations. Gaussian models are most widely used for estimating the impact of nonreactive pollutants, while numerical models are often employed for reactive pollutants in urban area-source applications. Gaussian models provide adequate spatial resolution near major sources, but are not appropriate for predicting the fate of pollutants more than 50 kilometers (about 31 miles) away from the source (U.S. EPA 1996b). The EPA recommends 0.1 and 50 km as the minimum and maximum distances, respectively, for application of the ISCLT2 model, a Gaussian model. In addition, Gaussian models do not provide adequate representation of certain geographical locations and meteorological conditions such as low wind speed, highly unstable or stable conditions, complex terrain, and areas near a shoreline.

These classes of models can be further categorized into two levels of sophistication: screening models and refined models. Screening models are simple techniques that provide conservative estimates of the air quality impacts of a source and demonstrate whether regulatory standards are exceeded because of the specific source. Refined models are more complex and more accurate than screening models, through a more detailed representation of the physical and chemical processes of pollution.

Some of these regulatory models have been used in modeling environmental risks in urban areas; for example, SHORTZ, an alternative air quality model according to EPA's classification, was used in the IEMD's Philadelphia study. In the Chicago study (Summerhays 1991), the Industrial Source Complex-Long Term (ISCLT) model was used to estimate impacts of point sources, while the Climatological Dispersion Model (CDM) was employed to model area sources. The Industrial Source Complex-Short Term (ISCST) model was used in estimating cancer risks from a power plant in Boston (Brown 1988). Multiple Point Gaussian Dispersion Algorithy with Terrain Adjustment (MPTER), which has been superseded by the Industrial Source Complex (ISC) model was used to calculate ground level concentrations from each utility source in Baltimore (Zankel, Brower, and Dunbar 1990). Most computer risk model packages incorporate ISCLT for simulating dispersion processes.

A model similar to the ISCLT2 was used in the EPA's Cumulative Exposure Project to estimate long-term, average ground level HAP concentrations for each grid receptor of each point source. Each point source has a radial grid system of 192 receptors, which are located in 12 concentric rings, each with 16 receptors (Rosenbaum, Axelrad, and Cohen 1999). For each grid receptor, annual average outdoor concentration estimates for each source/pollutant combination were obtained through a variety of meteorological condition combinations (such as atmospheric stability, wind speed, and wind direction categories) and the annual frequency of occurrence of each combination. These receptor concentrations were then interpolated to population centroids of census tracts, using log-log interpolation in the radial direction and linear interpolation in the azimuthal direction. For the resident tract where the source is located, the ambient concentration was estimated by means of spatial averaging of those receptors in the tract rather than interpolation.

Traditionally, and in all applications mentioned above, the lifetime exposure needed to estimate risk is generally found by multiplying the ambient concentration by the length of lifetime, e.g., 70 years. This is based on the assumption that people reside at a particular place and breathe the air with that pollutant concentration for 70 years. However, both ambient concentrations of pollutants and the time-activity patterns of people change substantially over the lifetime. This may introduce considerable uncertainties for calculation of the lifetime risks due to environmental pollution. Incorporating human time-activity patterns into estimating exposure was attempted recently to refine the exposure estimation and deserves further research efforts.

4.2.1.3　Time-Activity Patterns and Exposure Models

People's time-activity patterns and locations change daily and over a lifetime. In a day, an individual spends varying amounts of time in different microenvironments. A microenvironment is defined as a "location of homogeneous pollutant concentration that a person occupies for some definite period of time" (Duan 1982). Examples are homes, parking garages, automobiles, buses, workplace, and parks. Over a lifetime, an individual has considerably different activity patterns from childhood through early adulthood and middle age to old age. Some efforts have been made in modeling the variability of this exposure.

Total Human Exposure Study has developed two basic approaches (Ott 1990): (1) the direct approach using probability samples of populations and measuring pollutant concentrations in the food eaten, air breathed, water drunk, and skin contacted; and (2) the indirect approach using exposure models, as described below, to predict population exposure distributions. Studies of volatile organic compounds, carbon monoxides, pesticides, and particles in 15 cities in 12 states have been conducted for over a decade. Some very interesting and important findings have been discovered (Wallace 1993):

- For nearly all of 50 or so targeted pollutants, personal exposures exceed outdoor air concentration by a large margin and, for most chemicals, personal exposures exceed indoor air concentrations
- The major sources of exposure are personal activities and consumer products

The so-called *exposure models*, evolved by the school of Total Human Exposure, are based on the general assumptions of pollutant concentration distribution in different microenvironments, the activity patterns that determine how much time people spend in each microenvironment, and the representativeness of a sample to the population that might be exposed to a contaminant (National Research Council 1991).

An individual's total exposure can be obtained by summing the products of concentration and time spent in each microenvironment, a process labeled microenvironment decomposition (Duan 1981). Pollutant concentration in each microenvironment is measured or modeled, and time-activity patterns are employed to estimate the time spent in each microenvironment. Population exposure can be obtained through extrapolating the individual exposures through modeling.

Three types of models have been developed to estimate population exposure: a) simulation models, b) the convolution model, and c) the variance-component model (National Research Council 1991). The Simulation of Human Activities and Pollutant Exposures (SHAPE) model (Thomas et al. 1984; Ott, Thomas, and Mage 1988) is a computer simulation model that generates synthetic exposure profiles for a hypothetical sample of human subjects, which can be summed into compartments or integrated exposures to estimate the distribution of a contaminant of interest. For each individual in the hypothetical sample, the model generates a profile of activities and contaminant concentrations attributable to local sources over a given period. At the beginning of the profile, the model generates an initial microenvironment and duration of exposure according to a probability distribution. At the end of that duration, the model uses transition probabilities to simulate later periods and other microenvironments. This model was originally designed to predict CO exposure in urban areas. Similar models are the NAAQS Exposure Model (NEM) (Johnson and Paul 1982), and the Regional Human Exposure (REHEX) model designed for ozone exposure (Lurman et al. 1989).

The convolution model was developed to calculate distributions of exposure from distributions of concentration observed in defined microenvironments, and the distribution of time spent in those microenvironments (Duan 1981, 1982). The variance-component model assumes that short-term contaminant concentrations can be divided into components that vary in time and those that do not (National Research Council 1991). SHAPE deals mainly with the time-varying component, while the convolution model deals with the time-invariant exposure. The two components can be summed or multiplied to yield an estimated concentration value.

The three models differ in their assumptions (National Research Council 1991). SHAPE assumes that the short-term pollutant concentrations within the same microenvironment are stochastically independent, and independent of activity patterns. As a result, the microenvironmental concentrations are not correlated with activity time in that microenvironment and the variance of concentration decreases in inverse proportion to activity time. The convolution model assumes that microenvironmental concentrations are statistically independent of activity patterns. This implies that the variance of the concentration stays constant. In the variance-component model, the time-invariant components are assumed to be stochastically independent of the time-varying components. It is also assumed that the time-varying components have an autocorrelation structure, as is done in the variance-component model.

Although human exposure studies have received increasing attention, human exposure models have been used in few actual applications in assessing environmental risk.

4.2.2 MODELING DOSE-RESPONSE

"Dose-response assessment is the process of characterizing the relation between the dose of an agent administered or received and the incidence of an adverse health effect in exposed populations and estimating the incidence of the effect as a function of human exposure to the agent ..." (National Research Council 1983:19).

The dose is an exposure averaged over an entire lifetime, usually expressed as milligrams of substance per kilogram of body weight per day (mg/kg/day). The response is the probability (risk) that there will be some adverse health effect.

EPA typically assumes that the dose-response relationships for carcinogens and non-carcinogens are different, no threshold for the former and thresholds for the latter (U.S. EPA 1993a). That is, for carcinogens, health effects can occur at any dose, while for non-carcinogens, threshold levels exist below which no adverse health effects will occur. This dichotomy is not fully supportable by current scientific evidence and provides no common metric for comparison between carcinogenic and non-carcinogenic effects. The Presidential/Congressional Commission on Risk Assessment and Risk Management (1997b) recommended evaluations of two potentially useful common metrics: margin of exposure (MOE) and margin of protection (MOP). MOE is defined as "a dose derived from a tumor bioassay, epidemiologic study, or biologic marker study, such as the exposure associated with a 10% response rate, divided by an actual or projected human exposure" (Presidential/Congressional Commission on Risk Assessment and Risk Management 1997b:45). MOP is a safety factor that accounts for variability and uncertainty in the dose-response relationship for non-cancer effects. A NOAEL, a lowest-observed-adverse-effect level (LOADEL), or a benchmark dose is divided by MOP to derive estimates of acceptable daily intakes (ADI), reference doses (RfD), or reference concentrations (RfC).

Dose-response relationships can be established through either epidemiological data or animal-bioassay data. Epidemiological data are absent for most chemicals, and its accumulation generally requires a long time lag after release of chemicals to which humans are exposed. These limitations, therefore, necessitate reliance on the animal-bioassay data collected from experiments on rats or mice. The fundamental premise underlying experimental biology and medicine is that the results from animal experiments are applicable to humans. The standard protocol of a chronic carcinogenesis bioassay requires testing of two species of rodents, often mice and rats, testing of at least 50 males and 50 females of each species for each dose, and at least two doses administered (the maximum tolerated dose and half that dose) plus a no-dose control. For this protocol, the minimum number of animals required for a bioassay is 600 and with this number only relatively high risks can be detected. The detection of low risks requires an extremely large number of animals; the largest experiment on record involved 24,000 animals and was designed to detect a 1% risk of tumor (National Research Council 1983). However, lower risks, such as one in one million, are the major concern of regulatory agencies. Some extrapolation is inevitable.

Establishment of the dose-response relationship through either epidemiological or bioassay data requires some extrapolation models that can be used to estimate the response at environmental doses through extrapolating from high dose responses. A number of statistical cancer models have been developed for the extrapolation to low doses. The most commonly used are the one-hit model, the multi-hit model, the multi-stage model, the probit model, the logit model, and the multistage with two stages.

The one-hit model assumes that a single dose of a carcinogen can affect some biological phenomenon in the organism that will subsequently cause the development of cancer (White, Infante, and Chu 1982). As a direct extension of the one-hit model, the multi-hit model assumes that more than one hit is required to induce a tumor (Rai and van Ryzin 1979). It can be also viewed as a tolerance distribution model, where the tolerance distribution is gamma (Munro and Krewski 1981).

The multi-stage model is based on the assumption that tumors are the end result of a sequence of biological events (Crump 1984). This is a no-threshold model that implies that a tiny amount of a toxic substance which can affect DNA has some chance of inducing cancer.

The probit model assumes that the susceptibility of the population to a carcinogen has a normal distribution with respect to dose. And the log-probit model assumes that the logarithm of dose that produces a positive response is normally distributed. The logit model assumes a logistic response distribution of the population to a carcinogen with respect to dose.

Comparison of these models indicates systematic differences in the low-dose extrapolation. The one-hit and linearized multistage models will usually predict high risk, and the probit model will predict the lowest (Munro and Krewski 1981). The one-hit model is linear at low doses, and the multistage model is linear when the linear coefficient in the model is positive, and is sublinear otherwise. The logit and multi-hit models are linear at low doses only when the shape parameters are equal to one, and sublinear when these parameters are greater than one. The probit model is inherently sublinear at low doses and extremely flat in the low-dose region. EPA uses the linearized multistage model.

Statistical models are based on the notion that each individual in the population has his or her own tolerance to the test agent. Any level of exposure below this tolerance level will have no effect on the individual, but otherwise result in a positive response. These tolerance levels are presumed to vary with individuals in the population, and the lack of a population threshold is reflected in the fact that the minimum tolerance is allowed to be zero. Specification of a functional form of the distribution of tolerances determines the shape of the dose-response curve and thus defines a particular statistical model (Paustenbach 1989).

Critiques of the dose-response models lie in two major areas: the models themselves and interspecies conversion.

1. Most of these models can fit the observed data reasonably well, and it is impossible to distinguish their validity using the statistical goodness-of-fit criterion. Even with good fit to the experimental dose region, the models tend to diverge substantially in the low-dose region of interest to regulators. The results can have differences of five to eight orders of magnitude (Munro and Krewski 1981).
2. Most models have been based on statistical rather than biological methods and the biological mechanisms have not been considered in the models.
3. The extrapolation of the dose-response relationship from animal to human has been challenged based on two major aspects. First, animals and humans metabolize substances differently, and thus the level of the chemical reaching various parts of the animals and humans can vary widely. Consequently, different health effects may be produced for animals and humans. Second, the metabolism of chemicals differs at high and low doses (National Research Council 1983).

The Proposed Guidelines for Carcinogenic Risk Assessment consider dose-response assessment as a two-part process — range of observed data and range of extrapolation (U.S. EPA 1996a). In the range of observation, the dose and response relationship is modeled to determine the effective dose corresponding to the lower

95% confidence dose limit associated with an estimated 10% increased tumor or relevant non-tumor response (LED10). The LED10 would serve as the default point of departure for extrapolation to the origin (zero dose, zero response) as the linear default or for a margin of exposure (MOE) analysis as the nonlinear default. Whenever data are sufficient, a biologically based extrapolation model is preferred. Otherwise, three default approaches — linear, nonlinear, or both — are applied in accordance with the mode of action of the agent.

Computer models have been developed for quantitatively assessing environmental risks, e.g., RISKPRO, HEM-II, and AERAM. These computer models are based on the risk-modeling methodology described above. RISKPRO is a versatile modeling system for estimating human exposure to environmental contamination and environmental risk from various environmental media, e.g., air, soil, surface water, and ground water (McKone 1992). The Human Exposure Model II (HEM-II) was designed to evaluate potential human exposure and risks generated by sources of air pollutants (U.S. EPA 1991a). It can be used to either screen point sources for a single pollutant and rank the sources according to potential cancer risks, or to conduct a refined analysis of an entire urban area that includes multiple point sources, multiple pollutants, area sources, and dense population distributions.

4.3 MEASURING AND MODELING ECONOMIC IMPACTS

Measuring economic impacts of environmental pollution and programs has been a subject of inquiry by economists. This field of study is concerned about damages and environmental costs associated with deterioration of environmental quality caused by environmental pollution, benefits of environmental quality improvement as a result of environmental policies and programs, and costs associated with these policies and programs. Economic effects can be quantified for direct impacts on humans such as human health (morbidity and mortality) and non-health (odor, visibility, and visual aesthetic), for impacts on the ecosystem such as agricultural productivity, forestry, commercial fishery, recreational uses, ecological diversity and stability, and for impacts on non-living systems such as materials damage, soiling, production costs, weather, and climate (Freeman 1993). See Freeman (1993) for an excellent, comprehensive treatment of theory and methods for measuring environmental and resource values.

In the context of environmental justice analysis, we focus on evaluation of economic impacts from noxious facilities such as hazardous waste sites. Two methods are generally used for such evaluation: the contingent valuation method and the hedonic price method.

4.3.1 CONTINGENT VALUATION METHOD

The contingent valuation method elicits respondents' valuations of a hypothetical situation through direct questioning. It is typically used to elicit respondents' monetary values of goods, services, or environmental resources that do not have a market or for which researchers cannot infer an individual's values from direct observations.

A typical question asks respondents their maximum willingness to pay for improving environmental quality (which is interpreted as a measure of compensating surplus) or for avoiding a loss (interpreted as a measure of equivalent surplus) (Freeman 1993). Few studies use the contingent valuation method in the case of noxious facilities (Nieves 1993).

4.3.2 HEDONIC PRICE METHOD

Environmental quality can be considered as a qualitative characteristic of a differentiated market good (Freeman 1993). An individual chooses his/her consumption of environmental quality through his/her selection of a private goods consumption bundle. To the extent that environmental quality varies across inter-urban and intra-urban space, individuals, when making their decisions about which city and which location of a city to choose for their residential location, also choose their levels of exposure to environmental risks. The levels of environmental quality and risks are implicitly incorporated into land values in the land market. Households' demand for environmental amenities should be revealed through housing price differentials for different locations. On the supply side, environmental quality or risk affects land productivity, and land productivity differentials should be revealed through land rents or value differentials for different locations.

Therefore, the equilibrium price of housing should reflect the structural characteristics of a house, neighborhood characteristics, location characteristics, and environmental characteristics. A hedonic price function is used to represent the relationship between housing price and various characteristics. From a hedonic price function, we obtain the partial derivative with respect to any characteristic, which gives us the marginal implicit price for that characteristic. This marginal or hedonic price is the additional cost required to purchase an additional unit of a particular characteristic.

The hedonic price method has been used to study the impacts of nonresidential land uses on residential housing prices. The hypothesis is *nonresidential land uses have a significant negative impact on housing values.* Empirical tests have produced mixed results, which appear to depend on the unit of analysis and underlying assumptions about the impact scope of a negative externality due to nonresidential land uses. Studies using individual housing units as the unit of analysis tend to provide no support for this hypothesis (Maser, Riker, and Rosett 1977; Grether and Mieszkowski, 1980). Maser, Riker, and Rosett (1977) and Grether and Mieszkowski (1980) found insignificant effects on housing values of most land uses except for industrial land use. These studies indicate that most externalities have very localized effects. The neighborhood defined in these studies is a small homogeneous area, such as blocks or block groups, and this definition has an advantage for controlling such variables as the levels of taxes and public services. However, such a proximity to nonresidential land use may be too restrictive to measure non-localized externalities (Stull, 1975; Lafferty and Frech 1978; Burnell, 1985). Burnell (1985) pointed out two types of externalities associated with nonresidential land uses: localized negative impacts on adjacent residents and city-wide positive externality effects in terms of job opportunities and fiscal benefits.

Studies using a municipality as the unit of analysis have provided supporting evidence for this hypothesis. These studies suggested that there might be non-localized externalities associated with nonresidential land use. Stull (1975) took median housing values from suburban municipalities in Boston to be a function of physical accessibility, public sector, and environmental characteristics measured as the proportion of nonresidential land use area in a municipality. The findings indicated that housing values increased for small amounts of commercial land use and decreased for large amounts of commercial land use and for industrial and vacant agricultural land use. Lafferty and Frech (1978) largely confirmed the results derived by Stull (1975), but found the amount and dispersion of industrial land use did not affect housing values. Extending these two studies, Burnell (1985) found that concentrating industrial activity had a positive effect on housing values, but major air polluting industries had a significantly negative effect on housing values. This implies that not only the presence but also the type of industrial activity can affect residential location decisions.

Polluting industries are part of the noxious facilities that have received extensive hedonic price analyses. Some researchers believe that there is broad consistency in the findings of property value studies that noxious facilities depressed property values of real estate in proximity to those facilities (Nieves 1993; Dale et al. 1999). Nieves (1993) reviews 13 hedonic price studies of noxious facility impacts on property values and observes that these studies consistently find facility proximity to be associated with depressed property values. Nine of these 13 studies appear in peer-reviewed journals. Noxious facilities in these studies include hazardous waste facilities, solid waste facilities, industrial land use, nonresidential land use, electric utility power plant, nuclear power plants, feed materials production facilities, petrochemical refineries, chemical weapon storage sites, radioactive contaminated sites, and lique-fied natural gas storage facilities.

In reviewing literature, however, other researchers find inconsistent and mixed results (Zeiss 1991; Nelson, Genereux, and Genereux 1992). Of ten studies reviewed, Zeiss (1991) reported six cases that showed significant negative effects on nearby property values, eight cases that found no significant effects, and one study that indicated positive effects. Two of the ten studies are concerned about several municipal solid waste incinerators, and the others target landfills. Except for one study, these studies appear in unpublished reports.

The effects of any noxious facility have both spatial and temporal dimension. On the spatial dimension, there is little dispute that the effects decline over distance from the noxious facility. However, there is a lot to argue about concerning how far the effects need to diminish to reach an insignificant amount. The question being debated is: How far is far enough? Hedonic price studies evaluate the impacts of noxious facilities on property values in relation to distance from the subject facilities (Table 4.2). Some researchers reviewed hedonic price studies and observed that economic impacts of hazardous waste sites occurred mostly in an area within one-quarter mile (400 m) of the site (Greenberg, Schneider, and Martell 1994). Others find much larger impact areas (see Table 4.2); most of these studies report that impacts diminish to an insignificant amount between 2 and 4 miles from individual sites. There is also some evidence that the distance decay function is nonlinear (Kohlhase 1991).

TABLE 4.2
Hedonic Price Studies of Noxious Facility Impacts on Property Values

Facility Types	Location and Operation Period	Property Sale Records	Sale Period	Hedonic Function	Critical Distance (miles)[a]	Price Gradient with Respect to Distance	Source
Landfill (solid waste)	Ramsey, Minnesota (1969–1990s)	708 single-family sales between 0.35 and 1.95 mi	1979–89 during operation period	Linear regression	2–2.5	$4,896/mi or 6.6% mean value/mi	Nelson, Genereux and Genereux (1992)
Incinerator (solid waste)	Marion County, Oregon (1986—)	145 residential sales	1983–86 siting period, 1986–88 construction and operation	Linear regression		Insignificant for individual periods and entire period	Zeiss (1991)
Incinerator (solid wastes)	North Andover, Massachusetts (1985—)	2593 single-family home sales between 3,500 and 40,000 feet from the incinerator (sample sizes = 595, 302, 662, 711, 323 for five periods)	1974–78 pre-rumor, 1979–80 rumor, 1981–84 construction, 1985–88 Online, 1989–1992 ongoing operation	Log-log functional form (linear regression with the natural log of price index and of distance)	3.5	Insignificant for pre-rumor and rumor periods; $2,283/mi for construction period; $8,100/mi for online period; $6,607/mi for ongoing operation period	Kiel and McClain (1995)
Superfund sites (2 sites)	Woburn, Massachusetts	2209 single-family home sales in the town of Woburn (sample sizes = 106, 406, 362, 689, 463, and 183 for six periods, respectively)	1975–76 prior period, 1977–81 discovery phase, 1982–84 EPA Superfund NPL announcement phase, 1985–88 cleanup plan 1 phase, 1989–91 cleanup plan 2 phase, 1992 cleanup phase	Log-log functional form (linear regression with the natural log of price of distance)		Insignificant for prior period; $1,854/mi for discovery period; $1,377 for Superfund phase; $3,819 for cleanup plan 1; $4,077 for cleanup plan 2; $6,468 for cleanup phase	Kiel (1995)

Superfund sites (6 sites)	Houston, Texas	1969 single-family sales in 1976, 1083 sales in 1980, 1811 sales in 1985. 0.2 to 7 mi from the nearest site.	1976 pre-Superfund, 1980 Superfund program began, 1985 Post Superfund-NPL-announcement	Semi-log functional form with linear and quadratic terms of distance from site and distance from CBD	6.19 for pooled 5 sites 5.34 for pooled 6 sites 1.86–4.76 for individual sites	Insignificant in 1976 Negative on linear term and positive on quadratic term, indicating attraction to toxic sites up to 2.7 mi in 1980 $2,364 (2.2%)/mi at 3.62 mi from site for pooled 5 sites in 1985 (nonlinearly with distance—$4,940 more at 1 mi from site, $3,476 more at 2 mi, $2,606 more at 3 mi, $1,607 more at 4 mi, $690 more at 5 mi, and $100 more at 6 mi) $1,742 a mile at 3.67 mi from site for pooled 6 sites in 1985 $1006–$3310 at the average distance from site for 4 individual sites and negative for 2 sites in 1985.	Kohlhase (1991)
Superfund site (1 lead smelter)	Dallas, Texas (1934–1984)	203,353 single-family sales from 0.9 to 24 mi from site during 1979–95 (sample sizes = 18,180; 40,721; 26,156; 47,932; 70,328 for five periods, respectively)	1979–81 unpublicized period 1981–84 discovery and closure period 1985–1986 cleanup period 1986–90 post-cleanup period 1991–1995 new publicity period	Semi-log functional form (linear regression with a natural log of price)		The combined distance/neighborhood model results 2.27% per mi or $1,282/mi (in constant 1979 dollars) for unpublicized period 2.13% for discovery and closure period 0.978% for cleanup period –3.05% for post-cleanup period	Dale et al. (1999)

continued

TABLE 4.2 (CONTINUED)
Hedonic Price Studies of Noxious Facility Impacts on Property Values

					Distance[a]		
Hazardous waste sites (11)	Suburban Boston, MA	2182 single-family home sales	Semi-log functional form (linear regression with a natural log of price)	11/1977–3/81 pre-discovery short-term response (6-month after discovery) post-short-term response period		−4.42% for new publicity period positive for the 2-mi, high-income white neighborhood negative for the nearest 2-mi, lower income minority neighborhood	Dale et al. (1999)
						Full sample model 0.3% per mi for pre-discovery period 1.6% per mi for short-term response period 2.2% per mi for post-short-term period.	Michaels and Smith (1990)
Landfills						5–7% for urban homes	Reichert, Small, and Mohanty (1992)
Hazardous and non-hazardous waste sites					4	1% per mi for non-hazardous waste sites 2% per mi for hazardous waste sites	Thayer, Albers, and Rahmatian (1992)
Hazardous waste site	Pleasant Plains, NJ			1974 pre-contamination 1975 post-contamination	2.25	no significant effect for houses within 1.5 mi $2,700 per mi for homes between 1.5 and 2.25 mi	Adler et al. (1982)

[a] Distance at which the price effect diminishes to insignificance.

The temporal dimension is more complex than the conventional wisdom that a noxious facility decreases property value. Rather, the effects are dynamic and evolve over the life cycle of the facility. Market response may vary with different stages of the life cycle. Kiel and McClain (1995) classified five stages: pre-rumor, rumor, construction, on-line, and on-going operation. The pre-rumor stage is before any mention of the possibility of a noxious facility and reflects a pre-treatment equilibrium between supply and demand in the community. The rumor stage begins when news of the proposed project leaks or is announced to the community. The market responds to a probabilistic event. Homeowners and potential buyers will make their sale/buying decisions based on their perceived risks, damages, and benefits under uncertainties. Those risk-averse will relocate as quickly as possible. During the construction stage, households make their relocation decision based on expected damages and moving costs. During the on-line stage, the market continues to make price adjustments based on more and more information about the environmental and health effects of the facility. Finally, the market reaches a new equilibrium between supply and demand in the on-going operation stage.

There is evidence that the discovery of toxic and hazardous waste sites and the EPA announcing placement of those sites in the Superfund NPL negatively affected property values near those sites (Kohlhase 1991; Kiel 1995). The public usually does not perceive these sites as detriments prior to awareness of the risks. Hedonic price studies show that there is no significant location premium far away from these sites prior to discovery (Kohlhase 1991; Kiel 1995). The Love Canal and other events in late 1970s and the Superfund legislation in 1980 raised the public's awareness of the danger of hazardous waste sites. Price gradients with respect to distance to these waste sites increased substantially after the discovery and the EPA announcement (see Table 4.2). Negative impacts of hazardous waste sites are fully capitalized in real estate properties after the public receives information about potential risks from those sites. Further, evidence indicates that remediation programs help the market rebound; the housing prices rebound after the sites are cleaned up (Dale et al. 1999; Kohlhase 1991). This rebound is nonlinear with distance as the neighborhood closest to the site gains the most (Dale et al. 1999).

4.4 MEASURING ENVIRONMENTAL AND HUMAN IMPACTS FOR ENVIRONMENTAL JUSTICE ANALYSIS

For environmental justice studies, there is a spectrum of methods available for measuring environmental and human impacts from environmental hazards (Table 4.3). For human health impacts, we can use various actual monitoring or modeled measures to approximate health risks to humans. Proximity measures are used most often in environmental justice studies and are also the most controversial. One thing that surely accounts for their wide use: they are easy and economical to operationalize in studies. The easiest is perhaps to obtain facility address ZIP Code. For other census-geography-based proximity measures, the analyst needs to geocode the facility address and associate the facility's location with census geography, either manually or using GIS. For distance-based proximity measures, the analyst geocodes the facility location and uses GIS to delineate the boundary. All these measures assume that environmental impacts are constrained to the defined areas. With all sorts of geographic units, a debate focuses on which unit is the most representative of environmental impacts (see Chapter 6). This debate, however, does not address the magnitude of environmental impacts.

Proximity does explain, to a certain degree, some environmental impacts. Some epidemiological studies have shown a significant relationship between residential proximity to hazardous waste sites and increased health risk and disease incidence, especially among pregnant women and infants (Berry and Bove 1997; Croen et al. 1997; Goldman et al. 1985; Guthe et al. 1992; Knox and Gilman 1997). However, a few other studies did not find such a relationship (Bell et al. 1991; Polednak and Janerich 1989; Shaw et al. 1992).

Surveys have repeatedly reported the public's aversion to living near various noxious facilities. Using a national survey, Mitchell (1980) found nuclear power plants and hazardous waste sites to be the most undesirable land uses. Only 10 to 12% of the population would voluntarily live a mile or less from a nuclear power plant, and this figure is about 9% for a hazardous waste disposal site. It would take 100 miles from respondents' homes to the nuclear power plant or hazardous waste site for the majority of respondents (51%) to accept it voluntarily. In a survey of residents in suburban Boston, Smith and Desvousges (1986) found the threshold distance to be about 10 miles for a majority to accept a hazardous waste site and about 22 miles for the nuclear plant.

Studies have reported the inverse relationship between opposition to facility siting and distance (Furuseth 1990; Lindell and Earle 1983; Lober and Green 1994). Using in-person attitudinal surveys of Connecticut residents, Lober and Green (1994) developed a predictive model of the effect distance has on various waste disposal facilities. The chance of opposition is a negative function of distance; the odds of opposing a recycling center are 25% for someone living 0.5 miles (0.8 km) away from the facility.

As discussed in the last section on hedonic pricing, noxious facilities depressed property value in proximity to those facilities. Some studies find nonlinearity of these negative impacts; the nearest area bears the worst impact. The critical distance at which these impacts will diminish to zero is between 2 and 4 miles, as most studies report.

On the positive side, proximity captures some social, economic, psychological, and health impacts of noxious facilities. On the minus side, proximity is a very poor approximation of actual health risks imposed by noxious facilities.

On the other end of the measure spectrum is an exposure or risk measure. To establish the relationship between environmental health risks and various demographic groups, we can directly measure or estimate internal dose or health effects by various demographic groups. Lead exposure is the best example using this method. A few studies of lead pollution employ actual measurements of lead exposures such as pediatric blood lead level data (ATSDR 1988; Brody et al. 1994; Earickson and Billick 1988).

TABLE 4.3

Measuring Environmental and Human Impacts for Environmental Justice Studies

Measurement Method	Examples	Strength	Weakness
Proximity • Census geography within which emission sources are located • Distance from emission sources	ZIP code, census tracts, block groups; 0.5 or 1.0 mi from emission source	Easiest and most economical to use May capture impacts other than health	Poorest approximation to actual health risk
Emission • Emission monitoring • Emission models/methods	Species fraction, emission factor, and material and energy balance models	Data are widely available Easy and very economical to implement	Very poor approximation to actual health risk
Ambient environmental concentrations • Environmental monitoring • Environmental modeling	Criteria pollutant monitoring Air quality models, water quality model	Data are widely available Large geographic coverage	Poor substitute for human exposure and health risk
Micro-environmental concentrations • Micro-environmental monitoring • Micro-environmental modeling	Time-activity patterns	Good human exposure indicators	Difficult and costly
Internal dose • Direct measures • Exposure modeling	Personal monitoring, biological markers	Best approximation to health risk	Very difficult and costly
Effects • Epidemiology • Toxicology	Dose-response model for carcinogenic effects	Accurate measures of health risk	Most difficult and costly to implement
Contingent valuation	Willingness to pay survey	Easy implementation and interpretation	Potential biases
Hedonic pricing	Linear regression	Summarize environmental impacts in a single number	May not capture all impacts due to imperfect information

Exposure to lead occurs through multiple routes and pathways, such as inhalation and ingestion of lead in air, food, water, soil, or dust. Children are most susceptible to lead poisoning. The major pathway for them is ingestion through the normal and repetitive hand-to-mouth activity. Residential paint, household dust, and soil are the major sources of lead exposure in children. Lead can adversely affect the kidneys, liver, nervous system, and other organs. Children are a sensitive population; excessive exposure to lead may cause neurological impairments such as seizures, mental retardation, and/or behavioral disorders. The Centers for Disease Control and Prevention designates 10 µg/l in blood lead level as a threshold for any harmful health effect. Between 1976 and 1991, the percentage of children 1 to 5 years old with blood lead levels exceeding this threshold declined from 88.2 to 8.9% in the United States (Council on Environmental Quality 1996). However, disparity in lead exposure has remained; poor, urban, African-American and Hispanic children are still at the highest risk of lead poisoning (see Table 4.4).

The exposure or risk measure is the most accurate method for evaluating human health impacts. However, it is costly and difficult to implement. Partly because of its great cost and lack of suitable measurement methods, personal measurements of exposures to toxics are limited to a few case study areas and a few toxics. When we do not have actual measurements of exposure as in most cases, we have to rely on ambient monitoring or environmental modeling in the areas where people are likely to be exposed.

Ambient environmental quality data have been collected for the purpose of complying with environmental laws and regulations for some years. Monitoring networks have been operated at global, national, and local levels. They provide a rich database of environmental quality. These data permit longitudinal and trend analysis of environmental quality for a particular area, and inter-city comparisons of environmental quality. For some pollutants such as ozone, intra-city variations can also be analyzed by interpolating data at the monitoring stations (see Chapter 10). However, the monitoring system has also some limitations. The pollutants monitored are largely limited to those with environmental quality standards. For example, until recently, regular air quality monitoring has covered only criteria air pollutants. There is a paucity of data for toxics such as

---

TABLE 4.4

Percentage of U.S. Children 1 to 5 Years Old with Blood Lead Levels 10 µg/dl or Greater by Race/Ethnicity, Income Level, and Urban Status: 1988–1991

Income/ Urban Status	Total	Non-Hispanic White	Non-Hispanic Black	Mexican American
Low	16.3	9.8	28.4	8.8
Middle	5.4	4.8	8.9	5.6
High	4.0	4.3	5.8	0.0[a]
Central city ≥ 1 million	21.0	6.1[a]	36.7	17.0
Central city ≤ 1 million	16.4	8.1	22.5	9.5
Non-central city	5.8	5.2	11.2	7.0

[a] Estimates may be unstable because of small sample size.

Source: Council on Environmental Quality (1996).

hazardous air pollutants. All but a few previous environmental justice studies use some risk or exposure surrogates, such as ambient concentration or pollutant emission. For criteria air pollutants, ambient concentrations are used as a proxy for exposure/risk. For non-criteria pollutants, emission inventories are often the only choice. The approximations are economical but of limited accuracy in predicting or classifying risk or exposure (Perlin et al. 1995; Sexton et al. 1992; NRC 1991).

Where there are ambient monitoring programs, the spatial representation of the monitoring network is generally poor. The number of monitoring stations is often small, and the monitoring network is often geared toward specific pollution spots. As a result, the existing networks may not capture micro-scale variations of environmental quality, which are often the focus of environmental justice concerns. For example, site-specific impacts and transportation-related pollution are often localized and decay rapidly away from the sources. In these cases, site-specific monitoring programs are often required and can be costly. For environmental justice analysis, a particular concern is that the existing air quality network may not be representative of population characteristics for the study area, although it is supposed to be designed to represent the airshed (Liu 1996).

Environmental modeling is a very useful alternative and also complementary to ambient monitoring. The spatial dimensions in environmental modeling are especially appealing for environmental justice analysis. Environmental modeling has been widely used for assessing environmental impacts of existing and proposed facilities in regulatory settings. These applications are mostly site-specific. For site-based environmental justice analysis, environmental models can be used to project the plume footprint of ambient pollutant concentrations. Coupling environmental models and GIS would enable the analyst to delineate the impact boundary and the geographic units of analysis more accurately than simply relying on predefined census geography (see Chapter 8). Coupling environmental models and urban models would permit a better understanding of the relationship between urban activities and environmental quality (see Chapter 9). Of course, the outputs from environmental models are still only a substitute for human exposure.

Most environmental justice studies target a single type of environmentally risky facility or LULU in a city, county, region, or state. To what extent is this choice of LULUs relevant to better our understanding of the relationship between location of environmental risks and population distribution? The relevance is partial and could be distorted. Those who make the decisions about the locations of a LULU may take into account a series of variables including the host community's characteristics. For a single type of LULU, there are likely many common variables incorporated in the siting decision processes. Therefore, conducting a cross-sectional study of a single type of LULU can hold other things equal and better our understanding of the topic of interest: the association between siting and the host communities.

However, a LULU included in such studies is most likely to be only one of many environmental stressors in a community. There are also some other environmental indicators and environmental amenities in a community. Households make their residential choice based on a comprehensive appraisal of a residence and their surroundings. As a result, a residential location pattern in an area demonstrates a balance of many variables among all residence locators. Similarly, industrial locators make their location choices based on a balance of various factors and trade-offs. Not only do they consider what degree of externality they would produce but they might also take into account the current pollution and risk level resulting from the existing industries in the community. For environmental justice analysis in a metropolitan area, it appears to be more appropriate to incorporate major environmental risks and amenities.

4.5　CRITIQUE AND RESPONSE OF A RISK-BASED APPROACH TO EQUITY ANALYSIS

The EPA's 1992 report, "Environmental Equity: Reducing Risks for All Communities," calls for use of risk assessment and management in studying and dealing with environmental justice and equity issues. This recommendation immediately drew attacks from environmental justice advocates.

Opposition to risk assessment from the environmental movement has been evident and consistent since the 1980s (Tal 1997). The origins of opposition are linked to the Reagan administration, which used risk assessment, along with cost-benefit analysis, to undermine environmental protection policies and to support the deregulatory agenda. A national survey of environmental groups identified the following reasons for opposition to risk assessment:

- Misuse and manipulation of risk assessment for political purpose, particularly in the Reagan administration and in the 104th Congress
- Poor scientific basis
- Immorality
- Political disempowerment
- Asking the wrong question — emphasis on quantifying rather than reducing or eliminating risks

Every aspect of risk assessment has been scientifically controversial. Uncertainties associated with risk assessment have been acknowledged by the scientific community, environmental groups, and the business community. All agree that scientific basis for risk assessment is still inadequate. Not surprisingly, environmental groups believe risk assessment is biased toward underestimating risk, while the business community holds the opposite position. The scientific community recognizes the pervasive uncertainty in risk assessment and responds with new methods to take into account uncertainties.

Uncertainties in risk assessment can come from three basic sources (Suter 1993):

- The inherent randomness of the world (stochasticity)
- Imperfect or incomplete knowledge of things that could be known (ignorance)
- Mistakes in conducting assessment (error)

Stochasticity, characteristic of the natural systems, can be described and estimated but cannot be reduced. For example, wind, rainfall, and temperature are essentially stochastic. Recently, stochastic modeling of atmospheric dispersion has increased in popularity (National Research Council 1991). Recent research indicates that three-dimensional stochastic models will offer considerable predictive improvement over conventional Gaussian plume models. The third source, human error, is an inevitable attribute of all human activities. Such errors are primarily a quality assurance issue.

The current literature has focused on identifying, quantifying, and reducing the second type of uncertainty in risk assessment. This uncertainty results from our inability to accurately describe, count, or measure everything related to risk estimation. For example, the reasonable maximal exposure (RME), originally recommended by EPA, is an upper-bound point estimate based on a combination of conservative assumptions. This point estimate and the similar worst-case "maximally exposed individual" (MEI) estimate have been criticized as unrealistic and exaggerated. They provide little information for risk managers and may result in unreasonable clean-up goals. Taking the randomness of input variables into account, probabilistic risk assessment (PRA) produces quantitative distributions of modeled variables (such as exposure and risk) using Monte Carlo simulation. PRA is much more conducive to sensitivity analysis than the point estimate approach (such as RME or MEI) as the latter uses many variables at or near their maximums. PRA has enjoyed increasing popularity (Finley and Paustenbach 1994; Thompson, Burmaster, and Crouch 1992). The Presidential/Congressional Commission on Risk Assessment and Risk Management recommends that agencies move away from using MEI and use methods that "combine the many characteristics of probable exposure into an assessment of the overall population's exposures" (1997b:iv).

Critics also assail the narrow focus of risk assessment on a single pollutant in a single environmental medium for causing health effects through a single pathway. This narrow focus fails to assess health effects caused by environmental hazards in the real world, where multiple environmental agents coexist and interact in multiple environmental media, resulting in adverse health effects through multiple pathways. Over 60,000 chemicals are currently in use. Multiple exposures occur to people from any combinations of these chemicals and other environmental agents. Furthermore, these chemicals interact and lead to effects that are additive (whereby $1 + 1 = 2$), antagonistic (whereby $1 + 1 < 2$), or synergistic (whereby $1 + 1 > 2$). Synergistic effects are of particular concern because the additivity assumption that is usually used in risk assessment underestimates risk in the case of chemical mixtures where synergism occurs. Although the scientific basis for single chemicals is still limited, we know much less about the health effects of chemical mixtures. Even less is known about cumulative impacts.

EPA has recently shifted from the old narrow-focused paradigm of risk assessment to a new broadly based paradigm that takes into account multiple sources, chemicals, media, pathways, routes, and endpoints (U.S. EPA 1997a). The Presidential/Congressional Commission on Risk Assessment and Risk Management (1997a) recommended consideration of the risk's multisource, multimedia, multichemical, and multirisk context in assessing and managing risks.

A fundamental critique of risk assessment lies in its ethics and democracy. Environmental groups, particularly environmental justice groups, hold that risk assessment is fundamentally immoral and undemocratic (Tal 1997). Risk assessment often excludes affected communities from participating in the decision-making processes. Without financial and technical capabilities, these affected communities feel powerless when facing the technocratic nature of risk assessment. As discussed above, the current practice of risk assessment fails to consider multiple, cumulative, and synergistic effects. Environmental justice advocates argue that this failure has a disproportionate effect on the poor and minority communities, where multiple, cumulative, and synergistic exposures tend to be disproportionately high (Israel 1995). Environmental justice advocates also argue that risk assessment emphasizes the aggregate nature of risk and fails to incorporate "unusual exposure patterns" and "unusual susceptibility." To the extent that certain groups are highly exposed or susceptible to certain environmental risks, they suffer disproportionately from the failures of risk assessment. As a result, risk assessment de facto ignores risks experienced in the poor or minority communities. In this sense, some claim that risk assessment methodology as currently practiced is itself discriminatory (Israel 1995). Robert Bullard argues that the comparative risk approach helps institutionalize a system of unequal environmental protection across racial and class lines (Finkel and Golding 1993).

In response, the new blueprint Risk Assessment and Risk Management in Regulatory Decision-Making places stakeholders in the center of a new risk management framework and establishes guidelines for engaging stakeholder involvement. It recommends that exposure assessment include specific groups, such as infants, children, pregnant women, low-income groups, and minority groups. Furthermore, risk assessment should "characterize the scientific aspects of a risk and note its subjective, cultural, and comparative dimensions" (Presidential/Congressional Commission on Risk Assessment and Risk Management 1997b:21).

This call for incorporating stakeholder perception of a risk in risk characterization is also reflected in the recommendation made by the Presidential/Congressional Commission on Risk Assessment and Risk Management (1997b), for use of comparative risk assessment (CRA) as a tool for setting priority for risk management at the federal level. CRA, which has been conducted mostly by state, local, and tribal governments, emphasizes stakeholders' participation in the process of measuring, comparing, and ranking environmental risks. EPA's *A Guidebook to Comparing Risks and Setting Environmental Priority* recommends engaging stakeholder participation and addressing environmental justice issues during each phase of a comparative risk project (U.S. EPA 1993a). The comparative risk process includes establishing a project team of different stakeholders, making a comprehensive list of environmental problems, analyzing and characterizing risks associated with these problems in terms of human health, ecosystems, and quality of life, comparing and ranking risks posed by different problems, and developing priorities for risk-reducing and risk-preventing strategies. Ranking methods include negotiated consensus among stakeholders, voting by participants, formulas that integrate different quantitative data into a composite score, decision analysis, and other methods. Various state, local, and tribal comparative risk projects explicitly incorporate public values and perceptions of risks in a process of diverse stakeholder involvement.

Executive Order 12898 orders each Federal agency, whenever practicable and appropriate, to collect, maintain, and analyze information assessing and comparing environmental and human health risks borne by populations identified by race, national origin, or income. Particularly, environmental human health analyses, whenever practical and appropriate, shall identify multiple and cumulative exposures.

Sexton, Olden, and Johnson (1993) proposed a risk-based conceptual framework for studying the relationships between demographic variables and environmental health risks. This framework addresses three questions:

(1) How do important exposure- and susceptibility-related attributes affect environmental health risks?
(2) How do class and race affect important exposure- and susceptibility-related attributes?
(3) How do class and race differentially affect environmental health risks?" (Sexton, Olden, and Johnson 1993:715).

From this conceptual framework, we can see how environmental risk and location models contribute to the process. For exposure-related attributes (e.g., proximity to sources, occupation, and activity patterns), we can utilize residential and employment location models to simulate their roles and interactions; these models will be discussed in detail in the next chapter. For each component of environmental health risks, models are often necessary since direct measurement is expensive or simply unavailable. Some of the models have been discussed in this chapter.

Failure of risk assessment to address differential distribution impacts does not mean there is any intrinsic hindrance for preventing it from doing so. But rather, risk assessment, from the beginning, has been used to answer the question of how large a risk is, rather than how a risk is distributed. Its major concerns have been the worst cases such as "maximally exposed individual," rather than the spatial distribution of a risk. In fact, spatial distribution can be explicitly incorporated into risk assessment, although not without difficulties. The spatial dimension can be demonstrated in risk assessment processes, such as emission estimations, ambient concentrations modeling, and exposure modeling. The addition of a space dimension certainly increases requirements for data, data processing, and analysis. In the following discussion, we focus on a few areas where improvements can be readily made. These include spatial disaggregation, area source emission and dispersion models, mobile sources, micro-environment and human activity patterns, and use of GIS.

Spatial specification in emission estimation, dispersion model, and population exposure can affect model results significantly. Unfortunately, the spatial scale at which we can obtain emission data and the data to estimate emissions (e.g., at the county or census tract level), the grid system defined in dispersion models, and the scales at which census population data are available do not generally overlap each other. Some conversion has to be made and hence may introduce some degree of uncertainty. This issue is not well addressed in those urban risk analyses discussed previously. Ideally, grid resolution should be fine enough or the size of grid as small as possible, but it is difficult in practice. One way is to test the sensitivity of the results with different spatial specifications to find the best spatial resolution.

Since area sources may have considerably significant contributions to environmental risks, the performance of various atmospheric dispersion models used for evaluating area sources should be evaluated. Certainly, the performance is also linked to the spatial resolution of the study area. Area sources are closely related to population distribution. Distribution of emissions will change as the population distribution changes over time. The current practice of assuming constant emission rates certainly biases exposure and risk estimates. Population forecasting and residential location models can be incorporated to increase the accuracy of not only area emissions but also population exposure.

The Chicago study has demonstrated the relative contribution of mobile sources to environmental risks (causing 16% of the total cancer cases). EPA-recommended model MOBILE can be used to estimate mobile source emission factors in a specific urban area. This is being done by major metropolitan organizations around the nation to comply with the Clean Air Act Amendments (CAAA), the Intermodal Surface Transportation Efficiency Act (ISTEA), and TEA-21. Then, a transportation network model can be used to estimate the mobile source emission in the highway network. This modeling effort will improve the quality of mobile source emission estimates.

More accurate and realistic results can be obtained if human time-activity patterns and various microenvironments are incorporated in estimating lifetime exposure and risk. This may include an individual's daily activity patterns, change of such patterns over the lifetime, change of residence, the extrapolation of individuals to population using the distributions of these variables. To address environmental justice, dividing population into subpopulations by race, income, and age is important in estimating exposure and risks. As mentioned above, a population-forecasting residential location model can produce a better description of how many people will be exposed to air toxics than the current assumption of constant population at each site.

GIS can be used to store, manipulate, display, and analyze the results from risk model runs (Chapter 8). Many powerful functions are particularly suitable for urban risk analysis. For example, the spatial distributions of risks imposed by individual pollutants can be created and displayed separately to see their respective features, while the overall risk distribution can be obtained by overlaying them. This is, of course, based on the assumption of additivity of environmental risks. Another way to use GIS is to identify the hot spots of environmental risks in urban areas, and to relate these spots to their socioeconomic characteristics.

4.6 SUMMARY

As seen in this chapter, a wide range of impacts can occur from environmental hazards, such as human health, ecological health, and economic and psychological impacts. We discussed a spectrum of methods and tools for measuring these impacts. Our concerns for environmental justice have been largely focused on human health. However, there is still a knowledge gap in the causal linkage between environmental agents and health risks. If we want to investigate human health impacts alone, proximity-based measures are simply the poorest surrogates. Ambient environmental models, in conjunction with ambient monitoring and emissions models, can be easily adapted to provide more accurate measures of potential exposures. We still have a long way to go to measure actual exposure or risk. Economic methods such as hedonic pricing methods, although seldom used in environmental justice analysis, show potential for estimating economic impacts by different subpopulations. The proximity approach can capture some non-health-related impacts because evidence has shown that some economic and psychological impacts occur in the vicinity of environmentally risky facilities. However, some of the assumptions underlying the proximity approach are seldom realistic. Hedonic pricing methods can provide information about where economic impacts diminish and how they are spatially distributed. Hedonic pricing methods and GIS can be combined to generate a more accurate picture of economic burdens associated with different population groups. For evaluating health impacts, the debate on the risk-based approach illustrates some pitfalls in the current practice of risk assessment, but risk assessment can be and should be reformed to serve the needs of environmental justice analysis.

5 Quantifying and Projecting Population Distribution

In this chapter, we review the content of the census, the major source of population and housing data. We then examine various measures of population such as race/ethnicity, income, age, housing, and education, and look at their application to environmental justice analysis. Next, we discuss the spatial patterns of population distribution by race/ethnicity and income in 1990 and place the national, regional, and local patterns in a historical, social, and economic context. Finally, we review three categories of population forecasting techniques and discuss their advantages and disadvantages.

5.1 CENSUS

Many analysts take census data for granted. Few realize that taking the census derives from political need. The U.S. Constitution mandates it every ten years for the primary purpose of providing a basis for apportioning congressional representation among the states. Each state is guaranteed one seat, and 435 seats in the U.S. House of Representatives are distributed once every 10 years among the states in proportion to population size. Similarly, apportioning political power based on population size is done at the state and local level. Census data are the basis used to draw congressional, state, and local legislative districts. Another major use of census data has to do with economic power. Each year billions of dollars of federal funds (currently, over $150 billion annually) are allocated to the state and local governments according to formulas that rely on census data. Census data used in allocation formulas include population, per capita income, unemployment rates, and age of housing. These funds cover a wide range of social concerns from education and employment to health care, housing, and transportation.

Prior to 1970, population was counted by door-to-door enumeration. Since 1970, mail enumeration has been used: census questionnaires are mailed to each known residential address, and households are asked to complete and return them. For nonrespondents (25.9% of the households in 1990), enumerators were sent for door-to-door collection of census information.

Two types of census questionnaires have been used to collect data in most recent censuses. A short-form questionnaire has a brief list of questions and goes to the majority of all housing units (5 in 6 or 83% of housing units for Census 2000). A long-form questionnaire has a larger number of questions (including those in the short form) and goes to a sample of housing units (1 in 6 housing units for Census

TABLE 5.1
2000 Census Content

Short Form (asked of all housing units)

Population	Housing
Name	Tenure
Sex	(home owned or rented)
Age	
Relationship	
Hispanic Origin	
Race	

Long Form (asked of 1 in 6 housing units)

Population	Housing

Social Characteristics
Marital status
Place of birth, citizenship, and year of entry
Education, school enrollment and educational
  attainment
Ancestry
Residence 5 years ago (migration)
Language spoken at home
Veteran status
Disability
Grandparents as caregivers

Physical Characteristics
Units in structure
Number of rooms
Number of bedrooms
Plumbing and kitchen facilities
Year structure built
Year moved into unit
Housing heat fuel
Telephone
Vehicles available
Farm residence

Economic Characteristics
Labor force status (current)
Place of work and journey to work
Work status last year
Industry, occupation, and class of worker
Income (previous year)

Financial Characteristics
Value of home
Monthly rent (including congregate housing)
Shelter costs (selected monthly owner costs)

Notes:

Changes from the 1990 census

- Added grandparents as caregivers
- Deleted children ever born (fertility), year last worked, source of water, sewage disposal, condominium status
- Moved from short form to long form: marital status, units in structure, number of rooms, value of home, and monthly rent

Changes in the 1990 census from the 1980 census:

- Added congregate housing (meals included in rent), disability
- Added more detailed questions in shelter costs
- Moved from long form to short form: condominium status
- Moved from short form to long form: number of units in structure
- Deleted: number of bathrooms, air conditioning, stories in building, marital history

2000). The content of the questionnaires varies slightly from one census to another. Census 2000 covers 7 subjects in the short form and 34 subjects in the long form, compared with 12 and 38 subjects, respectively, for the short and long forms in 1990. The Census 2000 short form is the shortest form in 180 years. Table 5.1 shows the variables in the questionnaires for Census 2000 and the changes from 1980 and 1990 censuses.

Table 5.1 classifies census data into two categories: population and housing. For the housing universe, the fundamental unit is the housing unit, which can be vacant or occupied as separate living quarters. The occupied housing unit defines a household. Individual persons in a household are the fundamental population units. These individual persons are either working or not working and have their own economic status (Myers 1992). Several census variables can be confusing, such as household vs. family, household population vs. total population. Household population is equal to total population minus institutional population, which includes military personnel, college students, retirees in group homes, prisoners, homeless persons, and any others who do not live in households.

A family is a group of two or more persons related by birth, marriage, or adoption who live together. For example, if a married couple, their nephew, their daughter and her husband and two children all lived in the same house or apartment, they would all be considered members of a single family. On the other hand, a household consists of all the persons who occupy a housing unit (house or apartment), whether they are related to each other or not. If a family and an unrelated individual live in the same housing unit, they would constitute two family units, but only one household.

While decennial censuses are the most important source for socioeconomic data, it is a snapshot of the census year and soon becomes outdated. The usefulness of census data to represent current socioeconomic situations gradually diminishes between two censuses. Still, you will find many analysts using the 1990 census data at the end of the 1990s. For slowly changing areas, using previous census data will probably not result in many biases. It will be problematic for rapidly changing areas. In these cases, it is necessary to rely on the most recent estimates for non-census years. For non-census years, socioeconomic data available are limited in both data items and geographic levels.

For environmental justice analysts, the good news is that census reports devote a lot of space to data on disadvantaged groups of the society, who are subjects of federal programs. The bad news is that census data tend to be the least accurate for society's disadvantaged groups. This is where the most controversial issue in recent censuses arises: undercount, which will be discussed in detail later.

5.2   POPULATION MEASUREMENTS: WHO ARE DISADVANTAGED?

While measuring environmental risks in space is difficult, measuring the socioeconomic characteristics of population distribution is not without problems. Researchers are first confronted with the question of which subpopulation(s) in a society should be the focus for the purpose of environmental justice and equity analysis. Legally, several segments of the population are protected from discriminatory practices. Title VI of the Civil Rights Act and related regulations prohibit discrimination on the basis of **race, color, national origin, religion, sex, age, or disability**. Therefore, these legally protected populations should be considered for equity analysis. Specifically for environmental justice, Executive Order 12898 targets minority populations and low-income populations. The segment of the population that EPA and other federal agencies focus on includes only minority and low-income populations. These two subpopulations are also the subjects in most environmental justice and equity analyses. Greenberg (1993) argues that environmental justice and equity studies should include the subpopulation who is young and old because it is more vulnerable and susceptible. In some sense, they are socioeconomically disadvantaged groups.

The second issue is how to measure these socioeconomically disadvantaged groups. There are various measures, each of which has advantages and disadvantages. In the following, different variables and their measurements used in environmental justice and equity studies are reviewed, and their advantages and disadvantages are discussed.

5.2.1 RACE AND ETHNICITY

Race and ethnicity are used daily. However, concepts of race and ethnicity are becoming more difficult to define in modern times (Zimmerman 1994; Rios, Poje, and Detels 1993). Historically, physical features (e.g., skin color, hair characteristics, and facial features) were used to classify race. These features were believed to possess distinctive hereditary traits that allowed biologically relevant classifications (Rios, Poje, and Detels 1993). This classification is reflected in EPA's early definition of race. "'Race' differentiates among population groups based on physical characteristics of a genetic origin (i.e., skin color)" (U.S. EPA 1992a:10).

However, very complex combinations of genetic traits resulting from interracial marriages have rendered biological classification of race less relevant and useful (Rios, Poje, and Detels 1993). Concerns have been raised about the use of race as a variable for measuring social and economic disadvantage by health researchers and social scientists (Montgomery and Cater-Pokras 1993). Some demography scholars have argued against the use of race for classifying population. The United Nations recommended the use of the term "ethnic group" as a comprehensive descriptor for classifying culturally and socially allied populations (UNESCO 1975).

Ethnicity is not a concept without any practical difficulty in conceptualization and implementation. "*Ethnicity* usually refers to common or shared cultures, origins, and activities (originating within the culture)" (Zimmerman 1994). And similarly, according to EPA, "'ethnicity' refers to differences associated with cultural or geographic differences (i.e., Hispanic, Irish)" (U.S. EPA 1992a:10). However, cultures are subject to individual interpretations and identifications, and there are no universal criteria for defining the concept of ethnicity.

Race and ethnicity data are collected in two separate questions in the census. Race and ethnicity are determined through *self-identification* (Bureau of the Census 1992a; Myers 1992). "The data for race represent self-classification by people according to the race with which they most closely identify" (Bureau of the Census

1992a:B-28). Race categories used in the census do not reflect biological stock scientifically defined but "include both racial and national origin or socio-cultural groups." The census race and ethnicity categories reflect a "social-political construct" and are "not anthropologically or scientifically based."

The difficulties of this self-identification approach include possible confusion of race with national origin, language, and religion, possible lack of match with the standard categories provided in the census, and complication for multiracial families (Myers 1992). The race/ethnicity classification standards have been under attack, particularly since the 1990 census. Critics believe that the race/ethnicity classification standards do not reflect the increasing diversity of the nation's population.

In response to the criticisms, the Office of Management and Budget initiated a comprehensive review in 1993. As a result of this review, OMB decided to revise race and ethnicity standards: (1) the Asian or Pacific Islander category will be separated into two categories — "Asian" and "Native Hawaiian or Other Pacific Islander," and (2) the term "Hispanic" will be changed to "Hispanic or Latino." The revised standards will have five minimum categories for race: American Indian or Alaska Native, Asian, black or African-American, Native Hawaiian or Other Pacific Islander, and White. There will be two categories for ethnicity: "Hispanic or Latino" and "Not Hispanic or Latino." When self-identification is used, respondents will be given the choice of reporting more than one race. OMB decided that the method for respondents to report more than one race should take the form of multiple responses to a single question and not a "multiracial" category. The adoption of "Hispanic or Latino" is to better reflect regional differences in usage: Hispanic is commonly used in the eastern portion of the U.S., whereas Latino is commonly used in the western portion. The reason for a breakdown of the Asian or Pacific Islander category is to better "describe their social and economic situation and to monitor discrimination against Native Hawaiians in housing, education, employment, and other areas."

The new categories and definitions are

- American Indian or Alaska Native. A person having origins in any of the original peoples of North and South America (including Central America), and who maintains tribal affiliation or community attachment.
- Asian. A person having origins in any of the original peoples of the Far East, Southeast Asia, or the Indian subcontinent including, for example, Cambodia, China, India, Japan, Korea, Malaysia, Pakistan, the Philippine Islands, Thailand, and Vietnam.
- Black or African American. A person having origins in any of the black racial groups of Africa.
- Hispanic or Latino. A person of Cuban, Mexican, Puerto Rican, South or Central American, or other Spanish culture or origin, regardless of race. The term, "Spanish origin," can be used in addition to "Hispanic or Latino."
- Native Hawaiian or Other Pacific Islander. A person having origins in any of the original peoples of Hawaii, Guam, Samoa, or other Pacific Islands.
- White. A person having origins in any of the original peoples of Europe, the Middle East, or North Africa.

This change in racial categories and terms is not the only one, and various terms have previously been used in census questionnaires and reports (Myers 1992). Negro was used in the pre-1980 censuses. Instead of Hispanic, Hispanic/Spanish was used in the 1980 census, and Spanish was used in the 1970 census. To analyze demographic changes over time, the analyst needs to be careful about the changing definitions of the census data. In particular, changes in the definition of Hispanic greatly affect comparability over time of Hispanic and race data between 1970 and post-1970 censuses. In 1970, inconsistent definitions of Spanish origin were used across the country. The 1980 census reports higher counts of Hispanics through better coverage, but is not directly comparable with the 1970 census. The 1980 census also reports a much larger proportion of Hispanics identified as *other* races than the 1970 census. In 1970, only 1% of Spanish origin population identified themselves as *other* races, but 38% did so in 1980. As a result, the 1970 white population was inflated, while *other* races were deflated (Myers 1992).

Race and Hispanic origin are complete-count variables in the census, which implies that they are free of any random sampling errors. However, this does not mean that they are free of non-random errors. Since first conducted in 1790, each decennial census has striven to count each and every person in the country. How well has each census reached this goal? The goal has remained elusive. Recent evidence suggests *net undercounting* of the population; that is, the undercounting is greater than the overcounting. If this undercounting occurs evenly among different subpopulations and places, the impacts will be trivial in terms of allocating political power and financial resources. It is the differential net undercounting among different subpopulations and places that skews the allocation of political power and financial resources and that recently received great attention. To investigate the undercount, the Census Bureau conducted demographic analysis and the post-enumeration survey during the 1990 census (Wolter 1991).

The good news from these analyses is that net undercounting of the population declined steadily from 5.4% in 1940 to 1.2% in 1980. The bad news is that net undercounting rose to 1.8% in 1990 and minorities such as blacks have been consistently undercounted at a much higher rate. The differential net undercounting between blacks and nonblacks increased from 3.4% points in 1940 to 4.4% points in 1990. For the 1990 census, blacks had a net undercount of 5.7%, compared with 1.3% for nonblacks.

Furthermore, differential net undercount occurs at different degrees in different places. A post-enumeration survey (PES) was conducted to cross-check a sample of 170,000 housing units in approximately 5,400 block clusters (Hogan 1990). Through the capture–recapture method, the PES tried to estimate the number of persons missed by the census and those factors for allocating adjusted counts to small areas in the nation. Post-strata (1,392 in total) were defined for types of persons by four race groups, six age groups, two sexes, region of the country, type of location, and type of housing (rented or owned). New Mexico had the highest net undercount rate of 4.5%, followed by California with 3.7%.

In spite of these difficulties, the census data of race and ethnicity are used in almost all environmental justice studies. Table 5.2 lists a series of race/ethnicity variables used in environmental justice and equity studies. Percent black and percent

Hispanics are the two most frequently used variables, while few studies include percent Native American and percent Asian/Pacific Islander. The aggregated variables, such as percent nonwhite and percent minorities, are very helpful for providing a holistic picture of the aggregated groups as a whole and for making comparison with the white share. However, the aggregated variables mask the differences among various groups in terms of location choice, behaviors, and cultures. More detailed disaggregations are very helpful for detecting any differences among the minority groups.

Definitions in the environmental justice guidance of federal agencies generally follow census definitions. Furthermore, the CEQ Environmental Justice guidance provides the following for identifying minority population (CEQ 1997).

Minority populations should be identified where either: (a) the minority population of the affected area exceeds 50 percent or (b) the minority population percentage of the affected area is meaningfully greater than the minority population percentage in the general population or other appropriate unit of geographic analysis. In identifying minority communities, agencies may consider as a community either a group of individuals living in geographic proximity to one another, or a geographically dispersed/transient set of individuals (such as migrant workers or Native American), here either type of group experiences common conditions of environmental exposure or effect. The selection of the appropriate unit of geographic analysis may be a governing body's jurisdiction, a neighborhood, census tract, or other similar unit that is to be chosen so as to not artificially dilute or inflate the affected minority population. A minority population also exists if there is more than one minority group present and the minority percentage, as calculated by aggregating all minority persons, meets one of the above-stated thresholds.

5.2.2 INCOME

There are many measures of income that can be used to classify economically disadvantaged populations. In the census, income is defined as total money income received by persons in the calendar year preceding the census. The eight types of income reported in the census are wage or salary income; nonfarm self-employment income; farm self-employment income; interest, dividend, or net rental income; social security income; public assistance income; retirement or disability income; and all other income. The income information collected in the census clearly represents only current income before taxes, not wealth. Not represented in the current income measures are, for example, home ownership and car ownership, which may be an important factor in an individual's economic well-being. Therefore, we must recognize the discrepancy between wealth and income, which grows larger at older ages, and may vary with social groups and across places. Fundamentally, we want to ask the question: How good an indicator is the current income measure as collected in the census for classifying economically disadvantaged populations? Public health research has shown that home ownership and car ownership have inverse relationships to mortality (Montgomery and Cater-Pokras 1993). Housing-related measures have been used in environmental justice and equity studies (see Table 5.2), but car ownership has never been used.

TABLE 5.2
Examples of Population Measures Used in Environmental Justice Studies

Population Variables	Measures
Race/ethnicity measures	% black or African American, % Native American, % Asian/Pacific Islander, % other races, % Hispanic, % nonwhite, % minorities.
Income	% families below poverty level, % population below poverty level, per capita income, median family income, mean family income, family income distribution, median household income, mean household income, household income distribution, % households receiving public assistance, median black household income, % poor, % poor whites, % poor blacks, % poor blacks among all the poor, % poor blacks among blacks
Age	% population under 5 years old (% young)
	% population under 15 years old
	% population under 18 years old
	% population 65 years old or older (% elderly)
	% female age 15 to 44
	Median age
Housing	Median value of owner-occupied housing units (housing stock)
	Median rent
	Mean estimated house value
	Median % of income devoted to rent
	Mean age of housing units
	% housing units built before 1940
	Housing tenure (owner occupied or rent)
	% housing units occupied by owners
	% housing units vacant
Education	% population with 12 or more years of schooling
	% adults with 4 years of college
	Average years of school by persons age ≥ 25

Note: Native American = American Indian, Eskimo, and Aleut

Minority is often defined as the segment of population composed of (UCC, 1987; Glickman, Golding, and Hersh, 1995):

- Black population not of Hispanic origin
- Native American not of Hispanic origin
- Asian and Pacific Islander not of Hispanic origin
- Other races not of Hispanic origin
- Population of Hispanic origin

% poor = the number of persons living below the poverty level ($12,674 for a family of four in 1990) divided by the number of persons in the adjusted total population (i.e., total population less those held in institutions such as prisons and psychiatric hospitals).

Another question is: Given different measures of current income, which is the most appropriate one for the purpose of environmental justice and equity analysis? Or, is there a most appropriate single measure? As can been seen in Table 5.2, a number of income measures have been employed in environmental justice and equity analysis. Each one of them measures some aspect of current income. There are three units of analysis for income calculations: family, households, and population. For computing the family income measure, all members 15 years old and over in each family (family members and related persons) are included. Those unrelated persons living in the same household are excluded. Families are only a subset of households, which include the householder and all other persons 15 year old and over, whether related or not (Bureau of the Census 1992a; Myers 1992). Both family and household income measures reflect relative income (earning) levels in an area, and therefore are useful for cross-sectional comparisons. But total-population-based income measures, such as per capita income, are not well suited for comparing income across time and places because they include children and other nonworkers, which may also vary across time and places.

These income measures can be used via a point value (such as mean or median) or a distribution. As is well-known, the income distribution is highly skewed. Therefore, a median is a better measure for the actual income distribution than a mean, but not as good as the distribution measure itself. When aggregation of different areas has to be done, as we see in some environmental justice and equity analyses, mean values are more convenient. Used in cases of aggregations (Been 1994), a so-called weighted median is derived by multiplying each median by its base (e.g., the number of households or families), summing these products and then dividing the sum by the total base (e.g., total number of households or families). It must be pointed out that this weighted median is often a flawed measure for the median of the aggregated data unless the individual areas assume some unique distributions. A detailed discussion of this issue will be presented in Chapter 7.

Poverty measures are often used to represent the economically disadvantaged population. The federal governments use two slightly different versions of the poverty measure:

- The poverty thresholds
- The poverty guidelines

The poverty thresholds are the original version of the federal poverty measure. The thresholds are used mainly for statistical purposes; all official poverty population figures are calculated using the poverty thresholds, not the guidelines. They are based on a definition originated by the Social Security Administration in 1964 and subsequently modified in 1969 and 1980 (Bureau of the Census 1992a). This definition has as its core the 1961 economic food plan, the least costly of four nutritionally adequate plans designed by the Department of Agriculture. Poverty levels are set according to the cost of the economic food plan. The income cutoffs for determining poverty status include a set of thresholds taking into account the family size, number of children, and age of the family householder or unrelated person (see Table 5.3 for an example). The official poverty definition counts money income before taxes and excludes capital gains and noncash benefits (such as public housing, Medicaid, and food stamps). The poverty threshold line also makes some adjustment in the cost of living across years, based on the Consumer Price

TABLE 5.3
Weighted Average Poverty Thresholds Vary by Size of Family

Size of family unit	1980 ($)	1989 ($)	1998 ($)
One person	4,190	6,310	8,316
Two	5,363	8,076	10,634
Three	6,565	9,885	13,003
Four	8,414	12,674	16,660
Five	9,966	14,990	19,680
Six	11,269	16,921	22,228
Seven	12,761	19,162	25,257
Eight	14,199	21,328	28,166
Nine or more	16,896	25,480	33,339

Source: Bureau of the Census, Current Population Survey, Washington, D.C., 1999.

Index. However, it does not have an adjustment for regional differences in the cost of living, which varies considerably nationwide. Another problem is that the current definition may not catch up with the changes in the spending patterns of Americans (Montgomery and Carter-Pokras 1993). The Census Bureau is revising its definition of poverty with a formula that takes into account the changing spending patterns of what poor people spend on food, clothing, housing, and extras. Under the proposed new formula, for a family of four to be considered above the poverty line, its annual income would have to be $19,500 a year, instead of the current $16,660 per year. The change would make 46 million Americans, 17% of the population, poor. As of September 1999, only 12.7% were considered poor, the lowest level in almost a decade. This new formula would send more families below the poverty line.

In 1997, the poverty rate was 11.0% for whites, and 14.0% for Asians and Pacific Islanders, compared with 26.5% for blacks and 27.1% for Hispanics (Bureau of the Census 1998). Even though the poverty rates for whites (11.0%) and non-Hispanic whites (8.6%) were lower than those for the other racial and ethnic groups, the majority of poor people in 1997 were white. Among the poor, 69% were white and 46% were non-Hispanic white.

The poverty guidelines are issued each year in the Federal Register by the Department of Health and Human Services (HHS). The guidelines are a simplification of the poverty thresholds used for administrative purposes (see Table 5.4). For example, the guidelines or percentage multiples of the guidelines are used to determine financial eligibility for certain federal programs, such as Head Start, the Food Stamp Program, the National School Lunch Program, and the Low-Income Home Energy Assistance Program.

Unlike the poverty thresholds, the poverty guidelines are designated by the year in which they are issued. For example, the guidelines issued in March 1999 are designated the 1999 poverty guidelines. However, the 1999 HHS poverty guidelines only reflect price changes through calendar year 1998. Accordingly, they are approximately equal to the Census Bureau poverty thresholds for calendar year 1998.

TABLE 5.4
1999 HHS Poverty Guidelines

Size of family unit	48 Contiguous States and D.C. ($)	Alaska ($)	Hawaii ($)
One person	8,240	10,320	9,490
Two	11,060	13,840	12,730
Three	13,880	17,360	15,970
Four	16,700	20,880	19,210
Five	19,520	24,400	22,450
Six	22,340	27,920	25,690
Seven	25,160	31,440	28,930
Eight	27,980	34,960	32,170
For each additional person, add	2,820	3,520	3,240

Source: U.S. Department of Health and Human Services, Federal Register, 64, 52, 13428-13430, March 18, 1999.

Federal programs in some cases use administrative definitions that differ somewhat from the statistical definitions; the federal office that administers a program has the responsibility for making decisions about definitions. "Family unit" has been used in the poverty guidelines Federal Register notice since 1978, although it is not an official U.S. Bureau of the Census term. Either an unrelated individual or a family (as defined for statistical purposes) constitutes a family unit. In other words, a family unit of size one is an unrelated individual, while a family unit of two or more persons is the same as a family of two or more persons.

Both measures of poverty have been used in Federal agencies' guidelines on environmental justice. The CEQ Environmental Justice guidelines define low-income population using the annual statistical poverty thresholds from the Bureau of the Census' Current Population Reports, Series P-60 on Income and Poverty. The Department of Transportation Order on environmental justice uses the Department of Health and Human Services poverty guidelines.

To fully account for income status across space and time, household income measures (median or particularly distribution) appear to be a better choice. Household income data are not available at all geographic scales. The decennial census reports household income down to the block-group level. For non-census years, the county level is often the smallest geography for which income data are available, although you can find income estimates down to census tracts in some areas, particularly in metropolitan areas. For non-census years, household income estimates are the most widely used and come from different sources (Galper 1998). Estimates from private data companies are based on data from federal government agencies such as the Census Bureau, the Internal Revenue Service, the Bureau of Labor Statistics, and the Bureau of Economic Analysis (BEA).

For non-census years, the Census Bureau estimates median household income at the county and state levels. These estimates use the Current Population Survey, tax returns, BEA data, and 1990 census data. These estimates are considered robust and are widely used, but they are slightly outdated (Galper 1998). ES-202 is the most widely used data source for estimating household income and employment by place of work. ES202 is a quarterly report of employment submitted by non-exempt businesses and governments. Not covered by ES-202 are self-employed workers, some types of agricultural workers, domestic workers, some government employees, and members of the Armed Forces. Tax returns may be the most truthful source of income data, but only total money income as defined by the Internal Revenue Service is available at the county level. Total money income misses a significant amount of income such as tax-exempt interest, dividends, capital gains and losses, and others.

Private data firms also produce household income estimates at the county level. Market Statistics generates Median Household Effective Buying Income (EBI) and Average Household EBI at the county level (Galper 1998). This measure takes into account household income estimates made by the BEA, aggregate payments of income tax, and ES-202 data. It represents disposable income and is often used for marketing purposes. Woods & Poole Economics, Inc. estimates a Mean Household Income statistic, which is more inclusive. Its "total personal income" includes both earned and unearned income, and may more accurately represent total economic resources.

These county-level household income estimates may not help a fine-scale environmental justice analysis. For many metropolitan areas, analysts will be able to buy household income estimates at a smaller geographic level from private firms or metropolitan planning organizations (MPOs). As will be discussed in detail in Chapter 13, MPOs are responsible for making long-range transportation plans for metropolitan areas. The staff of MPOs or their forecasting committees usually estimate socioeconomic data including household income at the Transportation Analysis Zone (TAZ) level. As with other estimates, analysts must understand their underlying assumptions, definitions, and estimation methods.

5.2.3 HIGHLY SUSCEPTIBLE OR EXPOSED SUBPOPULATIONS

Susceptibility refers to an individual's biological sensitivity. "A 'sensitive' individual is one who shows an adverse effect to a toxic agent at lower doses than the general population or who shows more severe frequent adverse effects after exposure to similar amounts of a toxic agent as the general population" (U.S. EPA 1999a:1-4). Biological factors that affect susceptibility include genetic characteristics and disease frequencies, which vary with age, sex, race, and ethnicity (Rios, Poje, and Detels 1993). "Individuals are 'highly exposed' on the basis of their activities, preferences, and behavior patterns that differ from those established for the general population" (U.S. EPA 1999a:1-4). Exposure is often affected by nonbiological factors. These nonbiological factors include lifestyle factors such as smoking, diet and nutrition, substance abuse, activity patterns, residential proximity to waste facilities; socioeconomic status and social inequality such as access to education, employment, housing, and health care. These factors may also vary with age and gender. Individuals can be at a greater health risk when they are "more exposed" or "more susceptible." Although this distinction is necessary, EPA investigators also use the term "susceptible" to refer to those highly exposed individuals.

Health disparities by race/ethnicity and income have been extensively documented (Institute of Medicine 1999). The percentage of low-birth-weight was higher among African-American women (11.9%), American Indian (6.0%), Asian or Pacific Islander women (6.8%), and Hispanic women (6.0%) than white women (5.5%) with similar levels of education. Minorities also had higher infant mortality and overall mortality rates. A report from a committee of the Institute of Medicine finds that rates of certain types of cancer among the poor and certain ethnic minorities have remained high even though overall cancer rates in the U.S. have fallen in recent years (Haynes and Smedley 1999). In particular, the report finds:

- African-American males develop cancer 15% more frequently than white males.
- African-American men are more likely to develop prostate cancer.
- Asian-Americans are more likely to develop stomach and liver cancer than white Americans.
- Cervical cancer rates are higher among woman of Hispanic and Vietnamese descent.
- African-American women are less likely to develop breast cancer, but once detected, they are less likely than white women to survive.
- Native Americans have the lowest cancer survival rates.
- Poor individuals have high cancer incidence and mortality rates and low rates of survival from cancer. For example, in Appalachian Kentucky, the incidence of lung cancer among white men was 127 per 100,000 in 1992. The rate is higher than that for any ethnic minority group in the U.S. at that same time.

Table 5.5 shows some examples of subpopulations who are more susceptible or exposed. As discussed below, age is a salient factor in identifying both susceptible and exposed subpopulations. Children and seniors have the potential for being more susceptible and exposed than the general population. Evidence also shows gender-related differences in susceptibility and exposure. For example, pregnant women have the greater potential for being exposed to contaminants because of increased food consumption. We can also identify highly exposed populations on the basis of exposure pathways (Table 5.6).

5.2.4 AGE

It has been long recognized that people in different age cohorts have different health risks and different behaviors related to the life cycle. Age is useful to identify both highly susceptible and exposed populations. Both the young and the old are more vulnerable and susceptible because of immunological deficiencies, and are likely to be more exposed to environmental risks due to inactivity for the elderly or unique activity patterns for the young (Sexton et al. 1993). For example, children and fetuses are more sensitive to chemicals such as lead, which has neurotoxic effects. Infants and young children are more exposed to contaminants such as lead through hand-to-mouth behaviors. The elderly have a decreased capacity to detoxify chemicals and have a functional decline of the immune system (U.S. EPA 1999a).

TABLE 5.5
Examples of Highly Susceptible or Exposed Subpopulations

Subpopulation	Susceptible Factors	Subpopulation	Exposure Factors
Asthmatics	Increased airway responsiveness to allergens, respiratory irritants, and infectious agents	Industrial workers	Higher exposure to job-related hazardous chemicals through breathing and skin contact; more lung exposure due to physically demanding work
Fetuses	Sensitivity of developing organs to toxicants that cause birth defects	Farmers	Pesticide exposure
Infants and young children	Sensitivity of developing brain to neurotoxic agents such as lead	Infants and children	Higher consumption of fruit, vegetables, and fruit juices; higher inhalation rates
Elderly	Diminished detoxification and elimination mechanisms in kidney and liver	Subsistence and sports fishers	Higher fish consumption
Low income population	Nutritional deficiencies and poor access to health care	Low income and minority population	Higher exposure to lead, air pollution, and toxics.
α_1-Antitrypsin-deficient persons	Inherited deficiency of a protein that protects against chemical damage		
Gluthathione-S-transferase deficient persons	Diminished detoxification of some carcinogens and medicines		

Source: Adapted from the Presidential/Congressional Commission on Risk Assessment and Risk Management, 72, 76, 1997b.

Besides being a health risk factor, age pattern in a community reflects neighborhood characteristics and their dynamics, which are related to life cycle. As discussed in Chapter 2, the life-cycle model can help us understand the causal linkage in neighborhood changes involving environmentally risky facilities and LULUs. Greenberg (1993) noticed a lack of environmental justice research interest in the subpopulations who are young or old and called for more studies in this area.

TABLE 5.6
Identifying Potential Highly Exposed Populations on the Basis of Exposure Pathways

Exposure Pathway	Potential Highly Exposed Populations
Water ingestion	Athletes, residents of hot climates, outdoor activity participants in hot climates/weather
Soil ingestion	Children, pregnant women, migrant workers, outdoor activity participants (e.g., gardening, sports)
Inhalation	Athletes, children, outdoor sports participants, outdoor workers (e.g., farmers and construction workers)
Dermal contact with soil	Children, home gardeners, outdoor sports participants, outdoor workers (e.g., farmers and construction workers)
Fish ingestion	Fishers, Eskimos, Native Americans
Dermal contact with water	Fishers, occupational and recreational aquatic sportsmen (e.g., swimmers, boaters)

Source: Adapted from U.S. EPA, 1999a.

Age is measured in detail in the census. A lot of variables are cross-tabulated by age. Few have been used in environmental justice and equity studies (see Table 5.2). It is customary to define the elderly as those 65 years of age and older. It is less clear what the age cut-off is for the young. We need a more accurate biological definition of these two groups related to their susceptibility to environmental risks.

5.2.5 HOUSING

As mentioned earlier, housing is an indicator of households' wealth, reflecting not only earned income but also non-earned income. Housing is also a primary characteristic of communities. It signifies not only the economic well-being but also the social status of a community. To some extent, it can complement the income variable to identify socially and economically disadvantaged groups in environmental justice and equity analysis.

There are two broad types of housing measurements in the census: the physical and economic characteristics of the housing stock and the characteristics of the household's fit to the housing unit (Myers 1992). In the former are tenure (rent or own), number of units in the structure, number of rooms or bedrooms, age or year built, adequacy of plumbing, and its cost. The latter includes person per room (indicating the level of crowding), percentage of household income spent on the rent or mortgage (indicating affordability), and the length of time the household has occupied the structure.

With few exceptions (Earickson and Billick 1988), environmental justice and equity studies use the housing stock (rather than the fit) characteristics (see Table 5.2). These studies vary in their rationales in the selection of particular housing measures. In some multivariate exploratory data analyses, a wide range of housing characteristics was used (Napton and Day 1992; Earickson and Billick 1988). These housing variables represent the "characteristics of place" (Napton and Day 1992). Most studies chose a couple of housing variables. In confirmatory data analysis, housing values are used to represent the housing market (Zimmerman 1994) and potential compensation (Hamilton 1995). Home ownership has been taken as a substitute for political participation.

The age of housing can be a proxy measure for potential exposure to lead paint. Residential paint contained up to 40 to 60% lead by weight. EPA estimated that lead paint was used in 65% of the houses built before 1940, 32% of the houses built between 1940 and 1960, and 20% of the houses built between 1960 and 1975.

5.2.6 EDUCATION

Educational level is a frequently used measure of socioeconomic status in social science and health research. It has very limited use in environmental justice and equity studies (see Table 5.2). Education was used as a proxy measure for people's willingness to pay, and expected compensations (Hamilton 1995).

Using education as an indicator of social class has some major problems (Montgomery and Cater-Pokras 1993). Education is not linearly related with income, and socioeconomic returns on education change over time. Socioeconomic returns on education vary also by gender and race. There are also other confounding factors, such as regional variability, work experience, and professional certifications, which make education a noisy measure of socioeconomic status.

5.3 POPULATION DISTRIBUTION

Distribution of population has national, regional, and local patterns. Nationwide, minority distribution shows a U-shaped pattern. Minority concentration stretches from California on the West Coast, through the U.S.-Mexican Border States, then through the southeastern states, to the New York metropolitan area on the East Coast. The southern regions are called "the Deep South." In this U-shaped belt itself, each race/ethnicity group of minority population has its own spatial patterns. Hispanics reside predominantly in California and the southwest states, while African-Americans are especially concentrated in the southeast and mid-Atlantic states extending from Texas to the New York metropolitan area. Asians/Pacific Islanders are congregated in California and in the New York metropolitan area, and Native Americans are mostly scattered in the states throughout the West, particularly Arizona, New Mexico, and the Dakotas. Except for Native Americans, minority groups tend to have a predominant representation in the nation's largest metropolitan areas. In a metropolitan area, their concentrations are astonishingly extensive in the central cities.

The rich and poor are not geographically distributed evenly. Generally, urban counties tend to be richer than rural counties. America's wealthiest counties include exclusive suburbs of large metropolitan areas such as Fairfield, CT, financial centers like Manhattan, NY, retreat and retirement communities such as Palm Beach County, FL, and sparsely populated counties with huge natural resources such as Glasscock,

TX, which has oil wells and their owners. The poorest counties include rural counties without any economic engine such as Appalachia, the Deep South, and the Texas-Mexico border, large Indian reservations, and urban areas with large slums.

These distributional patterns have important implications for environmental justice analysis. Regional differentiation in population distribution indicates that the choice of study area is important; regional studies may generate results that differ substantially from nationwide studies. These spatial patterns contribute to the debate on which is an appropriate comparison area. If metropolitan areas are the universe of a study, the minority proportion of total population is certainly higher than the national average. Therefore, the findings may depend on which is chosen as a comparison, the metropolitan area as a whole or the nation as a whole (see Chapter 11). As shown in several previous studies, the urban/rural difference confounds the findings. It has been found that minorities have shouldered a disproportionate burden of potential exposure to a particular environmental pollution or risks, while the rich are also at the higher potential risk. This seemingly incompatible finding has to do with the income structure differentiation between urban and rural counties. Although there are concentrated poverty pockets in most central cities, urban counties are largely richer than rural counties, and the rural poor tend be far away from industrial facilities and their by-products, environmental pollution, and risks. Therefore, we need to look carefully at the regional differences, the urban vs. rural dichotomy, the sub-county differentiation, and other fine-grained comparisons.

These distributional patterns reflect the historical backdrop of population settlement and current and long-standing social and economic structures at the macro level. At the micro level, they show an aggregated nature of residential location choice in an imperfect world. As discussed in Chapter 2, households make their residential location choices on the basis of several factors such as space, accessibility, environmental amenities, and racial externality. Households' demand for housing is balanced by supply in the real estate market. This market is not perfect but full of barriers such as racial discrimination in the real estate market and redlining in mortgage financing. On a national and regional basis, international immigration and domestic migration shape population distribution. All these contribute to spatial patterns of population distribution.

We should understand geographic distribution of racial and ethnic groups in a historical context and as a distribution of social and economic relations (Pulido, Sidawi, and Vos 1996). International immigration by various race and ethnic groups into the U.S. has historical patterns, and immigrants' settlements depend on the social, economic, and political structure during different periods of the country's development. Europeans ventured into the New World after Columbus' discovery. Their initial entry was largely in the northeastern U.S., and moved outward from there. Africans were forced to migrate to the New World as slaves. From the first slaves brought to the colony of Virginia in 1619 to the abolishment of slave trade in 1808, approximately 400,000 Africans were transported into the New World. These slaves were mostly concentrated in rural plantations in the South.

The first peak of immigration occurred in 1854, with 428,000 immigrants entering the U.S. They were predominantly Irish and German. The second peak was composed predominantly of persons from Italy and Eastern Europe. They initially settled in the industrial base in the Northeast and migrated to the West. The east-to-west migration has been one of the most significant migration streams in the U.S. Another very significant migration stream has been the rural-to-urban migration, occurring in the wake of the Industrial Revolution and urbanization. Since the abolishment of slavery, there has been a very significant stream of migration northward and westward. In particular, during and after World War I, blacks moved out of the South in massive numbers.

These forces that shaped current population distribution patterns should be taken into account in an environmental justice analysis. They show us that environmental justice concern is more than a simple correlation, and it compels us go to the deepest roots.

5.4   POPULATION PROJECTION AND FORECAST

Censuses provide a snapshot of current and past population characteristics in an area. They are critical data sources for evaluating environmental justice issues for past and present policies and programs. When dealing with the potential distributional impacts of proposed policies and projects, analysts also resort to current census data. Certainly, present residents have a high stake in whether and how proposed policies and projects will affect their neighborhoods. Will the proposed policies and projects affect demographic composition of the existing neighborhoods? A few studies look at whether past projects affect neighborhood changes (see Chapter 12), but no known attempt has been made to examine the potential impacts of proposed projects on neighborhood characteristics. Even without the proposed policies or projects, population characteristics may change in the future. Will the proposed project change the locus of neighborhood changes in the future? What about future residents? Do they have a stake in the current decision making?

Both private and public sectors do short-term and long-term planning for the future. These plans may have distributional impacts. To prevent adverse distributional impacts from happening in the future, policy makers and analysts must assess proposed policies, plans, and programs from the equity perspective. This type of analysis entails population projection and forecast.

The Bureau of Economic Analysis (BEA) is the most widely known source of demographic and economic forecasts in the country. The BEA produces OBERS forecasts, the oldest and best known forecasts at the county level. These forecasts include population and personal income by state, by BEA economic areas, and by county for 50 years into the future, at 5-year intervals for the first 20 years and at 10-year intervals thereafter.

Almost all Metropolitan Planning Organizations (MPOs) use economic and/or demographic forecasts in the planning process (Lawrence and Tegenfeldt 1997). These forecasts usually include total population and employment at the Transportation Analysis Zone (TAZ) level. Some MPOs break them into sub-categories such as households by types of dwelling units (single-family and multi-family), retail vs. non-retail employment, or several aggregate categories of employment. A small number of MPOs (11% of 54 MPOs interviewed) use forecasts of household income.

5.4.1 METHODS

Demographers distinguish population projection and forecast clearly, while others use the two terms interchangeably. "Projections are conditional ('if, then') statements about the future" (Isserman 1984:208). For example, a county demographer may say that if current birth, death, and migration rates continue, the county's population will increase by 40,000 by 2020. This says nothing about the validity of the underlying assumptions, which are critical for the projection. In fact, if you change current rates, you will have different projections. Every projection is correct under its assumptions, but it is hypothetical. A projection provides useful information about what-if in the future but does not tell us what will likely happen in the future.

A forecast is a judgmental statement of the most likely future. To make such a statement, the demographer or analyst must evaluate alternative assumptions (the "ifs") and identify those that are most realistic and likely to occur. We use forecasts daily. Weather forecast is the most popular example. The public sector uses a variety of forecasts, both short term and long term. Planners and policy makers use population forecasts to plan future infrastructures such as roads, public water and sewer service, and public schools.

Three approaches are generally used for projection and forecast: mathematical trend extrapolation, cohort-component model, and demographic-economic models (Isserman 1984). The extrapolation technique quantitatively characterizes a past trend in the form of an equation and extends the trend to project or forecast the future. The analyst must first choose the historical database, decide which equation(s) to use, fit the data with equation(s), identify the most suited functional form(s), and apply the best equation to project or forecast the future. In making projections, the analyst can present the results under different assumptions: different years of historical data that are assumed to continue into the future and different functional forms that imply different growth rates. In making a forecast, the analyst must choose the most appropriate historical period that will best forecast the likely future. The analyst must also evaluate different equation forms and identify the one that will best describe the future. The choice of historical periods and equation forms matters greatly, but there is no simple decision rule to guide the analyst. Various forms of mathematical functions have their implicit assumptions, strengths, and weaknesses (Table 5.7).

The extrapolation technique assumes that past trends will continue into the future. If this assumption holds, the best fitting equation will produce the best forecast. As any investment brochure tells you, past performance is no guarantee for future performance. Therefore, the analyst must look at not only the goodness-of-fit of various equations but also the reasonableness of the forecast results. The analyst may choose the equation that forecasts a future consistent with his or her expectations. As a result, a forecast using the extrapolation technique strongly depends on the analyst's professional judgment. This is related to the extrapolation technique's failure to account for the underlying forces of population changes.

The cohort-component model is an accounting framework to trace the effects of future births, deaths, and migration on population (Isserman 1984). The population is disaggregated into cohorts based on age, sex, and race. Each cohort is traced into

TABLE 5.7
Different Functional Forms of Extrapolation Methods

Extrapolation Methods	Functional Form	Implied Growth Patterns	Strengths and Weaknesses	Applicable Areas
Linear	$Y = a + bX$	Constant amount of growth	Simple to use and easy to interpret No upper limit Rarely happen for demographic and economic phenomena	Small, slow-growing areas
Exponential	$Y = ae^{bx}$	Constant percentage rate of growth	Reasonable for demographic processes No upper limit Does not account for declining growth in the long run due to resource constraints	Rapidly growing areas Short-term
Parabolic	$Y = a + bX + cX^2$	Constantly changing (increasing or decreasing) rate of growth	No upper limit Does not account for resource constraints.	Rapidly growing or declining areas
Logistic	$Y = (c + ab^x)^{-1}$	Small initial growth increments, increasingly larger increments until a point of reflection, and then increasingly smaller growth	Has an upper limit Recognizes resource constraints	

Source: Klosterman (1990)

the future by taking into account the three components of population change: fertility, mortality, and migration. Demographic analysis is based on the simple equation:

$$\text{Population} = \text{births} - \text{death} + \text{immigrants} - \text{emigrants.}$$

Fertility, mortality, and migration rates vary with age, sex, and race. For example, mortality rates by age are generally higher for males than females, higher for non-whites than for whites, and higher for the poor than for the rich (Klosterman 1990). Fertility rates by age also vary with race/ethnicity. Accounting for these fundamental demographic processes for different segments of the population, the cohort-component model improves the accuracy of population projection. It provides particularly useful information by disaggregating population by age, sex, and race. These disaggregated forecasts are vital for environmental justice analysis, as well as for city and regional, environmental, health care, housing, and facility planning for schools and other public infrastructures.

Cohort-component models, however, do not forecast fertility, mortality, and migration rates but take them as external inputs to determine future populations. They are crucial for the end results. Unfortunately, there is no reliable procedure to determine these rates for the future. The analyst has to rely on past trends and expectations about the future to decide the likely rates for forecasting the future population. This certainly involves judgment.

Both the extrapolation technique and the cohort-component model have applicability limitations in terms of geographic scales. Both methods work best at the county or higher geographic level. Sub-county areas often contain diverse populations, which have dramatically different fertility, mortality, and migration rates (Klosterman 1990). Generally, these rates are statistically reliable for large geographic areas, less so for small areas, and even unavailable for sub-county areas. Furthermore, sub-county areas are also highly volatile and easily skewed by a single large development project.

The economic-demographic methods estimate population change based on economic change (Isserman 1984). Of the economic-demographic methods, the recursive models first determine economic activity (employment) in the future and then derive population using an expected ratio of labor-force age population to employment. To estimate the future population-employment ratio, the analyst must assume future unemployment, labor-force participation, and dependency rates such as the ratio of total population to the labor-force age population. For more sophisticated models, so-called integrated urban models, which will be discussed later in Chapter 8, forecast population in sub-county areas, among other things. A component of urban models is often referred to as land use models, which are used by regional planning agencies in transportation planning.

5.4.2 CHOOSING THE RIGHT METHOD

Choice of population forecast methodology depends on how the analyst views the causal linkage between population and the independent variables that drive the demographic processes. There are two general views of causal dynamics for population change and two analytical perspectives (Myers 1992). At the county or higher geographic levels, analysts believe that employment growth induces migration, which, in combination with fertility and mortality, results in population growth. With population growth comes growth in households, which require housing.

At the sub-county level, local analysts hold an exactly opposite view of this causal linkage. They believe that land is subdivided and building permits are issued for building houses, which are then occupied by households. Households consist of people of different ages and sexes, who engage in employment and other social activities.

As a result of these two views, analysts adopt two analytical perspectives for population forecast: top-down and bottom-up. In the top-down perspective, analysts start with a population forecast at a higher geographic level and allocate the total to lower geographic levels. The bottom-up approach begins with permitted, platted, or planned development in the smallest geographic level of analysis such as blocks or block groups. This usually entails geocoding of building permits issued and land subdivision approved. Then, analysts aggregate them up to higher geographic levels such as census tracts, TAZs, planning districts, Minor Civil Divisions (MCDs), municipalities, and counties. This so-called "pipeline" methodology is especially popular among local demographers for short-term forecasts. For longer terms such as 25 years, they usually take into account planned development, availability of developable land or land capacity, land use plan, and local development and land use policies. Local analysts rely heavily on local land use plans and regulations and are often influenced by the wishes of politicians. This forecast is sometimes known as "plancast."

Not surprisingly, the two approaches often produce inconsistent forecasts at the same geographic level. To reconcile the differences, planners or demographers often rely on a scaling method. They generally take the forecast at the higher geographic level (often the county level) as a control total, calculate a scalar as the ratio of the control total to the sum of the bottom-up results, and use the scalar to scale up or down proportionately the bottom-up results at the lower geographic level such as TAZ.

Myers (1992) proposes a housing-based allocation method to integrate the two perspectives. The two-tiered method links population and housing in two opposite orders at the metropolitan and local levels. At the metropolitan level, standard demographic methods are used to forecast regional population in the future, which is then converted to expected demand for housing. This conversion takes into account head of household rates by age-sex group, home ownership rate, vacancy rates for rental and owner units, and existing housing stock.

Rather than allocating population forecasts from a higher geographic level to a lower one in the top-down approach, the proposed method allocates shares of the expected housing construction to subareas. This allocation is based on the subareas' attractiveness and available resources. In the final step of the four-step processes, analysts convert the expected local housing to future population growth. This conversion takes into account residential mobility, including out-mover households, stayer households, and in-mover households.

When making the choice of forecasting methods, the analyst should also consider the accuracy of various methods, the type and quality of data available, the scale of analysis, the length of the forecast period, the purpose of the forecast, and time and budget constraints (Greenberg, Krueckeberg, and Michaelson 1978).

Most studies on forecast evaluation focus on the nation, states, or counties. They generally use post-hoc analysis to compare past forecasts to a known target year or employ stochastic models of population growth to measure the random error associated with demographic change (Tayman 1996). Lawrence and Tegenfeldt (1997) compared the OBERS forecasts with actual data reported in the Statistical Abstract for eight states in the Ohio River Basin. They found that population forecasts at the state and national level are generally close to the actual data for 5- or 10-year forecasts. The 5-year forecasts usually have errors of a couple of percentage points at the state level. The longer-term forecasts show larger degrees of error, reaching double digits for more than 10-year forecasts. Personal income forecasts have much larger errors than population forecasts. The good news is that the 5-year and 10-year forecasts have improved remarkably over time. The errors for recent 5-year and 10-year forecasts are reduced to a couple of percentage points at the state and national level, but the longer-term forecasts are still unacceptably high.

Evidence on the accuracy of these forecasts at a disaggregated sub-county level is considerably sketchy. Almost half of the 42 surveyed MPOs that use forecasts believe the forecasts to be accurate, but 41% found their forecasts to be inaccurate or unreliable in some way (Lawrence and Tegenfeldt 1997).

Isserman (1977) used eight extrapolation models and decennial census data from 1930 to 1960 to project the population of 1,777 townships in Illinois and Indiana for 1960 and 1970. Projections were compared with actual census data to evaluate the models' accuracy as measured by the mean absolute percentage error. He found that no model was superior for all townships but some models were more accurate for certain types of areas. Generally, errors tended to increase as population size decreased and as growth rates increased (Isserman 1977; Tayman 1996).

A few measures have been used to evaluate the accuracy of forecasts. Mean absolute percent error (MAPE) evaluates the difference between estimated values and observed values in percentage terms and ignores the direction of error. A range of MAPEs has been reported for small-area forecasts (Tayman 1996): 28% for 112 townships in northern Illinois; 11 to 17% for Illinois townships; 17% for areas in Texas and North Dakota with populations between 2,500 and 10,000; 17 to 28% in Florida's census tracts; 19% for census tracts in the Dallas–Fort Worth metropolitan area; 28% for TAZs in metropolitan Dallas; and 21% for census tracts in San Diego County.

The more complicated models do not necessarily produce significantly more accurate forecasts than simple techniques. Tayman (1996) evaluated the accuracy of forecasts using a spatial interaction modeling system at the census tract level in San Diego County. He found that a forecast produced by this modeling system has an accuracy comparable to that based on other techniques. Other studies also find that simple extrapolation techniques generate forecasts with an accuracy comparable to more complex techniques (Smith and Sincich 1992). Some researchers stress that the accuracy of a forecast depends on agreement of the forecast model's underlying assumptions with the actual course of events, which has nothing to do with the forecasting technique itself. Certainly, the forecasting techniques themselves have their own technical and theoretical strengths and weaknesses.

Extrapolation techniques have modest data requirements and are easy to apply to any geographic scale. What extrapolation techniques do not offer is the capability of structural models, such as spatial interaction models or other urban models, to forecast potential policy responses. The extrapolation models do not depict any relationships that determine population change, while structural models contain explicit representations of the demographic change processes. This type of representation offers structural models a considerable advantage in evaluating potential policies. In this regard, structural models enable better informed decision making. One major application of spatial interaction land-use models is to evaluate the impacts of transportation policies and projects on the future distribution of population, employment, and land consumption in a metropolitan region.

As alluded previously, the difficulty of achieving an accurate forecast increases with the length of the forecast period and decreases with an increasing geographic level of analysis. The smaller the geographic level of analysis, the greater variability in local characteristics and the greater need for various symptomatic data to forecast population change. Greenberg, Krueckeberg, and Michaelson (1978) believe that almost any forecasting method is suitable for short-term periods of up to 10 years. For longer-term projections, simple historical extrapolations would be the least effective, and a cohort-component model or spatial interaction model would be needed.

5.5   SUMMARY

Census is the best data source for current and past population distribution, but we need to be aware of its caveats for serving environmental justice analysis. As the society evolves, measures of race, income, and other socioeconomic variables also change. These changes have important implications for longitudinal or dynamics analysis of environmental justice concerns. The time dimension is also important because census is only taken every decade. We need both current estimates and future forecasts of population distribution. Population forecasts are essential for us to better understand the potential impacts of proposed actions and plans in the future. These tools are seldom used in environmental justice studies. We should examine population distribution in a broader historical, social, economic, and political context. All these factors operate in concert and lead to what we are and where we are.

6 Defining Units of Analysis

Any study starts with a unit of analysis. For environmental justice analysis, there are at least two crucial questions to be answered. To what extent is the unit of analysis chosen in a study relevant to environmental and human impacts of the phenomena analyzed? How sensitive are the results to the uncertainties in the choice of a unit of analysis? Early environmental equity studies usually ignored these two issues, and have been recently challenged.

In this chapter, we first discuss the unit-of-analysis debate and look at the geographic units of analysis used in environmental justice studies. Since most of the units are based on census geography, we review concepts, criteria, and hierarchy of census geography. Then, we examine three issues of using census geography as a unit of equity analysis: consistency, comparability, and availability. Since there are so many census units, we would like to know which one is most appropriate, if any. Finally, we explore alternative ways to define units of analysis for environmental justice studies.

6.1 THE DEBATE ON CHOICE OF UNIT OF ANALYSIS

Choice of unit of analysis is one of the most controversial and critical issues in environmental justice studies. As noted in Chapters 1 and 3, the landmark UCC study concluded that minorities, and to a lesser extent the poor, bear a disproportionate burden of commercial treatment, storage, and disposal facilities (TSDFs) (UCC 1987). The UCC study used the ZIP Code as a unit of analysis. Critics argue that ZIP code areas are so large a geographic unit of analysis that there is a danger of committing ecological fallacy (Anderton et al. 1994). Using the census tract as a unit of analysis, the UMass study reported no association between racial composition of census tracts and the presence of TSDFs (Anderton et al. 1994). Use of a 2.5-mi radius circle as a unit of analysis produced results similar to the UCC study. Therefore, the authors concluded that choice of units of analysis affects research findings. Critics of the UMass study argue that census tracts may be too small, particularly in the central city, to sufficiently cover the environmental impact boundary (Mohai 1995). This sparked a debate on census tracts vs. ZIP codes.

In an effort to explicitly examine the impact of using different units of analysis on the results of an equity analysis, Glickman, Golding, and Hersh (1995) compared five different units of analysis: block group, census tract, municipality, and 0.5- and 1.0-mi radius circles around a facility. In the context of manufacturing facilities releasing air toxics in Allegheny, Pennsylvania, they found that the choice of a unit of analysis had dramatic effects on findings related to race/ethnicity, but relatively little effect on findings related to poverty (see Table 6.1). To test the sensitivity of equity analysis

TABLE 6.1
Do Environmental Justice Analysis Results Depend on the Geographic Units of Analysis?

Geographic units of analysis	Glickman, Golding, and Hersh (1995)		Cutter, Holm, and Clark (1996)	
	Racial inequity	Income inequity	Racial inequity	Income inequity
Block group	No	Yes	No	Yes, small
Census tract	No	Yes	No	Yes, small
Half-mi buffer	Yes	Yes		
One-mi buffer	Yes	Yes		
Municipality	Yes for all	Yes		
	Yes for Pittsburgh			
	No for all except			
	Pittsburgh			
County			No (reverse	No (reverse)
			relationship)	
Study area	Allegheny County		South Carolina	
			MSAs	
Type of facilities	TRI		TRI sites, TSDFs,	
			CERCLIS sites	
Dependent variable			Number of facilities	
Statistics	Aggregate measures		Pearson correlation,	
	(population-		t-test, discriminant	
	weighted means)		analysis	

Source: Brody, D.J., et al., *Journal of the American Medical Association*, 272(4): 277–283. With permission.

findings, Cutter, Holm, and Clark (1996) used three geographic units of analysis (i.e., block groups, census tracts, and counties) and three types of waste facilities (i.e., Toxics Release Inventory (TRI), TSDFs, and National Priority List (NPL) sites). Their study found no disproportionate impacts of waste facilities on the minority population at three levels of analysis unit but a slight disparity by income at census tract and block group levels in the state of South Carolina (see Table 6.1). They concluded that census tracts and block groups were the most appropriate units of analysis.

That research findings vary with geographic units of analysis is not a new discovery and has long been known as "the modifiable areal unit problem" (MAUP) in geography. There are two types of MAUPs: the scale effect and the zoning effect (Wrigley et al. 1996). The scale effect occurs when different statistical findings are obtained at different levels of spatial resolution (e.g., census tracts, blocks, and counties). The zoning effect happens when different statistical findings are obtained from different zone structures at a given scale (e.g., for a given number of 100 TAZs, whose boundaries can be configured in different ways). It was found that the correlation between variables tends to increase as the zone size increases (Openshaw and Taylor 1979). In addition, scale and zoning effects may result in different degrees of goodness-of-fit, different regression coefficient estimates and t values, and Moran's I in the linear regression (Fotheringham and Wong 1991).

Sui and Giardino (1995) examined the impacts of different "scale" and "zoning" schemes on equity analysis results in the city of Houston. Block groups, census tracts, and ZIP codes were used to test the "scale dependency" hypothesis. To test whether different zoning schemes (different areal unit boundary) affect the results (the "zoning dependency" hypothesis), tract level data were regrouped into three sets of spatial units: (1) 1.5-mi buffers along major highways; (2) 1.5-, 3.0-, and 4.5-mi circular buffers around major population centers; (3) 45° sectoral patterns on four concentric rings for three major ethnic enclaves. The number of TRI sites was regressed on (1) minority population, per capita income, and population density; (2) percentage of black population, percentage of Hispanic population, percentage of Asian population. The results supported both hypotheses. As the geographic reso-lution became larger (from block groups to census tracts to ZIP codes), the impor-tance of the minority population variable increased in explaining the number of TRI sites, and per capita income and population density became less significant. As the zoning scheme changed from buffer zones along highways to circular buffers to sectoral radii, the minority population became substantially less important.

A variety of units of analysis have been used in environmental justice studies (see Table 6.2). These include legal units such as states, counties, MCDs, incorpo-rated places; administrative entities such as ZIP codes; statistical entities such as Metropolitan Areas (MAs), census tracts/block numbering areas, block groups, blocks; and GIS-based units such as a circle around a facility. Almost all of these units are based on census geography, to which we are turning next.

TABLE 6.2
Geographic Entities of the 1990 Census and Their Use in Environmental Justice Studies

Type of geographic entity	Status	Number	Used in EJ studies?
Nation (the United States)	Legal	1	
Regions (of the United States)	Statistical	4	
Divisions (of the United States)	Statistical	9	
States and Statistically Equivalent Entities[a]	Legal	57	Yes
Counties and Statistically Equivalent Entities	Legal	3,248	Yes
County Subdivisions and Places		60,228	
Minor Civil Divisions (MCDs)	Legal	30,386	Yes
Sub-MCDs	Legal	145	
Census County Divisions (CCDs)	Statistical	5,581	Yes
Unorganized Territories (UTs)	Statistical	282	
Other Statistically Equivalent Entities[b]	Statistical	40	
Incorporated Places[c]	Legal	19,365	Yes
Consolidated Cities	Legal	6	
Census Designated Places (CDPs)	Statistical	4,423	Yes
American Indian and Alaska Native Areas (AIANAs)		576	
American Indian Reservations (no trust lands)	Legal	259	

continued

TABLE 6.2 (CONTINUED)
Geographic Entities of the 1990 Census and Their Use in Environmental Justice Studies

Type of geographic entity	Status	Number	Used in EJ studies?
American Indian Entities with Trust Lands	Legal	52	
Tribal Jurisdiction Statistical Areas (TJSAs)	Statistical	19	
Tribal Designated Statistical Areas (TDSAs)	Statistical	17	
Alaska Native Village Statistical Areas (ANVSAs)	Statistical	217	
Alaska Native Regional Corporations (ANRCs)	Legal	12	
Metropolitan Areas (MAs)		362	Yes
Metropolitan Statistical Areas (MSAs)	Statistical	268	Yes
Consolidated Metropolitan Statistical Areas (CMSAs)	Statistical	21	Yes
Primary Metropolitan Statistical Areas (PMSAs)	Statistical	73	Yes
Urbanized Areas (UAs)	Statistical	405	
Special-Purpose Entities		404,583	
Congressional Districts	Legal	435	
Voting Districts (VTDs)[d]	Legal	148,872	
School Districts	Administrative	15,274	
Traffic Analysis Zones (TAZs)[e]	Administrative	200,000	Yes
ZIP Codes[e]	Administrative	40,000	Yes
Census Tracts and Block Numbering Areas (BNAs)		62,276	Yes
Census Tracts	Statistical	50,690	Yes
Block Numbering Areas	Statistical	11,586	Yes
Block Groups (BGs)	Statistical	229,192	Yes
Blocks	Statistical	7,017,427	Yes

[a] Officially, "the United States" consists of the 50 States and the District of Columbia. In addition, the 1990 decennial census includes American Samoa, Guam, the Northern Mariana Islands, Palau, Puerto Rico, and the Virgin Islands of the United States.

[b] The 40 entities include the 40 "census subareas" in Alaska.

[c] The city of Honolulu is included as an incorporated place for statistical presentation purposes.

[d] Include only those eligible entities participating under the provisions of Public Law 94-171.

[e] Estimated value.

Source: Bureau of the Census, Geographic Areas Reference Manual, 1994, 2-3 and 2-4.

6.2 CENSUS GEOGRAPHY: CONCEPTS, CRITERIA, AND HIERARCHY

6.2.1 BASIC HIERARCHY: STANDARD GEOGRAPHIC UNITS

Census-defined geography has a hierarchical structure that the Census Bureau uses to collect, process, and distribute census data. This structure shows the geographic entities in a superior/subordinate relationship. At the top of this pyramid is the U.S., while at the bottom is the unit of blocks (see Table 6.2 and Figure 6.1). The country is divided into four regions that are groupings of states: Northeast, Midwest, South

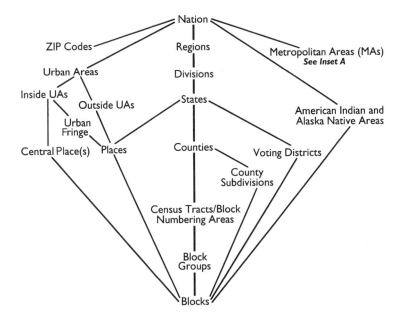

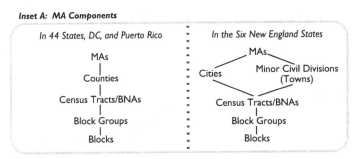

FIGURE 6.1 Geographic hierarchy of the 1990 Census. Bureau of the Census, Geographic Areas Reference Manual, 1994.

and West. Each of the four census regions is divided into two or more census divisions (also groupings of states); there are nine divisions.

Counties are the primary political divisions in most states; and some states have county equivalents such as "parishes" in Louisiana, "boroughs" and "census areas" in Alaska, and "independent cities" in Maryland, Missouri, Nevada, and Virginia. There are 3,141 counties and county equivalents in the nation.

County subdivisions are the primary subdivisions of counties and their equivalents. They include Minor Civil Division (MCD), Census County Division (CCD), Census Subarea, and Unorganized Territory. MCDs are defined in 28 states, and "represent many different kinds of legal entities with a wide variety of governmental and/or administrative functions"(Bureau of the Census 1992a:A-6). They are often known as towns and townships, and serve as general-purpose local governments in

12 states. In some states, they are variously designated as American Indian Reservations, assessment districts, boroughs, election districts, precincts, etc.

In 21 states that do not have legally established MCDs or have MCDs subject to frequent change, CCDs are defined. In contrast to MCDs, they have no legal, administrative, and governmental functions. "The primary goal of delineating CCDs is to establish and maintain a set of subcounty units that have stable boundaries and recognizable names. A CCD usually represents one or more communities, trading centers or, in some instances, major land uses. It usually consists of a single geographic piece that is relatively compact in shape" (Bureau of the Census 1997). In Census 2000, a CCD is delineated on the basis of census tracts and has a minimum population of 1,500 persons.

In Alaska, census subareas are statistical subdivisions of county equivalents. Unorganized territory is defined as a residual area of a county in the nine MCD states where there is some territory that is not covered in an MCD.

Places include incorporated places and census designated places (CDP). Incorporated places include cities, boroughs, towns, and villages. Exceptions are the towns in the New England States, New York, and Wisconsin, and the boroughs in New York, which are recognized as MCDs, and the boroughs in Alaska, which are county equivalents. Each state enacts laws and regulations for establishing incorporated places. As the statistical counterpart of incorporated places, CDPs are "closely settled, named, unincorporated communities that generally contain a mixture of residential, commercial, and retail areas similar to those found in incorporated places of similar sizes" (Bureau of the Census 1997:39). The 1990 census uses the criteria of total population size, population density, and geographic configuration for delineating CDPs. For Census 2000, a substantial change from all prior CDP criteria is that there are no minimum or maximum population thresholds for defining a CDP. Other criteria include presence of an identifiable core area, the surrounding closely settled territory, a reasonably compact and contiguous land area internally accessible to all points by road, not being coextensive with any higher-level geographic area recognized by the Census Bureau, and boundaries following visible and identifiable features. Figure 6.2 shows the Columbia CDP in Maryland and its relationship to census tracts and block groups.

Census tracts are small, relatively permanent statistical subdivisions of a county. When first delineated, they "are designed to be homogeneous with respect to population characteristics, economic status, and living conditions"(Bureau of the Census 1992a:A-5). Prior to Census 2000, Block Numbering Areas (BNA's) were delineated for non-metropolitan counties where local census statistical area committees have not established census tracts. Census 2000 combines BNA and census tracts into a single entity and retains the census tract name.

The goal of establishing census tracts is "to provide a small-area statistical unit with comparable boundaries between censuses" (Bureau of the Census 1994:16). The criteria for delineating census tracts for Census 2000 include the following (Bureau of the Census 1997).

1. A census tract must meet the population criteria (see Table 6.3). To provide meaningful tabulations, the Census Bureau maintains population size requirements for census tracts while allowing for some flexibility. With a few exceptions, census tracts must have between 1,500 and 8,000 persons

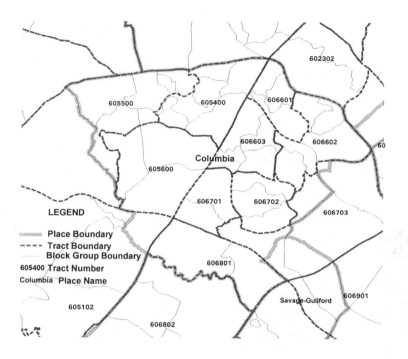

FIGURE 6.2 Columbia, Maryland CDP and its relationship to census tracts and block groups.

TABLE 6.3
Population Thresholds for Census 2000 Census Tracts and Block Groups

Area Description/Census Tracts	Optimum	Minimum	Maximum
United States, Puerto Rico, Virgin Islands of the U.S.	4000	1500	8000
American Samoa, Guam, Northern Mariana Islands	2500	1500	8000
American Indian reservation and Trust Lands	2500	1000	8000
Special Place Census Tract[a]	none	1000	none
Area Description/Block Groups	**Optimum**	**Minimum**	**Maximum**
Standard (most areas)	1500	600	3000
American Indian reservation and Trust Lands	1000	300	3000
Special Place Block Group[a]	none	300	1500

[a] Special places are correctional institutions, military installations, college campuses, workers' dormitories, hospitals, nursing homes, and group homes.

Source: Bureau of the Census, United States Census 2000 Participant Statistical Areas Program Guidelines: Census Tracts, Block Groups (BGs), Census Designated Places (CDPs), Census County Divisions (CCDs), FORM D1500 (10/97), U.S. Department of Commerce Economics and Statistics Administration, 1997.

(600 to 3,200 housing units), with an optimum (average) population of 4,000 (1,600 housing units). The minimum population threshold is lower than the 1990 census minimum threshold of 2,400 persons.

2. A census tract must meet the boundary feature criteria and be comprised of a reasonably compact, contiguous land area, all parts of which are accessible by road. A county boundary always must be a census tract boundary. Census tract boundaries should follow visible and identifiable features, such as roads, rivers, canals, railroads, and above-ground high-tension power lines. Some nonvisible, governmental unit boundaries are acceptable as census tract boundaries.

3. Census tracts must cover the entire land and inland water area of each county.

Block groups (BGs), made up of clusters of blocks, are a subdivision of census tracts or BNAs. The primary goal of establishing BGs is "to provide a geographic summary unit for census block data." Each census tract contains a minimum of one block group and a maximum of nine block groups. A block group consists of all census blocks whose numbers begin with the same digit, and is identified using the same first digit. The 1990 census used a three-digit block numbering system and reserved n00 and n98 for special uses and n99 for water areas. Therefore, each BG may include no more than 97 census blocks. This limitation has been lifted; Census 2000 uses a four-digit block numbering system.

The criteria for delineating block groups for Census 2000 include the following (Bureau of the Census 1997).

1. A BG must meet the population criteria (see Table 6.3). With a couple of exceptions, BGs in Census 2000 must have between 600 and 3,000 persons (240 to 1,200 housing units), with an optimum (average) population of 1,500 (600 housing units). The maximum population criterion is substantially increased compared with the 1990 housing unit criteria. The 1990 census guideline specified an optimum of 400 housing units for BGs, with a minimum of 250 and a maximum of 550 housing units.

2. A BG must meet boundary feature criteria and be comprised of a reasonably compact, contiguous land area internally accessible to all points by road. A census tract boundary must always be a BG boundary. BGs must cover the entire land and inland water area of a census tract. BG boundaries should follow visible and identifiable features, such as roads, rivers, canals, railroads, and above-ground high-tension power lines. Some nonvisible, governmental unit boundaries are acceptable as BG boundaries.

3. Each census tract must contain a minimum of one BG and may have a maximum of nine BGs.

4. A BG that is entirely within an American Indian reservation or trust land may extend across a state or county boundary for tabulation in the American Indian geographic hierarchy. For standard data tabulations, the portion of the BG in each state and county is treated as a separate BG.

As a subdivision of census tracts or BNAs, blocks are the smallest unit tabulated from the census. They are "bounded on all sides by visible features such as streets, roads, streams, and railroad tracks, and by invisible boundaries such as city, town, township, and county limits, property lines, and short, imaginary extensions of streets and roads"(Bureau of the Census 1992a:A-3). Census collection blocks generally do not cross other geography boundaries of states, counties, census tracts, BGs, CCDs, CDPs, and MAs. Incorporated places and MCDs may split a collection block. When this happens, alphabetic suffixes are assigned to all portions of the split collection blocks, which are referred to as census tabulation blocks. They may have zero population or thousands of residents in a high-rise building. For the first time, the 1990 census block-numbered the entire U.S. and its possessions. Figure 6.3 shows census tract 6054, its block groups, and blocks.

The Census Bureau used a computer routine to automatically assign census block numbers for the 1990 census. The goal was to maximize the number of census blocks within each BG. The computer routine analyzed the network of TIGER (Topologically Integrated Geographic Encoding and Referencing) database features that formed polygon areas within each 1990 BG and assigned a number to each block. It gave major consideration to the type of feature and the shape and minimum size of a potential census block.

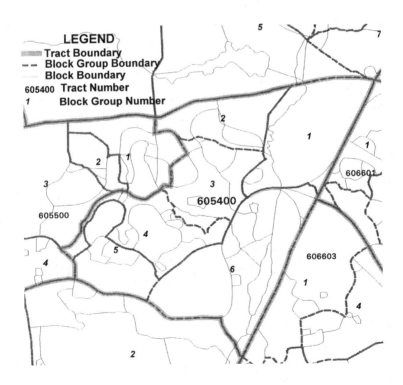

FIGURE 6.3 Census tract, block groups, blocks.

1. The minimum size of a census block was 30,000 square feet (0.69 acre) for polygons bounded entirely by roads, or 40,000 square feet (0.92 acres) for other polygons. There was no maximum size for a census block.
2. Based on polygon shape measurements, extremely narrow slivers were eliminated as potential census blocks.
3. Census features were ranked in terms of their importance as census block boundaries. The ranking criteria were (1) the type of boundary, (2) the feature with which it coincided, (3) the existence of special land use areas (such as military reservations), (4) the presence of governmental boundaries.
4. At least one side of a potential census block had to be a road feature.

6.2.2 NON-STANDARD GEOGRAPHIC UNITS

The Census Bureau also provides data for some supplementary geographic units. These units generally cut across the basic hierarchy of census geography.

An Urbanized Area (UA) "comprises one or more places ('central place') and the adjacent densely settled surrounding territory ('urban fringe') that together have a minimum of 50,000 persons" (Bureau of the Census 1992a:A-12). The urban fringe has to meet the census-defined population density criteria of at least 1,000 persons per square mile. UAs always follow the boundaries of tabulation census blocks. "Urban" for the 1990 census includes "all territory, population, and housing units in urbanized areas and in places of 2,500 or more persons outside urbanized areas" (Bureau of the Census 1992a:A-11).

A Metropolitan Area (MA) comprises "a large population nucleus, together with adjacent communities that have a high degree of economic and social integration with that nucleus" (Bureau of the Census 1992a:A-8). An MA must include (a) a city with a minimum population of 50,000, or (b) a census-defined urbanized area of at least 50,000 population and a total metropolitan population of at least 100,000 (75,000 in New England). An MA is comprised of one or more central counties (cities and towns in New England), and one or more outlying counties that have close economic and social relationships with the central county. An outlying county must meet certain standards such as the level of commuting to the central county, population density, urban population, and population growth. This definition indicates that an MA may include a suburban county that has both developed areas near the central city and an extensive rural hinterland. This is where an MA differs from an urbanized area, which includes only densely developed areas of counties.

MAs are classified as either Metropolitan Statistical Areas (MSAs) that are relatively freestanding or Consolidated Metropolitan Statistical Areas (CMSAs) that have at least 1 million people and two or more closely related components known as Primary Metropolitan Statistical Areas (PMSAs). The Office of Management and Budget issues the standards for defining metropolitan areas.

Congressional districts (CDs) are the 435 areas that state officials and courts define for the purpose of electing persons to the U.S. House of Representatives. "Each CD is to be as equal in population to all other CDs in the State as practicable, based on the decennial census counts"(Bureau of the Census 1992a:A-6). These CDs are often defined as groups of blocks, and may cross geographic units below the state level.

ZIP codes are administrative units established by the U.S. Postal Service for the most efficient distribution of mail. "The Zoning Improvement Plan (ZIP Code) was initiated July 1, 1963, to speed and improve mail handling. ZIP code is a five-digit system: the first digit represents one of the geographic areas; the second two numbers indicate a metropolitan area or sectional center; the last two represent a small town or delivery unit within a metropolitan area" (U.S. Postal Service 1982:21). They generally do not follow political or census statistical area boundaries, "usually do not have clearly identifiable boundaries, often serve a continually changing area, are changed periodically to meet postal requirements, and do not cover all the land area of the United States" (Bureau of the Census 1992a:A-13). ZIP Codes cut across various geographic boundaries of census geography. Figure 6.4 shows an example of the boundary relationship between ZIP codes and census tracts in downtown Baltimore.

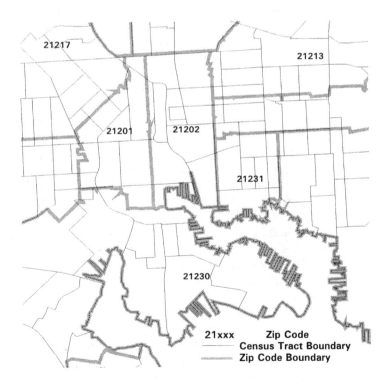

FIGURE 6.4 ZIP codes and census tracts in downtown Baltimore.

6.3   CENSUS GEOGRAPHY AS A UNIT OF EQUITY ANALYSIS: CONSISTENCY, COMPARABILITY, AND AVAILABILITY

The hierarchy principle of geography also applies to availability of census data. As a general rule, the smaller the geographic unit, the less detailed data that are available to users. The most detailed tabulations are available at the national and state level, while the least detailed data are at the block level. Block data are not released in printed documents, but rather in microfiche or computerized form.

This hierarchy principle helps preserve data confidentiality and reliability. The Census Bureau does not release data if users of the data could potentially identify individual households. The smaller the geography scale, the more likely individual households can be identified in the cross-tabulations, and the more sampling errors.

From the data availability perspective, environmental equity analysts need to choose a unit of geography that has data available for the variables they want to analyze.

6.3.1   HIERARCHICAL RELATIONSHIP AND GEOGRAPHIC BOUNDARY

The hierarchical structure of census geography does not guarantee an exact match of boundaries among different geographic units. Some geographic units have exact matches of boundaries. This is a many-to-one relationship between the lower and higher levels of geographic units. That is, many entities at the lower level (e.g., many block groups) belong to only one entity at the higher level (e.g., one tract). For example, for the boundaries of blocks, block groups, census tracts, and counties, the lower level of geography does not cross the higher level of geography. For computational purposes, you can aggregate the lower level data to the higher level by simply summing the individual components without worrying about any mismatch.

For some other geographic units, there is a many-to-many relationship, which complicates the analysis and warrants special attention. For example, between block groups and census tracts on the one side and places on the other, we often find that a municipality boundary divides a block group or census tract into two parts, one part in one municipality, the other in another municipality. For these split tracts or block groups, printed census documents report the data for split parts separately and again for the whole tract in a section called "Total for Split Tracts." In census tabulations, split tracts or block groups are generally indicated by a superscript. One common pitfall is that when finding a tract number in tabulations, people use it without realizing that it is a split tract.

The most complicated boundary relationship occurs when there is no match whatsoever between two geographic units. For example, the boundaries of census tracts and ZIP code areas usually do not match. Unlike basic census geography, "Zip Codes are not defined as enclosed spatial areas but rather as sets of postal carrier routes emanating from a local post office. More than one set of routes may penetrate the same spatial area" (Myers 1992:68). ZIP codes shown as enclosed spatial areas in maps actually represent the dominant ones in the area. There are cases where one building or complex has a single ZIP code. Figure 6.4

shows such a complicated boundary relationship in downtown Baltimore. Not shown in the map is ZIP code 21203, which is the main Post Office in Baltimore located inside ZIP code 21202.

Spatial configurations of census geography such as patterns, sizes, and shapes vary within and between areas. Factors that influence the overall configuration of census geography include "topography, the size and spacing of water features, the land survey system, and the extent, age, type, and density of urban and rural development" (Bureau of the Census 1994:11-9). We often see grid systems in old central cities and irregular shapes in modern suburban areas.

6.3.2 BOUNDARY COMPARABILITY OVER TIME

Any longitudinal study has to deal with the issue of the boundary comparability of census geography over time. This is a very important issue because some census geographic units do have different definitions and boundaries from one census to another. Even though you have the same name for a geographic unit for two censuses, you could have significantly different areas represented. Analysts should be very careful to avoid the pitfall of comparing apples and oranges.

Census geographic units can be arranged according to the degree of boundary stability (in decreasing order) as follows: states, counties, census tracts, blocks and block groups, MAs, UAs, county subdivisions, and places. States have fixed boundaries, and the boundaries of counties and county equivalents have rarely been changed. They are very stable, large geographic units for longitudinal analysis.

As indicated above, census tracts are designed to have a relatively stable, permanent boundary, and therefore are the most reliable small geographic units for longitudinal studies. However, analysts must be aware of and very careful of two types of census tract boundary changes to ensure tract comparability over time, particularly in growing and declining areas. When delineated, a tract has to maintain its population between the minimum and maximum thresholds. If tract population grows beyond the maximum threshold as often occurs in growing areas, a tract will be subdivided into two or more tracts in the next census. These subdivided tracts will have the same basic four-digit number as the original tract, and an extra two-digit suffix; for example, 6059.01 and 6059.02. To ensure comparability over two censuses, you have to aggregate the subdivided tracts in a later census. This is a one-to-many relationship between tracts in two censuses.

Conversely, in declining areas, a tract's population may fall below the threshold, and may have to be merged with adjacent tracts. In this case, you have to aggregate the merged tracts in the earlier census to achieve a comparable area. This is a many-to-one relationship.

A more difficult situation is when the tract boundary changes because of major new physical features such as new highways. This could be a many-to-many relationship, where you have to aggregate many tracts in both censuses to obtain a comparable area.

Approximately one quarter of all census tracts had some changes between 1970 and 1980 (White 1987), but fewer than 5% of all tracts had significant changes with over 100 people affected. Most changes resulted from tract splits.

For census tract boundary change information, the Census Bureau provides tables of tract comparability over two censuses. These tables list tracts with boundary changes and can be found in the front pages of the published census tabulations or in the machine-readable TIGER/Comparability file of the 1990 Census.

The Census Bureau has no intention of maintaining boundary stability in blocks and block groups. As indicated above, the Census Bureau has the goal of maintaining a certain number of persons or housing units in a block group. In growing and declining areas, block groups could have boundary changes when the population or housing units grow or decline beyond the census-established thresholds. Boundary changes for block groups are not reported in the printed format. Thus, for most census users, it is impossible to obtain BG comparability over past censuses. Fortunately, digital data for block group boundaries have become available for the 1990 census. Some local agencies that participated in delineating census boundaries have also digitized the 1980 block group boundaries. Similarly, ZIP code area boundaries, which have constant changes particularly in growing areas, are also becoming increasingly available in GIS digital format. Therefore, the analyst is able to identify boundary changes using these GIS data.

MAs have boundary changes because of definition changes and population changes. In growing areas, counties may be added, while counties may be subtracted in declining areas. Comparing MAs over time requires addition or subtraction of counties in different census data, to ensure comparability over time.

Boundaries of legal entities such as County Subdivisions and incorporated places may change because of (Bureau of the Census 1992a:A-4):

1. Annexations to or detachments from legally established government units.
2. Merge or consolidations of two or more governmental units.
3. Establishment of new governmental units.
4. Disincorporations or disorganizations of existing governmental units.
5. Changes in treaties and Executive Orders.

Between 1980 and 1990, nearly 40% of the incorporated places in the U.S. changed their boundaries. In some states, boundaries for incorporated places change frequently. In California, 80% of incorporated places changed their boundaries between 1980 and 1990. On the other hand, incorporated place boundaries seldom change in some states such as Maine, Massachusetts, New Hampshire, Rhode Island, New Jersey, Pennsylvania, Vermont, and Connecticut.

These boundary changes must be taken into account for longitudinal studies. Information on boundary changes between the 1980 and 1990 censuses is presented in the "User Notes" section of the technical documentation of Summary Tape Files 1 and 3, and in the 1990 CPH-2, *Population and Housing Unit Counts* printed reports. For previous censuses, see the *Number of Inhabitants* reports for each census. Boundary changes are not reported for census-designated places. The Census Bureau has conducted an annual Boundary and Annexation Survey (BAS) since 1972 and incorporated the BAS into the TIGER database.

6.3.3 Data Availability and Comparability Over Time

As indicated above, data detail varies with the hierarchical structure census geography. Data availability also changes over time. These changes happen because census data items and geographic coverage change over time. A longitudinal study should take into consideration these changes. Generally, recent censuses have increased area coverage in the smallest units of census geography.

In 1940, the Census Bureau first published census block data (housing statistics) for 191 cities that had a population of 50,000 or more at the time of the 1930 census (Bureau of the Census 1994). In 1950, the Census Bureau published census block data for 209 places. In 1960, the Census Bureau expanded the program to include the total population data and housing statistics for 295 cities and an additional 172 places. The 1960 census had a total of over 736,000 census blocks. In the 1970 census, mail enumeration was used for the first time for a large portion of the U.S. population, and as a result, census block coverage was dramatically expanded. The Census Bureau numbered approximately 1,618,000 census blocks in and adjacent to UAs and in areas that contracted for census block data, and published census block data by standard metropolitan statistical area (SMSA).

In the 1980 census, census block coverage expanded again to include all incorporated places of 10,000 or more persons, in addition to urbanized areas. With over 2.5 million census blocks, the coverage accounted for approximately 78% of the nation's population and 7% of its land area. Again, the Census Bureau published reports for tabulated block data by SMSA, and also issued digital tape files, Summary Tape File (STF) 1B, for census blocks and BGs.

The 1990 census was the first time the entire U.S. and its possessions were block numbered and block-group numbered. This was made possible by the development of the TIGER System, an automated geographic database. The automated delineation produced a total of 6,461,804 collection blocks for the nation (6,517,390 including Puerto Rico and the Outlying Areas) (Bureau of the Census 1994:11-13). Data were tabulated for a total of 6,961,150 census tabulation blocks in the U.S. (7,020,924 including Puerto Rico and the Outlying Areas). Nationwide, there were 234,078 water blocks, 864,423 census blocks with suffixes, and 2,023,109 tabulation blocks with zero population. The percentage of tabulation blocks without any population varied considerably from one state and region to another, from a low of 14.1% for Rhode Island to a high of 64.7% for Wyoming. The national median was 31.1% (the state of Washington). It should be emphasized here that nearly one third of blocks have zero population.

The Census Bureau first used BGs in data tabulations in the 1970 census. The coverage was limited to areas in and adjacent to UAs that had census block numbers. For the 1980 census, the Census Bureau published data for 154,456 BGs. The 1990 census delineated 224,691 collection BGs in the U.S., and a total of 228,202 BGs in all areas under U.S. jurisdiction. The average number of BGs was 3.7 per census tract for counties with census tracts, and 3.9 per BNA for counties with BNAs (Bureau of the Census 1994:11-9).

For Census 2000, census tracts are established for the whole country. In the 1990 census, the entire country was delineated into either census tracts or BNAs. Census tracts were delineated for all metropolitan areas and more than 3,000 census tracts were established in 221 densely populated counties outside MAs. Only six States (California, Connecticut, Delaware, Hawaii, New Jersey, and Rhode Island) and the District of Columbia were covered completely by census tracts. Prior to the 1990 census, coverage of census tracts was limited.

The 1980 census delineated tracts only for Standard Metropolitan Statistical Areas (SMSAs), which consisted of cities of at least 50,000 persons and their surrounding counties or urbanized areas. For the 1980 census, the Census Bureau changed the BNA delineation criteria, which made BNAs more comparable in size and shape to census tracts. The concept of BNA for 1980 is dramatically different from the one for 1990. The 1980 BNA was delineated for assigning census block numbers, while the 1990 BNA shared the same basic attributes as census tracts.

Obviously, the 1990 census has larger geographic coverage than the 1980 census. Any longitudinal study using these two censuses is constrained by the limited coverage of the 1980 census, and would have to omit those areas not covered in the 1980 census.

If you want to go back further, you will find increasingly smaller areas covered by census tracts. When the Census Bureau first collected data for census tracts in 1910, they were delineated in only eight cities with populations over 500,000 (Bureau of the Census 1994). In 1930, the coverage expanded to 18 cities. The Census Bureau adopted census tracts as an official geographic entity and published the first tabulations for them in the 1940 census. Also in 1940, the Census Bureau devised block areas to control block numbering in cities without census tracts. Block areas were renamed block-numbering area (BNAs) in 1960 and consisted of one or more enumeration districts and sometimes city wards. From 1956, the Census Bureau continued to expand the program to cover entire metropolitan areas.

Places as a unit of analysis show little comparability geographically. "Incorporated places vary greatly in population, in physical extent, in the stability of their boundaries, and in their usefulness as a measure of the urban population of an area. The largest incorporated place in the Nation has more than seven million inhabitants, the smallest, fewer than ten. The largest incorporated place, in areal measure, has more than 2,800 square miles; the smallest, a few acres" (Bureau of Census 1994:9-11).

The geographic coverage of places is limited. In 1950, 66% of the nation's population lived in CDPs and incorporated places. This percentage has increased gradually since then. In 1990, approximately 66 million people (26%) in the U.S. lived outside any place. Of a total of 23,435 places in 1990, 19,289 places were incorporated, and the remaining 4,146 were CDPs.

Criteria for qualification of CDPs have changed since CDPs' first official recognition in 1950. There are two types of criteria for CDPs: inside UAs and outside of UAs. Criteria for UA designation have also changed from one census to another since 1950. Therefore, it is difficult to ensure data comparability for places and Urbanized Areas over time.

The Census Bureau was first to officially recognize the metropolitan concept, and defined *metropolitan districts* for cities as at least 100,000 people in the 1910 census. For the 1930 and 1940 censuses, the criteria were modified and the population of cities was lowered to 50,000. There were 96 metropolitan districts for the 1930 census and 140 for the 1940 census. From 1910 to 1940, metropolitan districts were defined based on population density and the boundaries of MCDs. From 1950, the Census Bureau began to implement the metropolitan concept based on counties. The criteria for defining metropolitan areas have been slightly changed over the decades, and the standards were modified in 1958, 1971, 1975, 1980, and 1990. Although most of the criteria's changes have been minor, the collective term used for metropolitan areas has changed enough times to cause confusion. It was *standard metropolitan area (SMA)* in 1950, *standard metropolitan statistical area (SMSA)* in 1959, *metropolitan statistical area (MSA)* in 1983, and *metropolitan area (MA)* in 1990. The changes in standards have implications for comparability of metropolitan areas over time. Recent MA definitions were increasingly broader, and changes in MA definitions result in more coverage in population and to a larger extent, land area. For example, just going from a 1960 MA definition to a 1990 MA definition would increase the MA population by 24% and the MA land area by 88%.

For the 1990 census, ZIP Code data are tabulated for the five-digit codes, which do not cover all land areas of the U.S. The 1980 census was the first time in which every 5-digit ZIP Code area in the U.S. was tabulated. The 1970 census tabulated 5-digit ZIP code areas only for standard metropolitan statistical areas, and only 3-digit ZIP code areas for all other areas.

6.4   CENSUS GEOGRAPHY AS A UNIT OF EQUITY ANALYSIS: WHICH ONE?

There are advantages and disadvantages in using different census geographic units in environmental justice analysis. Both sides of the debate on census tracts vs. ZIP codes have articulated their justifications and attacked the other side. The following is a summary of the pros and cons for using census tracts and ZIP codes. Some of the arguments for using census tract as a unit of analysis include

1. Census tracts have a relatively permanent, clearly defined boundary; comparisons are thus possible over time.
2. Census tracts have a relatively homogenous population of about 4,000.
3. Census tracts are delimited by local persons and thus "reflect the structure of the metropolis as viewed by those most familiar with it" (Bogue 1985:137).
4. Using census tracts rather than larger units such as ZIP codes could reduce "the possibility of 'aggregation errors' and 'ecological fallacies;' that is, reaching conclusions from a larger unit of analysis that do not hold true in analyses of smaller, more refined units" (Anderton et al. 1994).
5. Using smaller units such as block or block groups is difficult to justify, because the impact often goes beyond the block or block group boundary and some data are unavailable for the block level because of the need to protect confidentiality (Been, 1994).

6. It is the most commonly used geographic unit of analysis (Anderton et al. 1994).
7. It is a reasonable approximation of the concept of a neighborhood (Denton and Massey 1991).

Some of the arguments against using census tract as a unit of analysis are as follows:

1. Census tracts do not always cover rural areas, where some serious environmental hazards exist.
2. Using census tracts runs the risk of too small units and making incorrect inferences; this happens because census tracts in metropolitan areas are small and the potential impact area "may very well extend beyond the boundaries of individual tracts" (Mohai 1995:634).
3. A national study using census tracts is expensive.
4. Census tract data have serious limitations in longitudinal studies. The data availability at the census tract level is very limited for older censuses, and if data are available, census tracts in pre-1960 censuses generally have much larger geographic areas than those in recent censuses. Because of the limited data in an older census, a longitudinal study may be forced to drop some facilities sited early in this century. Because of a large area covered by a tract in an old census, a census tract may be less representative of the impact area.

Some of the arguments for using a ZIP code as a unit of analysis are as follows:

1. It has been successfully used in marketing, "for appraising demographic and socioeconomic characteristics of potential customers" (UCC 1987:61).
2. It is "the smallest geographic unit that can be used for consistent and comprehensive database integration purposes" (UCC 1987:61).
3. ZIP codes are more inclusive than census tracts, covering rural areas.

Some of the arguments against using a ZIP code as a unit of analysis include

1. ZIP code populations vary highly in space; any comparison across space requires standardization.
2. ZIP code populations vary highly in time; any comparison across time is difficult.
3. ZIP codes are constructed for delivering postal services, and thus may not reflect the local neighborhoods.
4. Using ZIP codes that may be "too large a geographic unit invites the possibility of 'aggregation errors' and 'ecological fallacies'" (Anderton et al. 1994).
5. The unavailability of census data at the ZIP code level in the pre-1980 censuses makes a longitudinal study including pre-1980 events virtually impossible.

For a facility-based equity analysis, use of census-defined units has an implied assumption that people living in the chosen unit of analysis are equally affected by the facility and impacts vanish at the unit's boundary (Mohai 1995). This assumption is certainly questionable in some cases.

The relative homogeneity of census-defined units is now challenged. While a census tract is supposedly delineated to represent a relatively homogeneous small area by those most familiar with it, it can be found that some non-homogeneous components may exist in it. Pockets of minority or low-income communities that experience disproportionately high and adverse effects may be imbedded in a census tract that is predominantly non-minority (Bullard 1994; U.S. EPA 1998b).

Some have argued that because of the relatively homogeneous population averaging 4,000 in a census tract, comparisons are thus possible over space without adjusting for area or density. In fact, any cross-sectional study using census tracts but accounting for no area variation would produce misleading results. This is because, although the census controls the population size for defining census tracts, it does not control the area, which could and indeed does vary widely.

An analysis of recent ZIP codes and 1990 census tracts demonstrates dramatic differences in size between them. Table 6.4 shows summary descriptive statistics for some geographic units. Note the dramatic differences between mean and median values. Since both ZIP codes and census tracts have highly skewed distributions, it is more desirable to use median measures. A typical ZIP code is at least 8 times as large as a typical census tract. Excluding those very small ZIP codes (mostly in a single building), a typical ZIP code is 20 times as large as a typical census tract. Both ZIP codes and census tracts vary greatly in size. While census tracts have a

---

TABLE 6.4
Descriptive Statistics for Some Geographic Units

Geography	Number	Sum	Minimum	Maximum	Mean	Standard Deviation	Median
			Area (Sq. Miles)				
ZIP codes (all)	42,682	3,568,785	0	98,484	84	676	17
ZIP codes (areal)	29,483	3,268,020	0.011	18,555	111	380	40.5
Census tract	61,386	3,779,518	0	61,586	62	593	2.2
County	3141	3,560,536	1.8	156,741	1134	3777	619
State	51	3,596,102	69	580,435	67,920	86,127	55,942
			Population				
ZIP codes (areal)	29,466	248,709,873	0	112,167	8441	12,316	2785
Census tract	61,255	248,709,873	0	71,872	4060	2394	3755
County	3141	248,709,873	52	8,863,164	79,182	263,813	22,085
State	51	248,709,873	453,588	29,760,021	4,876,664	5,439,195	3,294,394

Note: ZIP codes (areal) include those that can be represented as spatial areas. Some ZIP codes cover only single buildings and thus are excluded.

Source: Caliper Corp., Geographic Data CD ROM, 1995.

very close mean and median population (4,060 and 3,755, respectively) and a small standard deviation, ZIP Codes have a large standard deviation and their mean and median population values differ dramatically (2,785 and 8,441, respectively).

The size distribution of both census tracts and ZIP codes is right-skewed (Figure 6.5). However, the size distribution of census tracts is relatively uni-modal and lepokurtic, while that of ZIP codes is multi-modal and platykurtic (Figure 6.5). Slightly more than half the census tracts cover an area less than or equal to 3.14 square mi (an area equivalent to a circle with a 1-mi radius), compared with only 7% for ZIP codes. Census tracts are predominantly concentrated in the size range of 0.03 to 7.07 square mi (an area equivalent to 0.1 to 1.5 mi in radius), accounting for 62.6% of all census tracts. Approximately 27% of census tracts have an area between 0.03 and 0.79 square mi, and 26% of census tracts have an area between 0.79 and 3.14 square mi.

If 0.8 square mi (approximately 2 square km), an area equivalent to a circle with a 0.5-mi (approximately 800 m) radius, is too small for an impact area, choosing census tracts as a unit of analysis has a 29% chance of being too small. Choosing ZIP codes as a unit of analysis would have a much smaller chance (1.5%) of committing the error of being too small.

If 50 square mi (approximately 130 square km), an area equivalent to a circle with a 4-mi (approximately 6.4 km) radius, is too large for an impact area, the chance for census tracts to err on being too large is 18% while the chance for a ZIP code is 44%.

If the largest possible size for an impact area has a 0.5- to 3-mi radius, then census tracts have a 47% chance of being the right choice while ZIP codes would have a 36.5% chance. That is, more than half of census tracts or ZIP codes are either

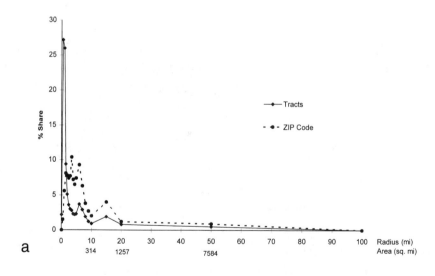

FIGURE 6.5 Size distribution of census tracts and ZIP codes. (a) Comparing distributions of census tracts and ZIP codes by area and radius of an equivalent-area circle.

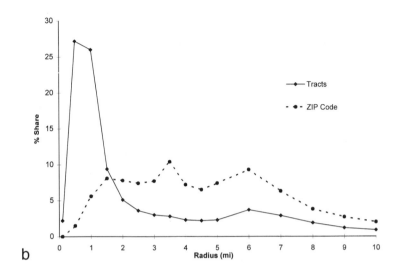

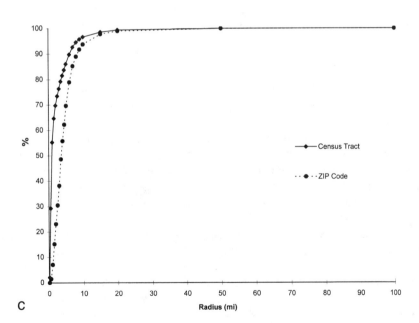

FIGURE 6.5 Size distribution of census tracts and ZIP codes. (b) Comparing size distributions of census tracts and ZIP codes by radius of an equivalent-area circle (enlarged). (c) Cumulative distributions of census tracts and ZIP codes by radius of an equivalent-area circle.

too small or too large to be appropriate impact areas. Obviously, both geographic units are not ideal for a typical environmental impact area of 0.8 to 28 square mi, although census tracts may have a better chance of being the right size.

For an environmental impact area with a 0.5- to 2-mi radius, census tracts have a 40.5% chance of being the right choice, compared with 21.6% for ZIP codes. If an environmental impact area has a 1- to 2.5-mi radius, then census tracts have only an 18% chance of being the right choice while ZIP codes have a 23% chance. For an environmental impact area with a 2- to 4-mi radius, census tracts have a 11.8% chance of being the right choice, compared with 32.7% for ZIP codes.

Furthermore, size distribution analysis does not account for the irregularity of census tracts and ZIP code configurations, let alone the relative location of a site in a census tract or ZIP codes. Even if the size of a chosen geographic unit is right for an impact area, irregularity could render it less representative. Irregularity in configurations is the rule rather than exception in census geography, particularly in post-WW II development areas. As will be demonstrated later, the relative location of a site in a unit of analysis is critical to determine the representativeness and research findings.

A further complication is the fact that environmental impacts do not radiate evenly in all directions from a pollution site. We cannot adequately judge which geographic unit is most appropriate without looking at the impacts of the environmental risks in question. What this calls for in an ideal situation is to delineate environmental impact areas case by case and then choose appropriate census geographic units to approximate the impact area. This may prove very difficult for a region-wide or nationwide study.

When knowledge about the scope of environmental impacts is inadequate or not taken into account in equity analysis, the choice of unit of analysis tends to be arbitrary. The choice of unit of analysis is "often dictated by expediency, determined by how existing data bases are aggregated and which level of aggregation provides the most data at the smallest geographic scale" (Zimmerman 1993:652). As a result, the unit of analysis could bear little relation to the actual impact area, and the results could be seriously distorted, as demonstrated by Zimmerman (1994). Especially controversial is the so-called "border issue," whereby environmentally risky facilities are located near the borders of two or more adjacent legal/administrative or statistical entities. Choosing only the entity where the facility is located can easily miss the real impact area across the border.

Zimmerman (1994) identified a number of NPL sites in two northeastern states that were within a few miles of county boundaries, and some of them were also within a few miles of state boundaries. It is unclear how representative the border phenomenon is nationally. If the border phenomenon is nationally or regionally widespread, then the validity of the findings from previous studies based on a legal or statistical area boundary could be seriously challenged. In more refined scales, Zimmerman (1994) illustrated the border and boundary issue with the Lipari Landfill in New Jersey, the EPA's top Superfund site. Located in Manua Township in Gloucester County, the site is within a mile of four townships or boroughs. Such a location could distort the results of an analysis that is based on the boundaries of political jurisdictions or statistical areas.

The border location phenomenon could be no accident. Ingberman (1995) dem-onstrates that a firm can win majority acceptance of a noxious facility by locating on political borders and using threat strategy. He illustrates the successful use of threat site strategy for expansion of two landfill sites along the border of two Pennsylvania townships. The result is market inefficiency in using economic instru-ments (e.g., compensation) to site noxious facilities. This hypothesis has important policy implications for facility siting and environmental justice. It would be inter-esting to see how this hypothesis bears out with national or regional data.

GAO (1995) conducted a survey of 500 metropolitan and 500 non-metropolitan landfills, of which 300 metropolitan and 150 non-metropolitan landfills were sub-sampled for identifying their exact locations on the U.S. Geological Survey 1:24,000 scale maps. Of the 450 landfills surveyed, 295 responses are usable. Two buffer areas were delineated: 1- and 3-mi radius. For 35 landfills, the 1-mi radius circle area extends into at least one other county. For 101 landfills, the 3-mi radius circle area extends into at least one other county.

In sum, the debate about which census geography is the most appropriate for environmental justice analysis has shown the limitations and constraints of census geography. We have seen the heterogeneity of census tracts in terms of size and shape, although they do have a relatively homogeneous population. Furthermore, environmentally risky facilities, or LULUs, may have impacts that may not easily match any census geography. Clearly, in order to choose the best unit of analysis, the analyst should consider a number of important factors: the size and shape of census geography, the impact boundary of environmental risks, the location of environmental risk sources relative to census geography, and types and magnitudes of potential impacts. We now turn to how we can best take into account these factors in defining appropriate units of analysis.

6.5   ALTERNATIVE UNITS OF ANALYSIS

One strategy to address the rigid census geography and border effect is use of GIS-delineated units. If a facility under study is located along the border of two or more census units, we can simply identify and aggregate these neighboring census units as one unit. For this purpose, we need to define the critical distance from the border to the point a facility has border effects. We also need to know the exact location of facilities and measure their distance to the boundaries of neighboring units. If the distance to an adjacent unit is under the critical distance, we can simply include that adjacent unit. To refine this method, we may further want to decide how much of that adjacent unit is under the influence of this facility and determine whether to include that adjacent unit based on some threshold for the sphere of influence, e.g., 50%. Without these data, we can still identify those census units that are adjacent to the census unit that hosts a facility under study. We can aggregate the adjacent units and the host unit as one or treat them separately as two groups of units — exposed unit and potentially exposed unit. Another commonly used method is a GIS-delineated buffer around a facility under study. Chapter 8 discusses both adja-cency analysis and buffer analysis in detail. Still, the actual environmental impact boundaries are not exactly accounted for.

While researchers and observers debate on which census geographic unit is the most appropriate for environmental justice analysis, they may share one view: Ideally, the unit of analysis should reflect the impact areas of environmental risks or pollution. As discussed in Chapter 4, the impacts are multi-dimensional: health, environmental, economic, social, and psychological. Although this makes defining a universal unit of analysis even more difficult than a single dimension, they provide a comprehensive perspective for examining environmental justice issues. In the following, we will examine the methods and their strengths and weaknesses of defining units for analysis based on these impact dimensions.

6.5.1 BASED ON THE BOUNDARY OF ENVIRONMENTAL IMPACTS

As shown in Chapter 2, experts and laypersons see environmental impacts differently. Environmental impacts can be measured as actual or perceived. Any difference between these two measures may lead to a difference in choice of units of analysis.

As discussed in Chapter 4, environmental modeling and monitoring have provided us with some data, methods, and models for delineating plume trajectory of pollutants and thus the impact boundary of single environmental pollutants. When an impact boundary associated with an environmental risk does not match one of the census-defined boundaries, we can use GIS to estimate the socioeconomic characteristics of the plume trajectory area. Chapter 8 examines the GIS-based plume trajectory method for delineating units of analysis in detail. This method is promising for improving the accuracy of defining units of analysis for environmental justice studies. Of course, the stochastic nature of environmental factors leads to uncertainties in the plume trajectory and thus impact boundary. We usually look at these boundaries under average environmental conditions for a certain period of time.

For aggregation of census units into the plume trajectory area, we should use the smallest census geography as a building block, if data permit. As noted earlier, the smaller the census units, the more limited the data that are available. Therefore, a compromise has to be made. A strategy is to use blocks for identifying "pockets" of minority or low-income neighborhoods at the first step, and if there is no such pocket, use block groups or census tracts that meet our needs for more variables. If these pockets are found, you need to devise a way to estimate data that are not available at the block level.

Even without accurate boundary data, we can still distinguish environmental risks with localized impacts from those with regional impacts, and make common-sense judgments about appropriate units of analysis. For those with localized impacts, it makes more sense to use fine-scale geographic units such as census tracts or block groups or even blocks, while it is hardly justifiable to use a county as a unit. In some cases, aggregations of census units may be needed.

While these considerations are essentially based on the "objective" aspect of environmental risks, it might be helpful to incorporate the public's risk perception into the choice of appropriate units of analysis. As discussed in the theories of risk, the public often disagree with the experts on the assessment of risks. Therefore, it can be expected that the impact boundary defined by the public will most likely diverge from that objectively determined by experts. For example, the psychological impact scope for a Superfund site might go beyond the objective, physical impact boundary. Which boundary should be more appropriate is, to some extent, dependent upon how risk is defined. Use of both boundaries is very helpful for better understanding the equity issue.

If an analyst would like to use the public perception in defining his/her unit of analysis, he or she should be careful in defining the public first. How close is too close or how far is far enough is a very subjective question. The answer to these questions may depend on to whom questions are addressed. In a landfill siting case in Pima County, Arizona, the proposed site is more than two miles away from the nearest residences (Clarke and Gerlak 1998). Two Hispanic county board members opposed the proposed site and argued that it was too close to residential areas. The proposed site was in the district of one of the two Hispanic members. Local residents in the district were mobilized and organized to fight the proposed site. They claimed that it was environmental racism. On the other hand, three non-Hispanic, white county board members supporting the proposed site argued that it was far way from the nearest population and it made no sense to claim environmental racism. Their constituents would not care where they put the landfill so long as it was not in their district. This case demonstrates how subjective it can be to define impact areas and a geographic unit of analysis based on public opinion.

Evidence also shows that actual and perceived proximity to hazardous waste sites or other LULUs may differ significantly. Studies of Three Mile Island and Memphis have found a positive association between actual residential proximity and public concern about exposure to environmental risks (Dohrenwend et al. 1981; Harris 1983). However, no association was found at Love Canal (Fowlkes and Miller 1983) and a New York county (Howe 1988). Instead, perceived residential proximity was a significant predictor of concern about environmental exposure to toxic waste disposal sites, while actual residential proximity was not (Howe 1988). Perceived distances to the closest waste sites bear no association with actual distances.

Clearly, the public has different opinions that vary from actual impacts or expert opinions. When it is difficult to establish a single best threshold for impact distance, use of multiple distances may provide extra insights. If possible, the analyst should evaluate the impacts on a case-by-case basis.

6.5.2 Based on the Boundary of Sociological Neighborhood

Neighborhood is a concept that is not easy to define precisely, "but we all know what they are and what they mean when we talk about them" (Hunter 1983: 5). Most definitions include the social and physical dimensions. Its basic elements include people, place, interaction system, shared identification, and public symbols (Schwirian 1983).

Sociologists have long debated what a unit of neighborhood is for study of neighborhoods and neighborhood changes (Hunter 1983). Although most sociologists view the neighborhood as shaped by physical features, such as streets, and cultural and symbolic structures, they disagree about the relative importance of physical and symbolic features in defining the neighborhood. While physical features are concrete and easily identifiable on the map, cultural and symbolic structures are subjective and reflect social interactions of residents within the neighborhood. Clearly, social interactions vary by individual residents, and this variation may be translated into wide differences in subjective boundary definitions among residents. Indeed, studies have shown that residents do not have a commonly perceived physical neighborhood (Guest and Lee 1984), and subjective boundaries of the neighborhood follow class, gender, and age lines (Haeberle 1988). Well-structured physical features, however, could constrain individual resident's social interactions and thus the subjective boundaries of their neighborhood. Pittsburgh is a city with many rivers, railroad lines, and mountainsides. The Pittsburgh Neighborhood Atlas Project developed a neighborhood map based on residents' perceptions (Ahlbrandt et al. 1977). These perceived neighborhoods matched those based on voting districts.

The relationship between census geography and sociological neighborhood is a subject of debate. Some sociologists believe that census tracts are a reasonable or closest approximation of the concept of a neighborhood (Denton and Massey 1991; Lee and Wood 1991). Others believe that census tracts may not be homogeneous and a single census tract may have multiple distinctive neighborhoods (Bullard 1996).

"Neighborhoods are spatial units where people have social and cultural attachments. These attachments may cross geographic and political boundaries of census tracts and ZIP Codes. Residents often define and defend their neighborhood along social, racial, ethnic, economic, and religious lines" (Bullard 1996:496).

Statistical units are usually established on distinct physical features and generally lack the symbolic dimension of the sociological perspective. This limitation led some sociologists and community activists to become dissatisfied with use of the census data for neighborhood analysis. The 1970s witnessed an increasing importance of neighborhoods as subunits of local governments and as a basic unit of planning and development (Fahsbender 1996). In response, the 1980 Census initiated the Neighborhood Statistics Program (NSP), and nearly 1,300 cities, counties, townships, and other areas participated in the program (Bureau of the Census 1984). These participants and local census officials worked together to define boundaries for the statistical neighborhoods. The NSP report includes statistics for 28,381 neighborhoods, covering more than half the total number of census tracts in 1980. However, the NSP program was "not as effective as expected" (Fahsbender 1996). Many neighborhoods are aggregates of census tracts. A more cost-effective user-defined program was introduced in the 1990 census.

6.5.3 BASED ON THE BOUNDARY OF ECONOMIC IMPACTS

As discussed in Chapter 4, environmentally risky facilities, or LULUs, have some economic impacts on surrounding areas. The externality associated with these facilities or other polluting activities translates into direct economic damage, decrease in property value, and reduction in neighborhood quality. Contingent evaluation and hedonic pricing models have been used to evaluate economic impacts. While some studies find little significant impact, most studies indicate the presence of significant negative economic impacts. The boundary of economic impacts varies with studies.

Some believe that economic impacts are limited to an area within 0.25 mi (400 m) of a site (Greenberg, Schneider, and Martell 1994). Others report a much larger impact area, as discussed in Chapter 4.

Economists have examined the impacts of different units of analysis on hedonic price analysis. Goodman (1977) compared intra-unit variation within census tracts and block group data in descriptive and analytical housing price models for the New Haven MSA. He found substantial intra-unit variation within the census tracts. Average home prices within block-group aggregations more accurately reflected neighborhood characteristics while census tract data masked behavior related to racial differences. Can (1992) measured neighborhood quality at tract and block-group level aggregations in Syracuse, New York, and concluded that tracts tended to obscure the underlying spatial patterns of neighborhood quality. Intra-unit heter-ogeneity within census tracts was also found in Dallas (Jargowsky 1994), where many integrated neighborhoods at the tract level were merely "two segregated neighborhoods lumped together in the data by virtue of tract boundaries that do not line up with the current pattern of racial segregation revealed at the block level" (Jargowsky 1994:291). The fixed neighborhood boundary also fails to take into account cross-boundary effects; namely, the value of a house at the neighborhood boundary will be affected by adjacent neighborhoods as well as its own neighbor-hood. These data issues affect the quality of hedonic price analysis. Dubin and Sung (1990) believe that the weakness of empirical data correlating neighborhood quality and housing prices can be at least partially attributed to errors in choosing neigh-borhood boundaries.

As discussed in Chapter 4, hedonic price methods have been used to evaluate economic impacts of environmentally risky facilities or LULUs. These studies also uncovered price gradients from facilities and identified the critical distance at which economic damage diminishes to an insignificant amount. These price gradients should be used to depict economic impact boundaries.

6.5.4 BASED ON THE ADMINISTRATIVE/POLITICAL BOUNDARY OR JUDICIAL OPINIONS

In some cases, a unit of analysis based on land use and zoning authority or other local political jurisdictions may be justified. These are the smallest geographic levels of decision making involving land uses and other social problems. These jurisdictions are responsible directly to the public, who may participate in the political decision-making processes. However, caution must be taken in interpreting results based on this type of unit of analysis. It is most likely that a legal or administrative boundary diverges from the actual or perceived impact boundary. Therefore, the results do not necessarily reflect "outcome" equity, but rather may have a "procedural" flavor.

Judicial opinions in environmental case law offer some insights into but incon-clusive guidance about what is legally a justifiable definition of units of analysis. In Bean v. Southwestern Waste Management Corporation, the court relied on census tract data in its decision that denies the Plaintiffs' motion for a preliminary injunction to revoke a solid waste landfill permit. However, the court acknowledged that (1) the possible "intra-tract variations could mask segregated conditions that might have given rise to a showing of discrimination," and (2) "the range of the facility's effects may transgress statistical boundaries and that the initial and determinative inquiry should be to determine the geographic reach of these effects" (Fahsbender 1996:156). In East Bibb Twiggs Neighborhood Association v. Macon-Bibb County Planning & Zoning Commission, the court focused on the census tract, without considering the extent of the landfill's effects. In R.I.S.E., Inc. v. Kay, the court used radii of one-half, one, and two miles to examine the racial composition of the areas surrounding the existing and proposed sites.

6.6   SUMMARY

Now we should be clear that census geography is not a perfect choice for environmental justice analysis. We cannot make a generalized characterization that one census unit is better than others. No single census unit can be well suited to a wide range of impact boundaries. A too large or too small census geography may lead to either "ecological fallacy" or "individualistic fallacy." The relative homogeneity of census units such as census tracts is only relative; intra-unit variations may mask the true relationship between population distribution and environmental risks. Use of multiple census units is helpful to detect the sensitivity of research findings. MAUP may change research findings; border effect can be significant. When using any census unit, the analyst should also be aware of the limitations associated with boundary comparability and census data availability and comparability over time.

All these limitations notwithstanding, census geography should be used as a building block for developing a more appropriate unit of analysis. To define such a unit of equity analysis, multiple dimensions such as environmental impacts, economic damage, health, social, and psychological impacts should be taken into account. GIS and environmental modeling tools are helpful for defining environmental impact boundaries. Hedonic pricing models can assist in delineating economic impact boundaries. Field surveys can be used for mapping social interaction, community structure, and sociologically defined neighborhoods.

7 Analyzing Data with Statistical Methods

This chapter is not Statistics 101, but rather it is intended to review potential use, actual use, and misuse of statistics in environmental justice analysis. Different statistical methods are applicable to different areas of environmental justice analysis. It is a matter of choosing the right method. Both descriptive and inferential statistics have been applied to environmental justice analysis. It has been shown that the type of statistics affects the results.

7.1 DESCRIPTIVE STATISTICS

Descriptive statistics are procedures for organizing, summarizing, and describing observations or data from measurements. Different types of measurements lead to different types of data. Four types of measurements or scales in decreasing order of levels are ratio, interval, ordinal, and nominal. A ratio scale is a scale of measure that has magnitude, equal intervals, and an absolute zero point. Some socioeconomic characteristics have a ratio scale such as income. Environmental data such as emission and ambient concentration are ratio measures. An interval scale has the attributes of magnitude and equal intervals but not an absolute zero point, for example, temperature in degrees Fahrenheit. An ordinal scale has only magnitude but does not have equal intervals or absolute zero point, for example, risk-ranking data are ordinal. A nominal scale is simply the classification of data into discrete groups that have no magnitude relationship to one another. For example, race can be classified into African American, Asian American/Pacific Islanders, Native American, Whites, and Other Races. A variable initially measured at a higher level such as a ratio scale may be reduced to a lower level of measure such as ordinal. However, the reverse cannot be done. For example, household income, initially a ratio measure can be classified as low, middle, and high income. This conversion helps us grasp large data concisely but results in loss of information initially contained in the higher level of measure.

Descriptive statistics include minimum, maximum, range, sum, mean, median, mode, quartiles, variance, standard deviation, coefficient of variation, skewness, and kurtosis. All these statistics are applicable to ratio measures, while some of them can be used to describe other measures. As can be found at the beginning of Statistics 101, a simple way to describe a univariate data set is to present summary measures of central tendency for the data set. Three representations of central tendency are

- Arithmetic mean, which is the sum of all observations divided by the number of observations
- Median, which is the middle point in an ordered set of observations; in other words, it is the number above and below which there is an equal number of observations
- Mode, which is the observation point that is most likely to happen in the data set

For a nominal variable such as race/ethnicity, mode can be used but mean and median cannot. An ordinal variable can use the median or mode as a central tendency measure.

Central tendency statistics are affected by the underlying probability distributions. Typical distribution shapes include bell-shaped, triangular, uniform, J-shaped, reverse J-shaped, left-skewed, right-skewed, bimodal, and multimodal (Weiss 1999). The bell-shaped, triangular, and uniform distributions are symmetric. An asymmetric unimodal distribution is either left-skewed or right-skewed. A left-skewed distribution has a longer left tail than right tail, so that most observations have high values and a few have low values. A right-skewed distribution has a longer right tail than left tail, so that most observations have low values and a few have high values. For a symmetrical unimodal distribution such as the bell-shaped normal distribution, the mean, median, and mode are identical. For a symmetrical distribution that has more than one mode such as a bimodal distribution, the mean and median values are the same but the mode may be different from them. For a skewed distribution, the mean, median, and mode may be different. Household size distribution in the U.S. is right-skewed. The proportion of minority or any specific race/ethnicity variables does not follow normal distribution, but rather is considerably skewed. They are often bimodal, reflecting residential segregation where most tracts are either all whites or all minorities (Bowen et al. 1995). Integrated tracts are still small in number in most places.

The mean is very sensitive to the extreme values, and thus the median is often preferred for data that have extreme values. However, the median measure also has its own limitations. We can perform all sorts of mathematical operations on a variable's mean but not necessarily on the median. Mathematical operations such as addition, subtraction, multiplication, or division on medians may not generate another median. In the environmental justice literature, some researchers treat the median measure like the mean and commit what I call the Median Fallacy. This often happens in estimating the median of a variable at the higher geographic level from the existing median data at the lower geographic level.

For example, median household income is often used in equity analysis. Census reports data on income from the census block-group level and does not report them at the census block level. In some cases, analysts need to aggregate a census geography unit such as a block group or census tract into a larger unit. For example, to correct the border effect, researchers aggregate census tracts or block groups surrounding the census tract or block group where a target facility is located. In this aggregation, researchers often use population or households in each census tract or block group as weights. In buffer analysis, the analyst has to aggregate the census units in the buffer and estimate buffer characteristics based on the characteristics of the census units (see Chapter 8). In these cases, most analysts use a proportionate weighting method for aggregation. The assumptions underlying this method will be discussed in Chapter 8. While this method could serve to approximate a simple variable such as population and average household income, it may generate wrong results for the median value of a variable such as median household income.

The following example illustrates the median fallacy. Suppose we have two census block groups. Each block group has five households and their household incomes are listed in Table 7.1. The median household incomes for BG1 and BG2 are, respectively, $20,000 and $60,000. The weighted median household income for BG1 and BG2 together is $40,000, with the number of households in each block group as the weights. The true median household income value of $60,000 is $20,000 higher than the estimated value based on a proportionate weight. In rare cases, such as a uniform distribution for each block group like those in Table 7.2, the estimated median value via the proportionate weighting method is the same as the true value.

Besides the median value, household income data are also reported as the number of households in each household income class in the U.S. census. An approximation

---

TABLE 7.1
True Median Household Income is not Equal to the Weighted Estimate

Block Group	BG1 ($)	BG2 ($)
Household income	20,000; 20,000; 20,000; 30,000; 60,000	60,000; 60,000; 60,000; 70,000; 70,000
Median household income at the BG Level	20,000	60,000
Weighted estimate of median household income for BG1+BG2	40,000	
True median household income for BG1+BG2	60,000	

TABLE 7.2
True Median Household Income is Equal to the Weighted Estimate

Block Group	BG1 ($)	BG2 ($)
Household income	20,000; 20,000; 20,000; 20,000; 20,000	60,000; 60,000; 60,000; 60,000; 60,000
Median household income at the BG level	20,000	60,000
Weighted estimate of median household income for BG1+BG2	40,000	
True median household income for BG1+BG2	40,000	

of median household income for the aggregated study areas can be calculated from the frequency distribution of household income through interpolation as follows:

$$Md \cong L + I\,(n_1/n_2)$$

where Md = median; L = the lower limit of the median class; n_1 = the number of observations that must be covered in the median class to reach the median; n_2 = the frequency of the median class; I = the width of the median class. This approximation is based on the assumption that the observations in the median class are spread evenly throughout that class. This assumption is adequate for a study with a large number of observations.

Like other inquiries, environmental justice analysis deals with two types of data: population and sample. Population data consist of all observations for the entire set of subjects under study (which is called a population in a statistical sense). A sample is a subset of population, which is often collected to make generalization to the population. Descriptive statistics are used in the context of a sample or a population. The mean of a variable x from a sample is called a sample mean, $\bar{x}$. A sample standard deviation measures the degree of variation around the sample mean and indicates how far, on average, individual observations are from the sample mean. A sample is often used to generalize to the population. Because of sampling errors, different samples from the same population may have different values for the same statistic, which show a certain distribution. The Central Limit Theorem states that for a relatively large sample size, the sample mean of a variable is approximately normally distributed, regardless of the distribution of the variable under investigation (Weiss 1999).

As discussed in Chapter 5, some census variables are based on the complete count and thus are population data, while others are based on a sample. Data based on the entire population are free from sampling errors but have non-sampling errors such as undercount. Sample data have both sampling and non-sampling errors. The sampling error is the deviation of a sample estimate from the average of all possible samples (Bureau of the Census 1992a). Non-sampling errors are introduced in the operations used to collect and process census data. The major non-sampling sources include undercount, respondent and enumerator error, processing error, and nonresponse. Non-sampling errors are either random or non-random. Random errors will increase the variability of the data, while the non-random portion of non-sampling errors such as consistent underreporting of household income will bias the data in one direction. During the collection and processing operations, the Bureau of the Census attempted to control the non-sampling errors such as undercoverage. Differential net undercounting is discussed in Chapter 5.

Sampling errors and the random portion of non-sampling errors can be captured in the standard error. Published census reports provide procedures and formulas for estimating standard errors and confidence intervals in their appendix (Bureau of the Census 1992a). The error estimation is based on the basic unadjusted standard error for a particular variable, the adjusting design factor for that variable, and the number of persons or housing units in the tabulation area and the percent of these in the sample (percent-in-sample). The unadjusted standard error would occur under a simple random sample design and estimation technique and does not vary by tabulation areas. "The design factors reflect the effects of the actual sample design and complex ratio estimation procedure used" and vary by census variables and the percent-in-sample, which vary by tabulation areas (Bureau of the Census 1992a:C-2). In other words, the standard error estimation is the universal unadjusted standard error adjusted by an area- and variable-specific sampling factor that accounts for variability in actual sampling. The percent-in-sample data for the person and housing unit are provided with census data, and design factors are available in an appendix of printed census reports. The unadjusted standard errors are available from census appendix tables or can be estimated from the following formulas.

For an estimated total, the unadjusted standard error is

$$SE(Y) = \sqrt{5Y(1 - Y/N)}$$

where N is the size of area (total persons for a person characteristic or total housing units for a housing characteristic) and Y is the estimate of characteristic total.

For an estimated percentage, the unadjusted standard error is

$$SE(p) = \sqrt{5p(100 - p)/B}$$

where B is the base of the estimated percentage and p is the estimated percentage.

These procedures are designed for a sample estimate of an individual variable in the census and are not applicable to sums of and differences between two sample estimates. For the sum of or difference between a sample estimate and a 100% count value, the standard error is the same as that for the sample estimate. For the sum of or difference between two sample estimates, the standard error is approximately the square root of the sum of the two individual standard errors squared. This method leads to approximate standard errors when the two samples are independent and will result in bias for the two items that is highly correlated. Census reports also provide estimation procedures for the ratio of two variables where the numerator is not a subset of the denominator and for the median of a variable (Bureau of the Census 1992a).

For Census 2000, the complete-count variables include sex, age, relationship, Hispanic origin, race, and tenure. Other census variables such as household income are estimates based on a sample and thus are subject to sampling and non-sampling errors.

7.2   INFERENTIAL STATISTICS

Inferential statistics are the methods used to make inferences to a population based on the observations made on a sample. Univariate inferences estimate the single-variable characteristics of the population, based on the variable characteristics of a sample. Bivariate and multivariate inferences evaluate the statistical significance of the relationship between two or more variables in the population.

As noted above, the purpose of using the sampling technique to collect data and derive estimates is to make assertions about the population from which the samples are taken. To make inferences about the population, we need to know how well a sample estimate represents the true population value. Standard deviation of the sample mean accounts for sampling errors, and the confidence interval tells us the accuracy of the sample mean estimate. The confidence interval is a range of the dispersion around the sample mean at a confidence level, indicating how confident we are that the true population mean lies in this interval. The length of the confidence interval indicates the accuracy of the estimate; the longer the interval, the poorer the estimate.

Like other inquiries, environmental justice analysis starts with a hypothesis. The null hypothesis is usually a uniform risk distribution. If risk is uniformly distributed, we should find no difference in the (potential) exposure or risk among different population groups. In a proximity-based study, the percentage of a particular subpopulation such as blacks in the vicinity of a noxious facility would be the same as that far away. The alternative hypothesis is typically a specific risk distribution pattern that we wish to infer if we reject the null hypothesis. Usually, it is the alleged claim that minority and the poor bear a disproportionate burden of real or potential exposure to environmental risks. Specifically for a proximity-based study, we consider the alternative hypothesis that the percentage of a particular subpopulation such as blacks near a noxious facility is greater than that far away. Here, we are dealing with two populations, one that is close to a facility and the other that is far away.

Hypothesis test for one population mean concerns whether a population mean is different from a specified value. We use the sample mean to make an inference about the population mean. Different hypothesis-testing procedures have different assumptions about the distribution of a variable in the population. We should choose the procedure that is designed for the distribution type under investigation. The z-test assumes that the variable has a normal distribution and the population standard deviation is known, while the t-test assumes a normal distribution and an unknown population standard deviation. The Wilcoxon signed-rank test does not require the normality assumption and only assumes a symmetric distribution (Hollander and Wolfe 1973). Given that many variables do not follow a normal distribution, the Wilcoxon signed-rank test has a clear advantage over the z-test and t-test. Outliers and extreme values will not affect the Wilcoxon signed-rank test, unlike the z-test and t-test. However, when normality is met, the t-test is more powerful than the Wilcoxon signed-rank test and should be used.

As noted above, most environmental justice analyses concern inferences about two population means or proportions. Here, we use two samples to make inferences about their respective populations. To compare two population means, we can collect two independent random samples, each from its corresponding population. Alternatively, we can use a paired sample, which includes matched pairs in two populations. For either sampling method, subjects are randomly and independently sampled; that is, each subject or pair is equally likely to be selected. The t-test again assumes a normal distribution, while the Wilcoxon rank-sum test (also known as the Mann-Whitney test) assumes only the same shape for two distributions, which do not have to be symmetric. Again, when normality is met, the t-test is more powerful than the Wilcoxon rank-sum test and should be used when you are reasonably sure that the two distributions are normal. For a paired sample, the paired t-test assumes a normal distribution for the paired difference variable, while the paired Wilcoxon signed-rank test assumes that the paired difference has a symmetric but not necessarily normal shape (Weiss 1999). The Wilcoxon rank-sum test for two groups may be generalized to several independent groups. The most commonly used Kruskal-Wallis statistic is independent of any statistical distribution assumptions.

A population proportion, p, is the percentage of a population that has a specified attribute, and its corresponding sampling proportion is $\bar{p}$ (Weiss 1999). This type of measure is often used in environmental justice studies for the percentage of population who are minority, or any race/ethnicity category, or in poverty. The sampling distribution of the proportion, $\bar{p}$, is approximately normally distributed for a large sample size, n. If p is near 0.5, the normal approximation is quite accurate even for a moderate sample size. When np and $n(1-p)$ are both no less than 5 as a rule of thumb (10 is also commonly used), the normal approximation can be used. For example, for a population proportion of 1%, the sample size needs to be no less than 500 for a normal approximation. For a large sample, the one-sample z-test is used to test whether a population proportion is equal to a specified value. For testing the difference between two proportions for large and independent samples, we can use the two-sample z-test. The assumptions require that the samples are independent and the number of members with the specified attribute x and its opposite set $n - x$ in each sample is no less than 5.

In general, parametric tests such as the t-test require more stringent assumptions such as normality, whereas nonparametric tests such as the Wilcoxon rank-sum test do not. In addition, nonparametric tests can be conducted on ordinal or higher scale data, while parametric tests require at least the interval scale. On the other hand, parametric tests are more efficient than nonparametric tests. Therefore, to choose appropriate methods, we need to examine the distribution of sample data. Normal probability plots and histograms will provide us with a visual display of a distribution, and boxplots are also useful. It is a normal distribution if the normal probability plot is roughly linear. In addition to these plots, statistical packages such as SAS provide formal diagnostics for normality. For example, if the sample size is no more than 2,000, the Shapiro-Wilk statistic, W, is computed to test normality in SAS, and the Kolomogorov D statistic is used otherwise (SAS Institute 1990).

Hypothesis testing entails errors. The null hypothesis is rejected when the test statistic falls into the rejection region. Type I error occurs when we reject the null hypothesis when it is true. Type II error occurs when we fail to reject the null hypothesis when it is false; that is, the test statistic falls in the nonrejection region when, in fact, the null hypothesis is false. The probability of making a Type I error is the significant level of a hypothesis test, α. This is the chance for the test statistic to fall in the rejection region when in fact the null hypothesis is true. The probability of making a Type II error, β, depends on the true value of μ, the sample size, and the significance level. For a fixed sample size, the smaller the significance level α we specify, the larger the probability of making Type II error, β. The power of a hypothesis test is the probability of not making a Type II error; that is, $1 - \beta$. The P-value of a hypothesis test is the smallest significant level at which the null hypothesis can be rejected, and it can be used to assess the strength of the evidence against the null hypothesis. The smaller the P-value, the stronger the evidence. Generally, if the P-value ≤ 0.01, the evidence is very strong. For $0.01 < P \leq 0.05$, it is strong. For $0.05 < P \leq 0.10$, it is weak or none (Weiss 1999).

7.3 CORRELATION AND REGRESSION

Correlation and regression are often used to detect an association between environmental risk measures and population distribution by race/ethnicity and income measures in environmental justice analysis. The most commonly used linear correlation coefficient, r, also known as the Pearson product moment correlation coefficient, measures the degree of the linear association between two variables. The correlation coefficient, r, has several nice properties, with easy and interesting interpretations. It is always between -1 and 1. Positive values imply positive correlation between the two variables, while negative values imply negative correlation. The larger the absolute value, the higher the degree of correlation. Its square is the coefficient of determination, or r^2. In the multiple linear regression, r^2 represents the proportion of variation in the dependent variable that can be explained by independent variables in the regression equation. It is an indicator of the predictive power of the regression equation. The higher the r^2, the more powerful the model.

Of course, we also use sample data to estimate r, and want to know whether this estimate really represents the true population correlation coefficient or is simply attributed to sampling errors. Usually, the t-test is used to determine the statistical significance of the correlation coefficients. However, like regression, the t-test requires very strong assumptions such as linearity, equal standard deviation, normal populations, and independent observations. When the distribution is markedly skewed, it is more appropriate to use Spearman's rank-order correlation (Hollander and Wolfe 1973). This is a distribution-free test statistic, which requires only random samples and is applicable to ordinal and higher scales of measurements. Interval and ratio data have to be converted to ordinal data to use this test. Another nonparametric statistic test is Kendall's rank correlation coefficient, which is based on the concordance and discordance of two variables. Values of paired observations either vary together (in concord) or differently (in discord). That is, the pairs are concordant if $(X_i - X_j)(Y_i - Y_j) > 0$, and discordant if $(X_i - X_j)(Y_i - Y_j) < 0$. The sum K is then the difference between the number of concordant pairs and the number of discordant pairs. The Kendall's rank correlation coefficient represents the average agreement between the X and Y values.

The classical linear regression model (CLR) represents the dependent (response) variable as a linear function of independent (predictor or explanatory) variables and a disturbance (error) term. It has five basic assumptions (Kennedy 1992): linearity, zero expected value of disturbance, homogeneity and independence of disturbance, nonstochasticity of independent variables, and adequate number of observations relative to the number of variables and no multicollinearity (Table 7.3). If these five assumptions are met, the ordinary least square (OLS) estimator of the CLR model is the best linear unbiased estimator (BLUE), which is often referred to as the Gauss-Markov Theorem. If we assume additionally that the disturbance term is normally distributed, then the OLS is the best unbiased estimator

TABLE 7.3
Assumptions Underlying the Classical Linear Regression Model

Assumptions	Definitions	Examples of Violations
Linearity	The dependent variable is a linear function of independent variables	Wrong independent variables, nonlinearity, changing parameters
Zero expected disturbance	The expected value of the disturbance term is zero	
Homogeneity and independence	The disturbance terms have the same variance and are independent of each other	Heteroskedasticity, autocorrelated errors
Nonstochasticity	The observations on independent variables can be considered fixed in repeated samples	Errors in variables, autoregressions, simultaneous equation estimation
No multicollinearity	There is no exact linear relationship between independent variables and the number of observations should be larger than the number of independent variables	Multicollinearity

Source: Kennedy, P., *A Guide to Econometrics*, 3rd ed., MIT Press, 1992.

among all unbiased estimators. Typically, normality is assumed for making inferences about the statistical significance of coefficient estimates on the basis of the t-test. However, the normality assumption can be relaxed, for example, to asymptotic normality, which can be reached in large samples. If there is serious doubt of the normality assumption for a data set, we can use three asymptotically equivalent tests: the likelihood ratio (LR) test, the Wald (W) test, and the Lagrange multiplier (LM) test. Inferential procedures are also quite robust to moderate violations of linearity and homogeneity. However, serious violations of the five basic assumptions may render the estimated model unreliable and useless.

Violations of the five assumptions can easily happen in the real world and take different forms (Tables 7.3 and 7.4). As shown in Chapter 4, the relationship between proximity to a noxious facility and the facility's impacts can be nonlinear, and the nearby area takes the greatest brunt of risks. Spatial association among observations, which will be discussed in detail later, can jeopardize the independence assumption. Indeed, in environmental justice literature, we see frequent violations of some assumptions.

One common violation is misspecification, which includes omitting relevant variables, including irrelevant variables, specifying a linear relationship when it is nonlinear, and assuming a constant parameter when, in fact, it changes during the study period (Kennedy 1992). As a result of omitting relevant independent variables, the OLS estimates for coefficients for the included variables are biased and any inference about the coefficients is inaccurate, unless the omitted variables are unrelated to the included variables. In particular, the estimated variance of the error term

TABLE 7.4

Consequences, Diagnostics, and Remedies of Violating CLR Assumptions

Violating Assumptions	Consequences	Diagnostics	Remedies
Omission of a relevant independent variable	Biased estimates for parameters,[1] inaccurate inference	RESET, F and t tests, Hausman test	Theories, testing down from a general to a more specific model
Inclusion of an irrelevant variable	The OLS estimator is not as efficient	F and t tests	Theories, testing down from a general to a more specific model
Nonlinearity	Biased estimates of parameters; inaccurate inference	RESET, Recursive residuals; general functional forms such as the Box-Cox transformation; non-nested tests such as the no-nested F test, structural change tests such as the Chow test	Transformations such as the Box-Cox transformation
Inconstant parameters	Biased parameter estimates	The Chow test	Separate models, maximum likelihood estimation
Heteroskedasticity	Biased estimates of variance; unreliable inference about parameters	Visual inspection of residuals, Goldfeld-Quandt test, Breusch-Pagan test, White test	Generalized least square estimator, data transformation
Autocorrelated errors	Biased estimates of variance; unreliable inference about parameters	Visual inspection of residuals, Durbin-Watson test, Moran's I; Geary's c	Generalized least square estimator
Measurement errors in independent variables	Biased even asymptotically	Hausman test	Weighted regression, instrumental variables
Simultaneous equations	Biased even asymptotically	Hausman test	Two-stage least squares, three-stage least squares, maximum likelihood, instrumental variables
Multicollinearity	Increased variance and unreliable inference about parameters, specification errors	Correlation coefficient matrix, variance inflation factors, condition index	Construct a principal component composite index of collinear variables, simultaneous equations

[1] This is true unless the omitted variables are unrelated with the included variables.

Source: Kennedy, P., *A Guide to Econometrics*, 3rd ed., MIT Press, 1992.

is overestimated, resulting in the overestimation of the variance-covariance matrix parameters. Consequently, inference about these parameters is unreliable. As a result, t-statistic values may be biased downward, and coefficients that are, in fact, statistically significant may become insignificant. Similarly, nonlinearity biases the parameter estimates and any inference about the parameters in the OLS estimation. Since a nonlinear function can be transformed into a polynomial one through a Taylor series expansion, the biases are approximately equivalent to omitting relevant variables that are higher-order polynomial terms.

These biases cast doubt on a couple of published studies that use only race and income variables in their regression models. A widely cited study of the Detroit area estimated two multiple linear regressions with only two independent variables — race (measured as 1 for white and 0 for minority) and income (Mohai and Bryant 1992). The dependent variable for one regression model is ordinal — a trichotomy of distance of the respondent resident to a commercial hazardous waste site, i.e., 1 = within 1 mile, 2 = between 1 mile and 1.5 mile, and 3 = more than 1.5 miles. This is a qualitative dependent variable, for which the CLR has its limitations (see Section 7.5). The dependent variable for the other regression is the distance of the respondent resident to the center of a facility. Clearly, the regressions failed to take into account some relevant independent variables other than race and income and resulted in a great deal of unexplained variance. With adjusted R-square values of 0.04 and 0.06, at least 94% of variations in the dependent variable were unaccounted for. In addition, the linear model assumes a linear relationship between exposure and distance from the source: exposure at 1 mile is exactly 10 times that at 0.1 mile. This assumption is hardly plausible and may misrepresent the true relationship (Pollock and Vittas 1995). These misspecifications lead to biased estimates of parameters for race and income and unreliable inferences about these parameter estimates. Consequently, these biases thwart the validity of the authors' objective and conclusion about the relative strength of race and income in explaining the distribution of commercial hazardous waste facilities in the Detroit area. Similar misspecifications can be found in a study of TRI release quantities in relation with race and income at the ZIP code level in the State of Michigan, its urban areas, and the Detroit Metropolitan Area (Downey 1998). No diagnostics of specification errors were reported in these studies.

Table 7.4 shows the consequences of violating assumptions underlying the CLR model, common diagnostic techniques used to detect these violations, and possible remedies for correcting these violations. For correcting specification errors, researchers have agreed that we should first and foremost consider relevant theories that would provide us with guidance on what variables to include and what functional forms to use. Unfortunately, for most social science disciplines, theories do not help us a lot. Social theories are particularly inadequate for prescribing appropriate functional forms. Recognizing this deficiency, researchers should consider various criteria for choosing a functional form such as theoretical consistency, domain of applicability, flexibility, computational facility, and factual conformity (Lau 1986).

Multicollinearity, measurement errors, and spatial autocorrelations deserve greater attention in environmental justice analysis. It is well-known that race and income are highly correlated. If these two variables are put in the same regression, multicollinearity may result. Although the OLS still maintains the BLUE, the variances of the OLS-parameter estimates for collinear variables are biased upward (Kennedy 1992). In this case, we do not know which collinear variables should be given credit for explaining variation in dependent variables and have thus less confidence on the parameter estimates of the collinear variables. If race is defined as the percentage of minority, we can see a negative correlation between race and income. Because of the high correlation, they share a large proportion of the variation (in the dependent variable) that can be attributed to them aggregately. Accordingly, the contribution unique to each variable is small. We do not know how to allocate the shared contribution. Because of the large variance caused by collinearity, we may fail to reject the individual hypothesis that race or income has no significant role in explaining environmental risk distribution. But the joint hypothesis that both race and income have a zero parameter may be rejected. This means that one of them is relevant, but we do not know which.

7.4   PROBABILITY AND DISCRETE CHOICE MODELS

A large proportion of environmental justice studies have treated environmental impact as a dichotomous or trichotomous dependent variable. That is, there is either presence or absence of a particular (potentially or actually) environmentally risky facility (such as TSDFs, Superfund sites, or TRI facilities) in a geographic unit of analysis (such as ZIP code, census tracts, or 1-mile radius). These facilities can be further classified as a trichotomous or polychotomous dependent variable, based on the degree of potential or real environmental risk. For example, we can have "clean" tracts without TRI facilities in them or in adjacent tracts, "potentially exposed" tracts with TRI facilities in adjacent tracts only, and "dirty" tracts with one or more TRI facilities (Bowen et al. 1995). The oft-cited UCC study classified 5-digit ZIP code areas into four groups according to the presence, type, and magnitudes of TSDFs in residential ZIP code areas. (See Chapter 4 for a discussion of the strengths and weaknesses of this proximity-based approach.) These dependent variables are qualitative or discrete, with which the CLR has some difficulties dealing.

Probability or discrete choice models are often used for qualitative or discrete dependent variables. These models really concern the probability of an event or making a discrete choice based on the decision-makers' characteristics and the attributes of alternatives. Some commonly used models include logit, probit, and Poisson models. Since the independent variables are often continuous, logit models are also referred to as logistic regressions. The simplest is a binomial logit model, where the dependent variable has two categories or choices. For example, a noxious facility is present in or absent from a census tract. Underlying logit and probit models is the random utility theory (see Chapter 9). The utility is formulated as a function of individuals' characteristics and attributes of alternatives, plus an error term. Probit models assume that the error term is normally distributed.

Logit models assume that the error term is independently and identically distributed (IID) as a Gumbel (log Weibull) distribution. Independence from irrelevant alternatives (IIA) property is crucial for logit modeling. What this property calls for is that alternatives cannot be very similar substitutes. "The choice probabilities from a subset of alternatives is dependent only on the alternatives included in this subset and is independent of any other alternatives that may exist" (Ben-Akiva and Lerman 1985:51). Otherwise, we will have the red bus/blue bus paradox. When the existing alternative is car and the only difference between the two buses is the color, they are not three alternatives meeting the assumptions of IIAs. It is wrong to expect that the three alternatives share the probability, but rather it should be that the blue bus and the red bus as a subset share the probability with the car alternative. One method of testing the IIA assumption is to compare the estimates of parameters from a logit model estimated with a full choice set and those from a logit model estimated with a reduced choice set. If the IIA holds true, the estimated coefficients should not change. Various procedures are also used to test for nonlinearity, heteroskedasticity, and outliers (Ben-Akiva and Lerman 1985).

The maximum likelihood method is used to estimate logit models. Asymptotic t-test is used to test the statistical significance of parameters (whether they are statistically significantly different from zero). Similar to the F-test in the CLR, the likelihood ratio test is used to test the joint hypothesis that all parameters are equal to zero. The goodness-of-fit is the likelihood ratio index (rho-squared), similar to R-square in the CLR.

If the dependent variable is ordered or ranked, an ordered logit model can be used. For example, clean tracts, potentially exposed tracts, and dirty tracts form a set of ranked alternatives. If the choice-making process follows a sequence with different stages and the decision in a later stage is nested and preconditioned on an earlier stage, a nested logit model can be used. Take car ownership for an example. A household first decides whether to buy a car or none at all. Then, if the household decides to buy, it needs to choose whether to buy one or more than one car. This process continues on and on until a desired level is reached.

In some cases, dependent variables are limited in the sense that they are censored and not observable at some known values of independent variables or they are truncated. In these cases, the OLS estimators are biased, and the maximum likelihood method is used for estimation. The Tobit model is used for the censored sample.

7.5 SPATIAL STATISTICS

Spatial statistics are based on the first law of geography and include spatial association, pattern analysis, scale and zoning, geostatistics, classification, spatial sampling, and spatial econometrics (Getis 1999). The First Law of Geography refers to the inverse relationship between value association and distance (Tobler 1979). Neighbors are more alike than points far apart. This means that if we have data for a point in space, it is possible to infer values for its neighbors. The questions are then what constitutes a neighbor or neighborhood and how we derive a value for a location from its neighbors. Two areal units (grid cells and polygons) are neighbors or contiguous if they share a common segment of their boundaries. A square contiguity or spatial weights matrix (W) is generally used to represent the neighborhood association for N locations or observations (Anselin 1993), where its element w_{ij} has a nonzero value when observations i and j are neighbors and a zero value otherwise. The spatial weights can be binary based on whether the pair has a common border or interval values that are based on inverse distance or inverse distance squared, or on the length or relative length of the shared border. Therefore, contiguity can generally mean not only whether or how much the two areal observations share a part of their borders but also how far they are from each other. In other words, two locations are considered to be contiguous if the distance between them is within a critical value. This weights matrix is often standardized by rows so that the elements sum to one across the row. Multiplying this weights matrix by an attribute vector, we obtain a product vector that consists of weighted averages of neighboring values. The resulting variable is called a spatial lag, by analogy of the time lag in time-series analysis.

Spatial autocorrelation is a measure of interdependence among spatially distributed data or the degree of correlation between a location and its neighbors. It is sometimes referred to as spatial dependence or spatial association. A positive spatial autocorrelation occurs when a large value for a location is surrounded by large values of its neighbors or when a small value is surrounded by small values of its neighbors. A negative spatial autocorrelation occurs when large values are surrounded by small values or vice versa. A positive spatial autocorrelation signifies spatial similarity or clustering, while a negative spatial autocorrelation means spatial dissimilarity.

Moran's I statistic and Geary's c statistic are the two most commonly used measures for testing spatial autocorrelation. Both measures use the individual values and mean of a variable and the spatial weights matrix for calculation. Their formulations differ in that Moran's I uses cross-products (covariance) to measure association while Geary's c uses the square of differences between associated locations. Like Pearson's correlation r, Moran's I takes values with a range between -1 and $+1$. Positive values mean a similar values cluster and negative values mean a dissimilar values cluster, while 0 indicates values are randomly distributed spatially. The Geary's c has a range from 0 to 2. In contrast to Moran's I, the Geary's c value that is smaller than its mean of 1 indicates positive spatial autocorrelation. When the Geary's c is 2, dissimilar values cluster.

Spatial autocorrelation is one form of autocorrelation, and another is temporal autocorrelation in time-series data. Autocorrelation leads to biased estimates of parameters and the variance-covariance matrix. The inference about the statistical significance of the parameter estimates is unrealizable. The Durbin-Watson test is the most popular test for non-spatial autocorrelation, while Moran's I and Geary'c are the two most popular tests for spatial autocorrelation. When Moran's I statistics are significant, first spatial differences can be used to provide a reasonable way to eliminate spatial autocorrelation problems (Martin 1974). When autocorrelation occurs, the estimated generalized least square (EGLS) method is often used to estimate a regression model. Another approach is to filter out the spatial autocorrelation using the Getis-Ord statistics and then use the OLS (Getis 1999).

7.6 APPLICATIONS OF STATISTICAL METHODS IN ENVIRONMENTAL JUSTICE STUDIES

In environmental justice analyses, the analyst is really concerned with the relationship between the distribution of an environmental impact such as environmental risks from Superfund sites and the distribution of the disadvantaged subpopulations such as minority and the poor. To this end, the analyst first estimates the two distributions and then identifies their associations. Chapters 4 and 5 presented the methods for measuring and modeling environmental impact and population distributions. Various statistical methods have been used to uncover their relationships, including univariate statistics, bivariate analyses, and multivariate analyses.

The analytical procedures usually proceed in two steps. The first step involves use of univariate statistics and bivariate analysis for independent variables. If treating the dependent variable as a discrete or categorical variable, the analyst divides the statistical population under investigation into two or more groups with one of them as the comparison (control) group and then summarizes the characteristics of these groups using univariate statistics. A few examples of group classification have been provided above, and Chapter 6 discussed various geographic units of analysis that can be used as the geographic basis for classifying these groups. As another example, census tracts can be categorized into three groups in a ranked order: tracts without any TRI release, tracts with a TRI release that does not contain a carcinogen or USEPA33/50 chemical, and tracts with TRI releases that have a carcinogen or USEPA33/50 chemical (Sadd et al. 1999a). Univariate statistics such as mean and median are then used to characterize these groups in terms of independent variables. Using the t-test or Wilcoxon test, the analyst examines whether these groups are statistically significant different from one another. Alternatively, the analyst can use the correlation coefficients to detect the association between an environmental impact measure or proxy and the percentage minority or the poor. For interval or ratio dependent and independent variables, the Pearson correlation coefficients can be used if t-test assumptions can be met. For an ordinal variable and the occasions where t-test assumptions cannot be met, nonparametric statistics such as Spearman rank-order correlation or Kendall's rank correlation coefficient should be employed. If the test statistic value indicates that the null hypothesis of no association can be rejected at a specified significance level, an association is established between environmental impact and minority or the poor. GIS, as will be discussed in Chapter 8, particularly helps visualize the geographic patterns of any association. Of course, the association between two variables may be spurious because of a third variable that affects both of them.

In the second stage, as most analysts agree, multivariate analyses should be used to control for the effects of multiple independent variables. Another purpose of using multivariate analyses is to determine the relative importance of various independent variables, particularly race and income, in explaining the dependent variables. For evaluating relative importance, the analyst can resort to the standardized parameter estimates and their statistical significance, or examine the marginal effect of a one standard deviation increase of each independent variable on the expected value of the dependent variable. For a dependent variable with an interval or ratio measure, the CLR can be used if their assumptions are reasonably met. For an ordinal or nominal dependent variable, logit and probit models are more appropriate. In the environmental justice literature, multivariate analyses have employed a variety of methods, including linear regression (Pollock and Vittes 1995; Brooks and Sethi 1997; Jerrett et al. 1997), very popular logit models (Anderton et al. 1994; Anderton, Oakes, and Egan 1997; Been 1995; Boer et al. 1997; Sadd et al. 1999a), probit models (Zimmerman 1993; Ringquist 1997), Tobit model (Hird 1993; Sadd et al. 1999a), discriminant analysis (Cutter, Holm, and Clark 1996), and others.

Greenberg (1993) illustrates that different statistics could lead to different findings about equity (see Table 7.5). Different statistics have different assumptions, advantages, and disadvantages. Table 7.5 shows a comparison of three measures: proportion, arithmetic mean, and population-weighted mean. In Greenberg (1993), the noxious facilities are the Waste-to-Energy Facilities in the United States in general and New Jersey in particular. In this case, we have the burden areas of the facility-hosting neighborhood (defined as towns) as the target (experiment) group and benefit areas (service area) as the control (comparison) group. This is essentially a paired sample. If we are only concerned about whether the burden areas are larger or smaller than the benefit areas in terms of minority or low-income population proportions, the dependent variable is essentially reduced to a dichotomy. If randomly distributed, the burden areas would have a 50% chance of being larger (smaller) than the benefit areas in terms of minority or low-income population proportions. Obviously, the conversion of a ratio or interval variable into a nominal or ordinal variable reduces the information available from the original data. Arithmetic mean and population-weighted mean take full advantage of the information in the original data. While arithmetic mean treats each observation in the sample equally, population-weighted mean treats each unequal observation in the sample unequally. Population-weighted mean reduces the sample into a single aggregate measure, and favors large population centers. Given the fact that minorities tend to be concentrated in large cities, it is not surprising to find the disproportionate burden on them when using population-weighted means as shown in Greenberg (1993) and Zimmerman (1993). To minimize these biases, Greenberg (1993) proposed use of segmentation by population size to make comparisons in each segment.

As discussed earlier, the proximity-based approach to defining the dependent variable tends to be arbitrary in classifying exposed vs. non-exposed observations and in using a certain radius such as 0.5 or 1 mile. Consequently, the findings may depend upon these arbitrary measurements. To correct for the arbitrariness, Waller, Louis, and Carlin (1997) proposed use of a cumulative distribution of exposure potential by subpopulations, which is then translated to risks (disease incidence) through a dose-response function. Exposure potential is measured as the inverse distance from the centroid of a census tract to the nearest TRI facility. Their combination leads to an injustice function, which shows the degree of injustice in relation to exposure potential (in this case, inverse distance). Further, they presented a Bayesian inferential approach to account for uncertainty in both exposure and response. This approach was implemented using Markov-Chain Monte Carlo methods. The proposed methodology is particularly appealing because of its ability to account for uncertainty, which is prevalent in environmental justice issues.

Most studies in the environmental justice literature report few diagnostics of the assumptions underlying their statistical methods, with a few exceptions. Jerrett and co-workers (1997) have reported the most comprehensive diagnostics so far in the environmental justice literature. Their statistical model regresses the total pollution emissions on median income, educational location quotient, average dwelling value, population density, total population, manufacturing employment location quotient,

TABLE 7.5
Different Statistics Lead to Different Findings in Environmental Justice

Statistic	Definition	Example	Strength	Weakness
Proportions	The percentages of communities in the test group (such as TSDF-hosting tracts) that have higher values than those in the control group (such as non-TSDF tracts)	Only 28.6% of towns hosting WTEFs are more affluent and only 38% have a larger percentage of African- and Hispanic-Americans than the U.S. as a whole	Each community treated equally	If communities have a wide range of population size, it is biased against the communities with a lot of people. It ignores the magnitude of difference between the two groups and thus loses important equity information
Arithmetic mean	Average of a variable across the communities in the test group, compared with that for the control group	Average of the percentage of blacks is 9.1% for Census Places or MCDs with NPL sites, compared with 12.1% for the nation as a whole. Percentage difference between these two numbers is −25%, while the absolute difference is −3%	Each community treated equally. Takes into account magnitude of a variable	If the variable does not have a normal distribution, it is a biased estimate of the central tendency because of extreme values
Population-weighted mean	Weighting each community in the test group by community population, compared with that for the control group	NPL sites as a whole have 18.7% blacks, compared with 12.1% for the nation. Percentage difference is 55%, while the absolute difference is 6.6%	Each person treated equally. Takes into account magnitude of a variable	Favors large population centers and is biased against small communities. Commits the median fallacy for median household income, as discussed above
Arithmetic mean, segmented by community population	Averaging a variable across each type of communities (segmented by population) in the test group, compared with that for the control group	Compared with the U.S., the average of the percentage African- and Hispanic-Americans for WTEF towns with at least 100,000 people is 62.9 higher, while that statistic for towns with less than 100,000 residents is 22.9 lower	Each community treated equally in each segmentation. Takes into account magnitude of a variable	Segmentation line is arbitrary and may affect the result

Sources: Greenberg, M. R., *Risk: Issues Health Safety,* 4(3), 235–252, 1993. Zimmerman, *Risk Anal.,* 13(6), 649–666, 1993.

and primary industry employment location quotient at the county level in Ontario, Canada. Extensive and unique transformations were conducted to the closest approximation of the normal or Gaussian distribution. In particular, total emissions were raised to 0.2 power, and the influence of high values was thus reduced. Use of location quotients for manufacturing and primary industry and educational level is useful (see another application of location quotients in Chapter 10). Except for educational location quotient (which was raised to 0.25 power) and manufacturing location quotient, five other independent variables were transformed in natural log. In addition, transformed independent variables were subtracted by their means. Based on criteria such as theoretical considerations, a measure of bias (Mallow's Cp statistic), adjusted R^2, and the standard error of the model prediction, the authors selected the best model that used 4 predictor variables among 128 possible combinations of the 7 variables. The four variables are median household income (in log), average dwelling (in log), total population (in log), and manufacturing location quotient. Although pollution emission is among the worst substitutes for environmental risks and the county level is too large a geographic unit of analysis, the selected model managed to explain 63% of the variation in the dependent variable. Diagnostics include normality of residuals, homoscedasticity of residuals, independence of residuals (spatial autocorrelation), linearity, multicollinearity, specification errors, outliers and influential cases, and cross-validation.

8 Integrating, Analyzing, and Mapping Data with GIS

A Geographic Information System (GIS) is defined in a variety of ways. "GIS technology is a data (information) integration and analysis engine which produces results that can be rendered using map displays at various levels of information resolution" (Nyerges, Robkin, Moore 1997:124). A GIS's functions include data storage, data retrieval, data analysis, and data display.

Spatial database and analytical tools, particularly GIS, have very important roles in equity analysis and environmental justice discourse. As part of its environmental justice strategy, EPA (1995a) established four sets of objectives for data collection, analysis, and access: (1) addressing data gaps; (2) improving quality and reducing burdens of data reporting; (3) data integration and analysis; and (4) improving public access. GIS can contribute to each of these four areas. EPA has undertaken major efforts to increase the accuracy of its locational data for its regulated facilities or sites and environmental quality monitoring points. EPA has been promoting the use of GIS to enhance identification of disproportionately affected communities.

The Final Guidance for Incorporating Environmental Justice Concerns in EPA's NEPA Compliance Analyses specifically recognizes how GIS can help in environmental justice analysis (U.S. EPA 1998b). "GIS technologies are useful for characterizing environmental justice issues by identifying the locations of minority communities that potentially may be affected by proposed actions and providing a visual understanding of how potential impacts may be distributed within a geographical area. GIS provides the technology for displaying and overlaying locational information and population and site characterization information on one or more maps. GIS allows for the visual display of vast amounts of spatially oriented information. In addition, GIS systems can be used to display alternative "what if" scenarios and provide for relatively quick and easy general comparisons of the potential impacts presented by alternative locations" (U.S. EPA 1998b:46).

GIS can serve as a bridge for positivism and participatory research. On one hand, it can be analytical and used for identifying patterns and associations; on the other hand, it is communicative and descriptive.

This chapter begins with a review of several basic concepts. Next, we discuss various spatial interpolation methods, including point-based and area-based interpolations. Using these methods, we estimate the spatial distribution of a variable under investigation at the geographic unit of analysis that we desire. As mentioned in Chapter 6, GIS can help in defining geographic units of analysis. In Section 8.3, we examine techniques that can be used to derive geographic units of analysis for

163

environmental justice research. These include adjacency analysis and buffer analysis, with a variety of refinements. Section 8.4 discusses overlay, a popular GIS function, and its origin and application in suitability analysis. In Chapter 2, we reviewed several theories of justice and equity. In Section 8.5, we examine how to operationalize these equity criteria in a GIS environment. Finally, we explore the strategies for integrating GIS with urban and environmental models for the purpose of environmental justice analysis.

8.1   SPATIAL MEASURES AND CONCEPTS

8.1.1   SPATIAL DATA

Spatial data consists of points, lines, and polygons (areas) in an increasing order. They are fundamental building blocks to characterize the Earth's surface. Points are the most basic component of a spatial database and all higher-order spatial data are composed of points. A point has neither length nor area. Examples of points are an intersection of two streets and the center of an area (centroid). A toxics-releasing facility may be treated as a point if it is small. A point is spatially represented by its coordinates, either two- or three-dimensional. A line consists of a string of points with beginning and end points. A line has length but no area; examples are the centerline of a road or stream, the boundary of a polygon (area), and the shoreline. An area, or polygon, such as a census tract or a parcel is an enclosed region that consists of a string of lines with the same beginning and end point. A polygon can be described in terms of its area, shape, and perimeter. These three entities can adequately represent a two-dimensional plane.

8.1.2   SPATIAL DATA STRUCTURE

Two basic data structures are used to store spatial data: raster and vector (Star and Estes 1990). In a raster data structure, the plane is divided into a system of cells (pixels), often regular square cells or possibly irregular cells. Each cell has a series of attributes. If the cell is a grid, the data are stored in a matrix with rows and columns representing the location of each cell. In a vector data structure, spatial features are modeled precisely by a series of points (coordinates) that comprise the features. In the case of a straight line, a vector data structure requires only the coordinates of the beginning point and the end point, whereas the raster data structure requires the coordinates of a series of pixels along the line depending on the resolution. The most widely used GIS database, TIGER™, is vector based. The two types of data structures correspond to raster- or vector-based GIS. An example of raster-based GIS is IDRISI®, and examples of a vector-based GIS are Arc/Info®, ArcView®, Atlas®, MapInfo®, and Maptitude®.

Both vector and raster data structures have advantages and disadvantages. Traditionally, raster structure is known for simple representation, high overlay efficiency, enhanced spatial modeling capabilities, and image processing capability. Vector structure has an edge in efficient data collection, compact and efficient representation, and precise topology. With advancement of high technology, these differences are gradually diminishing. For example, some raster-based GIS software can handle vector data, and some vector-based GIS packages incorporate raster-based capabilities.

8.1.3 DISTANCE

Distance is a critical measure for proximity-based environmental justice analyses. As shown in Chapter 4, environmental and economic impacts generally decline with distance. Almost all GIS packages provide a scale which allows the user to measure the airline distance from point a to point b. If necessary, we can calculate distance using the x, y coordinates of two points:

$$d = \sqrt{[(x_2 - x_1)^2 + (y_2 - y_1)^2]}$$

Whether using this simple formula or a handy tool in GIS, the analyst must keep in mind that the GIS layer should be appropriately projected before measuring the distance or obtaining the x, y coordinates for distance calculation. Equidistance projection methods avoid distortion of distance. If a GIS software has a network analysis component, the user can estimate the distance traveled along a street network.

8.1.4 CENTROID

Literally, a centroid is the center of an area (polygon) such as a census tract. Most GIS packages are equipped with a tool to calculate the centroid. In most cases, the centroid means the geometric or the areal unit centroid and is relative to the boundary of a polygon. When a polygon is irregularly shaped, its centroid could be found outside the polygon's boundary. Labels in a polygon layer are often attached to or near the centroids of polygons. Areal unit centroids are calculated based on a series of representative points around the perimeter of a polygon (Griffith and Amrhein 1991).

The areal population centroid, also referred to as the spatial mean, has been used to represent the geographic center of the U.S. It is the mean longitude and latitude of all population points (Garson and Biggs 1992). It is estimated through weighting the known x, y coordinates of the areal unit centroids for small geographic areas by these areas' populations. Similarly, we can choose other variables of interest for weighting such as household income. This type of centroid is sometimes called the "center of gravity."

8.2 SPATIAL INTERPOLATION

Spatial interpolation refers to the estimation of values for a variable at unknown locations or units from known locations or units. Usually, it is used in the context of transforming data from spatially scattered data to spatially continuous data or raster representation. For our purpose, this type of spatial interpolation is referred to as point interpolation. Another type is areal interpolation, which translates data from one spatial unit (area) to another.

8.2.1 POINT INTERPOLATION

An environment monitoring network consists of scattered stations (points) in space and provides discrete spatial data. We are interested in locating other points that do not have data, identifying spatial patterns, and visualizing the statistical surface. This requires the use of interpolation methods to make inferences about unknown points from known points. For example, air quality stations are distributed in certain places throughout a metropolitan area and provide ambient air quality data representing the station and surrounding areas. For air quality planning and environmental justice analysis, we would like to know the spatial distribution of air quality such as ozone (see Chapter 10 for an interpolation application).

Interpolation methods include the local neighborhood approach, the geostatistical approach, and the variational approach (Mitas and Mitasova 1999). The local neighborhood approach assumes those points nearby are more important than those far away. This approach includes inverse distance weighting, natural neighborhood, triangulated irregular network (TIN), and rectangle-based methods. Distance weighting is one of the most commonly used interpolation methods. The value of a variable at a point is estimated as the weighted average of the values at the sampled points, with the weight being an inverse function of distance raised to a power. A power or an exponent of 1 or 2 is usually used. As noted above, d^2 is simply the sum of the squares of x and y differences between two points. This method is intuitively very appealing and computationally efficient. However, it also has some limitations; in particular, the estimated values always fall within the range of the sample's minimum and maximum, and accordingly create local extrema at the sampled data points.

Tessellation refers to the partition of an Euclidean space into non-overlapping regions (Boots 1999). A planar tessellation has two dimensions. A region could be square cells (raster), triangles, hexagons, and other polygons. The resulting polygon is defined as a neighborhood for points within that polygon. Natural neighborhood interpolation is based on the concept of natural neighborhood coordinates derived from Thiessen (also called Dirichlet or Veronoi) polygons for two dimensions and from Thiessen polyhedra for three dimensions. Here, the weight is related to area or volume instead of distance. TIN is one of two dominant forms for terrain representation, the other being regular grids. Interpolation based on TIN uses a triangular tessellation to estimate a bivariate function for each triangle and uses this function to estimate the values at other locations. It is often used in dynamic visualization and visibility analysis. Although fast and commonly used, the two tessellation-based methods described above are among the least accurate.

Kriging is a popular method that uses the geostatistical approach. Kriging is actually a method of distance weighting with the weight being a stochastically derived distance function called semi-variogram (Declercq 1996). A spline interpolation method is based on a variational approach and emphasizes that the interpolation function should pass through observed data points and be as smooth as possible (Mitas and Mitasova 1999).

Different interpolation methods have been evaluated in terms of their predictive accuracy and the efficacy of portraying apparent spatial patterns (Declercq 1996). Studies have consistently given high marks to Kriging and distance weighting (d^{-2})

methods for their predictive accuracy. Declercq (1996) evaluated interpolation routines based on polynomials, splines, linear triangulation, proximation, distance weighting, and Kriging. Results showed that the effectiveness of distance weighting and Kriging methods depended largely on the number of neighbors used. For both gradually and abruptly changing data, the squared inverse distance-weighting method produced the most satisfactory spatial patterns when using few (4 to 8) and many (16 to 24) neighbors.

8.2.2 AREAL INTERPOLATION

Areal interpolation (cross-area estimation) refers to the transformation of attribute data from one type of zone (called source zones) to another (target zones) (Fisher and Langford 1996). Many areal interpretation methods have been proposed (Goodchild, Anselim, and Diechman 1993). They can be classified into cartographic, regression, and surface methods (Fisher and Langford 1995).

The simplest and most commonly used is the areal weighting method. A GIS package can be used to compute the area and proportion of a predefined census geography that falls in a target zone. This proportion is used as a weighting factor to estimate the population characteristics of the partially contained census units. Aggregating the partially and wholly contained census units provides the population characteristics in the target zone; for example, total population in a target zone can be estimated as

$$P = \Sigma_i P_i + \Sigma_j [P_j(a_{je}/a_j)]$$

where P_i is the population of a census unit entirely enclosed by a target zone; P_j is the population of a census unit partially contained by the target zone; a_j is the total area of a partially contained unit; and a_{je} is the area of a partially contained unit that is enclosed within the target zone.

The assumption underlying this method is that population is uniformly distributed in a pre-defined census geography. This assumption is seldom realistic. While some areas are relatively homogeneous with respect to residential land use, density, and socioeconomic characteristics, most others have residential land uses of different density and socioeconomic characteristics, which are interspersed with non-residential land uses such as agriculture, forest, and commercial and industrial land uses. Indeed, pockets of minority or low-income communities may be missed in such an analysis.

A number of methods have been introduced to correct this problem. A land-use/land-cover GIS layer provides information about the spatial distribution of different types of land uses, including residential land use. This additional information can be used to help allocate population from the source zone to the target zone. Like the areal weighting method, the dasymetric method is a weighting method, but it uses the residential land-use acreage and/or density as the weight, instead of the areal figures. We can simply assume that the residential land use has the same density in a zone. This requires only a binary classification of residential and non-residential land use in the GIS layer. Further refinement can be made if areas of different residential density classes are identified. Fisher and Langford (1995, 1996) evaluated the accuracy of five methods and how classification errors in land-use/land-cover layers (derived from the Landsat image) affect the accuracy of population estimates. The dasymetric method is not only the most accurate but is also robust; it produces better estimates than other methods after incorporating up to 40% error in the classified Landsat image.

8.3 GIS-BASED UNITS OF ANALYSIS FOR EQUITY ANALYSIS

As discussed in Chapter 6, a boundary of a pre-defined census unit seldom matches the boundary or boundaries of environmental risk impacts. For a site-based equity analysis, a single unit of census geography such as a census tract may be a poor unit of analysis, and could be misleading because of the border effect. GIS offers analytical capability for identifying alternative units of analysis.

8.3.1 ADJACENCY ANALYSIS

The simplest method for providing a remedy to the border effect is to identify those polygons that are adjacent to the facility-hosting polygon. For example, if we have the address of a site and geocode the site, we will be able to visualize the location of the site relative to a census tract or block group boundary. If it is near the border between two or three census tracts, we aggregate the adjacent tracts as a whole. If the number of sites for study is large, visual inspection will be time consuming. We need to calculate the shortest distance between a site and the boundary of the polygon that contains this site. Next, we need to decide the distance threshold for determining whether a site is near a border. A GIS script may be needed to complete this operation because existing low-end GIS packages may not have the capability.

Alternatively, we can identify a polygon that contains the site. Next, we complete an adjacency analysis that identifies those polygons that are adjacent to the site polygon. This operation is readily available in most GIS software packages.

This adjacency approach is intuitively better than the single census-unit approach. It works well in situations where polygons are well shaped and relatively homogeneous in size. However, this approach fails to take into account the actual or perceived impact boundary. Essentially, the adjacency analysis fails to control the size of the included area. If we study many sites in urban and rural areas, we may end up with a wide range of sizes. The irregular shape and great variation in polygon size, as often found in census geography, render this adjacency approach less accurate.

8.3.2 BUFFER ANALYSIS

Buffer analysis is a popular GIS function. It generates polygons around a point with a specified radius (circular buffer), along a line (such as a highway), or inside or outside a polygon. A circular buffer is often used in environmental justice analyses. It has several assumptions:

- A site is small enough to be treated as a point
- The impacts are confined in the specified circular area
- The impacts are equal and uniform in all directions

These are strong assumptions, and we will see later that they should be relaxed. If a site is large, such as some Superfund sites, we may still treat the site as a point as represented by the centroid of the site. However, a generated circular buffer does not accurately depict the area surrounding the site. If the radius is small and the site is large, the buffer may fall inside the facility's boundary. In this case, it is better to delineate the site as a polygon and make a buffer around the polygon. The analyst should collect and examine available data about the shape and size of the facility before deciding which type of buffer is appropriate.

Existing data indicate a wide range of sizes for waste facilities. In a GAO survey of 301 metropolitan and 322 nonmetropolitan landfills in operation in 1992, the average metropolitan landfill covered an area of 191 acres, with a range of 1 acre to 2000 acres (GAO 1995). Nonmetropolitan landfills were 98 acres on average, ranging from 1 to 1200 acres. Therefore, polygon buffers are needed for some large landfills.

Aggregation is usually needed to estimate the socioeconomic variables for the generated buffer based on the predefined census geography. The generated buffer for equity analysis seldom matches the predefined census geography. In almost all cases, the generated buffer includes portions of some census polygons as well as some whole census polygons. There are several methods for aggregating these polygons, namely, polygon containment, centroid containment, and buffer containment (Chakraborty and Armstrong 1997).

The simplest method is polygon containment, which includes all census polygons that are touched or entirely enclosed by the buffer (see Figure 8.1). Socioeconomic attributes for the generated buffer are a simple aggregation of all the census units. This operation relaxes the second and third assumptions described earlier. The generated buffer area is no longer circular. Variation in the shape of census geography may defeat the purpose of a circular buffer. A census unit, although having only barely touched the circle with a corner, will be included and bias will result. A variation of this method is to use some cutoff criteria to limit inclusion of polygons that are partially contained in the circle. We can estimate the area of a polygon intersected by the circle, and calculate the proportion of that intersected polygon. Then, we choose a decision rule for including polygons with a significant presence. For example, if a polygon has more than half of its area fall inside the circle, then we include it.

The second method is centroid containment. If a polygon has its centroid inside the buffer, it is included. This limits the intersected polygons that can be included in analysis. Again, this method relaxes the second and third assumptions. It assumes that the centroid really represents the polygon in terms of population characteristics.

The third method is buffer containment. Strictly speaking, buffer analysis chooses and retains a certain shape (a circle, a buffer along a line, a donut) and determines the characteristics of the buffer based on a predefined census geography. Usually, an areal interpolation method, as discussed earlier, is used to estimate the characteristics of those portions in the buffer that are partially contained in a census geography.

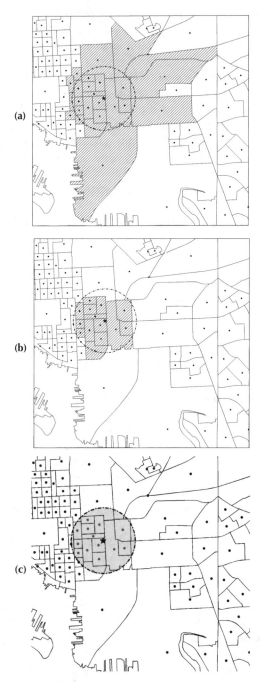

FIGURE 8.1 Methods for delineating units of analysis based on buffering: (a) polygon containment, (b) centroid containment, (c) buffer containment.

The choice of a radius is often arbitrary. The 1995 GAO report uses 1- and 3-mi radius buffers around the boundary of the landfill. Glickman, Golding, and Hersh (1995) chose 0.5- and 1-mi radii. A circular area of 0.8 sq. mi resulting from a 0.5-mi radius approximates the average area of a census tract (1.5 sq. mi) or a block group (0.5 sq. mi) in Allegheny County, the study area. "Half a mile is a distance that intuitively connotes close proximity" (Glickman, Golding, and Hersh 1995:103). A circular area with a 1-mi radius (3.14 sq. mi) is close to the average size of a municipality in the county (5.2 sq. mi), which is equivalent to a circle with a 1.3-mi radius. These justifications for choice of a circular radius do not take into account actual impacts of toxic or hazardous waste facilities.

Containment-based buffer analysis offers a fixed-size impact area for comparing many sites. However, a fixed size does not fit all facilities, which impose different magnitudes and spatial extents of environmental risks. To account for the differences among facilities, we can use a variable buffer whose radius varies with the amount of pollutant emissions. This variable buffer radius can be based on the range of TRI emissions in a study area divided by a constant that would give a maximum buffer radius of a chosen distance (Werner 1997). For example, if TRI emissions range from 10 to 1000 lbs and we assume a maximum impact distance of 5 mi, then these emissions should be divided by 200, and we should also assume a minimum impact distance such as 0.5 mi. Of course, these numbers are arbitrary and may not reflect the actual impact boundary.

Circular buffer analysis has clear advantages when census tracts may be too small and ZIP code areas may be too large. It works well in situations when facility impacts are uniformly and homogeneously distributed in a predefined circle. This seldom happens. Environmental risks are not usually distributed uniformly in all directions. More often than not, they have directional biases because of natural and/or environmental conditions such as wind directions and groundwater flow directions. Moreover, one fixed-size buffer does not fit all toxics in a site. The impact areas of different toxics vary with their types, release quantity, and environmental transport and fate.

Delineating impact areas by means of environmental dispersion models offers more accurate and realistic results. Geographic plume analysis (Chakraborty and Armstrong 1997) generates a composite plume footprint using dispersion models and historical weather patterns and estimates the composition of the population in the plume based on the aggregation methods discussed above. Air emission and dispersion models have been discussed in Chapter 4. These models project ambient pollutant concentrations downwind of a source. Therefore, it is possible to construct contour lines of pollutant concentrations (termed isolines) and identify the impact areas that are at risk based on some environmental standards. The impact area delineated is called a plume footprint. By overlaying a plume footprint layer (a polygon) over a census geography layer (a polygon) as shown in Figure 8.1, we can estimate the population characteristics of the areas at risk. The plume footprint is an irregular buffer, and the methods discussed above can be used for delineation and aggregation of a plume footprint.

In a case study of TRI facilities in Des Moines, Iowa, Chakraborty and Armstrong (1997) compared the results between circles of 0.5- and 1-mi radii and a composite plume footprint. Using the 1990 census block-group level data, they found that the results were sensitive to the shape and size of the buffer and the buffer delineation methods. The composite plume footprint has the smallest area, followed by the 0.5-mi circle and 1-mi circle buffers. The plume footprint method reports the highest percentages of nonwhite residents and individuals below the poverty line. The larger unit of analysis becomes similar to the city as a whole. Among the three buffer delineation methods, polygon containment generally includes the largest number of census units and thus the largest area. When combining the types of buffer and buffer delineation methods, they found that the 1-mile circular buffer delineated by the polygon containment method had the largest area and the smallest proportions of nonwhites and the poor. On the other hand, the composite plume footprint delineated by the centroid or buffer containment method had the smallest area and the highest proportions of disadvantaged groups. This finding that larger units of analysis tend to lower the proportions of disadvantaged groups is contrary to some previous findings such as Glickman, Golding, and Hersh (1995), who found that a 1-mile buffer led to a larger inequity than a 0.5-mile buffer.

This is a good example of integrating GIS and environmental modeling to conduct rigorous environmental justice analysis, as discussed in Chapter 4. Although the above application is for a potential spill or emission during chemical accidents, the general methodology is applicable to routine air pollutant emissions, property value impacts from a noxious facility, noise impacts from airport and highway, and impacts on groundwater aquifer.

8.4   OVERLAY AND SUITABILITY ANALYSIS

The overlay operation combines two or more layers according to the Boolean conditions. GIS layers can be point, line, or polygon layers. Point-in-polygon operation overlays a point layer over a polygon layer and identifies the polygon within which a particular point is located. Suppose we have a layer of points representing TRI facilities and a census block group boundary layer. We want to know within which census block-group each TRI facility is located. In other words, we want to associate each TRI facility with a census block group. This is one of the most commonly used GIS operations.

The Boolean operators include "AND," "OR," "XOR," and others (Star and Estes 1990). The "AND" operator merges two layers when two conditions are simultaneously met for the two layers, respectively. For example, we have two layers — a layer of 0.25-mi buffers around transit stations and a zoning layer. The "AND" operator can be used to identify parcels that are zoned for residential purposes and within 0.25 mi of a transit station. The "OR" operator combines two layers in a way that either one of two conditions are met. The "XOR" operator is an exclusive "OR," where only one of the two conditions (not both) is met. In a raster-based GIS, the overlay operation is conducted on a cell-by-cell basis. Complex conditions and scenarios can be readily calculated in a raster-based GIS, and this type of operation is commonly called cartographic modeling (Tomlin 1990). In a vector-based GIS, the overlay operation involves intersecting two layers, which generates a new geometry.

GIS originates from the overlay approach, originally proposed by Ian McHarg in his well-celebrated book *Design with Nature* (1969). His ecological approach to planning, particularly site suitability analysis, emphasizes integration of various social and environmental factors in planning. These factors are displayed in different map layers, which are then overlaid to generate the final map that shows all excluded areas as well as potentially developable areas. The relationships among various layers are governed by a variety of decision rules. The exclusionary rule identifies the areas that are inappropriate for a particular use. The discretionary or "weighting-and-rating" rule gives different weights to different layers before combining them. After overlaying, areas with the highest ratings are the best locations.

Lober (1995) expanded McHarg's representation of social values and explicitly incorporated public opposition in the siting analysis. Public opposition was estimated as a distance-decay function. Thus, the likelihood of public opposition was spatially differentiated and translated into a GIS layer, which represented the degree of acceptability or unacceptability of each location in space. Overlaying this social value layer with other environmental layers resulted in a GIS layer with locations that were both environmentally and socially acceptable. He put forth a GIS-based siting framework that considered three conflicting goals: equity, effectiveness, and efficiency.

Although the land suitability analysis framework is intended to be used from a prospective vantage, it can be used for a retrospective analysis. Using historical GIS layers such as land use/land cover, we can create GIS layers that show potentially suitable areas for industrial siting. These areas could be compared with the past and current spatial distributions of population by race and income, with the distribution of existing industrial and other environmentally risky facilities, and with the current zoning. These analyses will help us build an empirical GIS-based industrial location model, which takes into account various environmental, social, physical, and technological constraints.

Although never referencing McHarg and land suitability analysis, Wildgen (1998) used some land suitability concepts in deriving a distance-weighted race surface. In this GIS application, he considered physical limitations for siting, the distribution of protected population, and the location of economical choice sites. The industrial-compatible land-cover layer takes into account environmental constraints such as wetland. The distance-weighted 5-km buffer along the Mississippi River offers a cost surface of industrial development, particularly the petrochemical industry. Since the river has been both a major water supply source and a transport means, business costs increase with distance from the river. The black neighborhood layer is based on the neighborhood or focal operation, where each cell takes a value equal to the sum of blacks in the 2-km neighborhood. This operation smoothes the lumpy distribution of race/ethnicity that results from the arbitrary census boundary. Then, the black neighborhood surface layer is multiplied by the distance from the river. The generated layer reflects the distribution of aggregated economic and social (racial) cost. From this layer, it would be possible to delineate the minimum cost path, which indicates areas where both economic and social costs are minimum. This minimum cost path can serve as both a planning tool for future siting and an evaluation tool for examining past practices. The methodology proposed in this

GIS application takes full advantage of spatial modeling capability, such as the weighted-distance model and site-suitability model (ESRI 1997). It shows us that raster-based GIS applications to environmental justice analyses can be rich and thought provoking.

8.5 GIS-BASED OPERATIONALIZATION OF EQUITY CRITERIA

Lober's GIS-based operationalization of equity in the context of siting noxious facilities is interesting (1995). The utilitarianism notion of equity is cartographically expressed as locating a site as far as possible from areas of highest population density. This would protect most residents' rights of avoiding potential exposure to risks from the facility and satisfy the majority in society. In a GIS, identifying areas of highest population density is relatively straightforward. In a vector-based GIS, an attribute of population density (population divided by area) can be easily displayed through thematic mapping and highest density areas can be detected visually or through query. In raster-based GIS, a filter function can be used to reclassify a cell (pixel) based on its previous value and those of eight neighbors. As a result, the high density areas remain and low density areas become invisible. Rawls' notion of justice was expressed cartographically by "treating each house as having equal rights and identifying the regions that are farthest from all houses" (Lober 1995:488). In GIS, a distance function can be used to measure "the airline distance" between each cell and the nearest target cell (houses). It seems that this operationalization satisfies Rawls' first principle (see Chapter 2). However, the least advantaged group of society is not identified and incorporated in this method. It is not clear how the resulting site is to the greatest benefit of the least advantaged. To correct this, we certainly would also measure the distance between each cell and area of the least advantaged groups and give them more weight in the final decisions.

Less controversial but not less important is the provision of public facilities such as schools, libraries, parks, recreational facilities, and health care centers. Contrary to noxious facilities, planners want the public to have an easy access to these facilities. Although the notions of justice and equity as discussed in Chapter 2 are equally applicable to the provision of public facilities, special attention is needed in defining and measuring equity criteria in this specific context. For provision of public facilities, equity criteria can be equality based, need based, demand based, or market based (Lucy 1981). According to the equality-based standard, provision of public facilities is equitable if every one receives the same public benefit, regardless of socioeconomic status, willingness to pay, or need. The need-based criteria provide the largest public benefit to those who need the public facility the most and is an "unequal treatment of unequals" (Lucy 1981). The demand-based criteria give the largest weight to those who demand the services the most, either politically or economically. The market-based criteria subject provision of public facilities to how much people are willing to pay.

Accessibility measures can be used to operationalize the need-based equity criteria. Talen (1998) used four access measures in her equity mapping of parks. Gravity models, which will be discussed in detail in Chapters 9 and 13, describe the spatial interaction among human activities in the same way as Newtonian physics. The degree of access to parks from a resident location (e.g., a census block or tract) is proportional to the attractiveness of the parks and inversely proportional to the square of the distance between them. This measure emphasizes that distance is a major obstacle to access to parks, and the parks far away are much less important than those nearby.

Second, the minimum travel cost approach measures the average distance between an origin (e.g., a resident location) and each destination (e.g., every park in the city), and the goal is to minimize travel cost. This measure treats each park as equally important in a city and takes all parks into consideration. In the third method, covering objectives measure the number of facilities that is within a critical distance or covering radius for each resident location. This is similar to what is called an accumulative opportunity approach to measuring accessibility (Chapter 13). The goal here is to maximize the number of people covered in a critical distance. Only those parks within the critical distance are relevant, and this critical distance may vary with the type of facilities. For example, a critical distance of 1 mi was used to characterize access to parks. Finally, equity based on the minimum distance between a resident location and the nearest park would seek to minimize access inequality. In this case, only the nearest park really matters, and the goal is to minimize the distance to the nearest park.

Critical to these measures are the distance between a resident location and a public facility and the set of public facilities to include in the analysis. In GIS, distance is often measured in terms of the airline or straight-line distance, as discussed earlier. In this case, it is more accurate to use the street network distance. Resident location is represented by the centroid of a census unit such as block, block group, and tract. After these measures are computed for the chosen census unit of analysis, their spatial patterns can be visualized on maps and compared with those of socioeconomic characteristics. Thus, any spatial association can be detected visually or through statistical methods.

8.6 INTEGRATING GIS AND URBAN AND ENVIRONMENTAL MODELS

As argued in this book, the integration of GIS and urban and environmental models is very important for rigorous environmental justice analysis. This integration has been increasing gradually over the years and has moved faster due to the urgency of the environmental justice issues.

Strategies for linking models and GIS can take a form in the continuous spectrum from loosely coupled to strongly coupled (Batty and Xie 1994). Many practitioners have been using the loosely coupled approach to transfer data back and forth between GIS and various models such as air quality and land use models. The strongly coupled approach adds the functionality of one system to the other, such as embedding models within GIS or embedding GIS within models.

The loosely coupled approach can be accomplished with only the addition of interface programs or scripts between GIS and models. This approach is readily applicable to environmental justice analysis without formidable obstacles due to technical, budget, and time constraints. Based on this loosely coupled approach, we can derive an environmental justice analysis framework that runs a series of models and GIS operations in a sequential and iterative order. This framework starts with an urban model, which forecasts population and employment distribution, land-use consumption, and travel demand in a future year. The outputs from this model are taken as inputs to an emission/environmental dispersion model system (such as air quality or watershed model), which projects ambient environmental quality. The outputs from dispersion models are taken as inputs to exposure models, which are then linked to the risk (dose-response) model. Any interface between these models may be facilitated by scripts and GIS. Ultimately, the estimated environmental risk distribution and its relationship to the distribution of the disadvantaged subpopulations are displayed and analyzed in the GIS framework.

This loosely coupled approach has been tested recently in environmental justice analysis. In the context of a single source and two sources of pollution, Balagopalan (1999) illustrated the procedures and interfaces between environmental models and GIS. The models used include EPA's Industrial Source Complex Short Term (ISCST3) and California's Assessment of Chemical Exposure for AB2588 (ACE2588). ArcView Spatial Analyst and scripts were used to create risk isopleths and overlay isopleths with minority population distribution at the block group level.

Strongly coupled applications have been gradually increasing in the work of academic researchers and GIS software vendors. Batty and Xie (1994) showed a framework that embedded monocentric urban models in a GIS environment. METROPOLUS embeds a land use/transportation model in the ArcView environment. TransCAD is a software that strongly couples GIS and travel demand models in one platform. ESRI products have incorporated extensions that deal with location allocation, network analysis, and travel demand models.

The linkage between GIS and statistical methods also ranges from loosely coupled to strongly coupled forms in the continuous spectrum. Typically, exploratory data analysis techniques are strongly coupled with GIS (Anselin 1999c). They are used to identify usual spatial patterns and visualize and detect spatial association. Almost all GIS software packages provide descriptive statistics. Some GIS packages are equipped with tools for estimating correlation, linear regression models, and even logit models (for example, TransCAD). Some spatial statistics such as spatial autocorrelation are also available in some GIS packages. In statistical packages, extensions and interfaces have been created to use GIS-generated data to conduct spatial statistical analysis. For a strongly coupled bi-directional example, S+Gislink links S+ SpatialStats with ArcInfo. TransCAD provides procedures for calculating an adjacency matrix and Moran's I statistics (Caliper 1996). The adjacency matrix identifies areas that have common borders. Adjacency is measured in three ways: (1) the dichotomy of whether the two areas have a common border; (2) the length of the common border divided by the average of the perimeter of the two areas involved; (3) the length of the common boundary. One of the adjacency matrices can be used to compute Moran I, its expected value and the 95% confidence intervals.

9 Modeling Urban Systems

Urban environmental justice issues are salient. Some examples are lead-paint poisoning, multiple and cumulative exposures to toxics, and locally unwanted land uses. Some notable environmental justice cases have occurred in cities such as Los Angeles, Houston, Atlanta, New York, and Chester, Pennsylvania.

A city is a system, which is composed of many subsystems. Wegener's model of urban models consists of eight subsystems: land use, transportation networks, housing, population, workplaces, employment, travel, and goods movement (see Figure 9.1). Land use and the transportation network "represent the basic structure of urban settlements" (Wegener 1999:6). Land use includes where we live (residence or housing) and where we work (workplaces such as office and commercial buildings, and manufacturing plants). We wear different hats (population or employment), depending on where we are — residence or workplace. The spatial separation of where we live, where we work, where we shop, and where we play necessitates the need for travel and goods movement. These and other human activities have impacts on the urban environment (see Chapter 4).

As mentioned in Chapter 5, the use of urban models enables us to better understand the roles and interactions of location variables in current and future population distributions. More specifically, we want to know the role of environmental risks in these distributions.

Generally, there have been two fundamental approaches to urban modeling: economics and non-economics (Bertuglia, Leonardi, and Wilson 1990). The economics approach is based on neoclassical economics principles explaining the urban spatial structure in terms of a market process. The major principle is utility maximization as the basis on which, as this approach assumes, individuals make their choices of location and travel.

The non-economics approach did not initially have economics interpretation of its formulations and results. It studied social–spatial interaction and interdependency in space through a physical analogy, including so-called social physics, and later a family of spatial interaction models. The earliest conceptualization and formulation were generally credited to Batten and Boyce (1986). It greatly advanced in the mid-twentieth century thanks to Reilly (1931), Stewart (1948), Zipf (1946), Stouffer (1940), and Anderson (1955). Lowry (1964) contributed significantly to the widespread use of this approach in practice through his well-celebrated model. Introducing entropy maximization principles, Wilson (1970) made a monumental theoretical contribution to this approach by establishing its theoretical foundation and linking it to the economics approach. Ever since then, the relationship between the two fundamental approaches has been studied extensively.

While the economics approach focuses its attention on individual behaviors, the non-economics approach explores the statistical characteristics of individual behaviors. In this respect, the non-economics approach has usually been regarded as

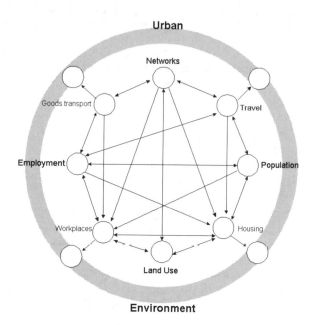

FIGURE 9.1 A model of urban models. (From Wegener, M., *J. Am. Planning Assoc.*, 60, 1, 22, 1994. With permission.)

macroscopic (aggregate) and the economics approach as microscopic (disaggregate) (Batten and Boyce 1986). In spite of significant differences, the two approaches are fundamentally identical, under certain conditions, in terms of their interpretations and results.

In the following, various approaches used to develop urban models are discussed. Section 9.1 reviews the background and development of location models using the non-economics approach. Section 9.2 covers the economics approach. Their advantages and disadvantages, particularly in practical application and evaluation measures, are discussed. Operational models and their use of these two approaches are then reviewed. Finally, the integration of urban and environmental models is discussed and an integration framework is presented based on GIS and a systems modeling approach for evaluating equity, efficiency, and sustainability goals.

9.1 GRAVITY MODELS, SPATIAL INTERACTION, AND ENTROPY MAXIMIZATION

Newton's law of universal gravitation inspired the creation of gravity and spatial interaction models for describing, explaining, and predicting geographic patterns of social and economic interactions (Batten and Boyce 1986). Newton's law states that the force of attraction between two bodies relates proportionally to their respective masses and inversely to the square of the distance between them (see Table 9.1).

Applying this law to the context of retail trade among two cities and an intermediate town, we would expect that the two cities' retail-trade attraction for the intermediate town would be directly proportional to the populations of the two cities and inversely proportional to the squares of the distances between the intermediate town and one of the two cities (see Table 9.1).

Of all spatial interaction models developed so far, the Lowry family of models using the gravity approach is perhaps the most well-known and widely used. Lowry's "Model of Metropolis" (Lowry 1964) was the first time that household and employment allocation, land use, and transportation were linked through an operational analytical model. A regionally constrained gravity model was used to simulate allocation of households to residential locations as a function of travel costs and employment locations and to simulate allocation of service employment locations as a function of household locations and transportation costs.

TABLE 9.1
Gravity Models

Model Names	Model Structure	Modelers
Law of Universal Gravitation	$F_{ij} = g \dfrac{m_i m_j}{d_{ij}^2}$	Newton (1686)
Law of Retail Gravitation	$\dfrac{P_{ij}}{P_{ir}} = (p_j/p_r)^\mu (d_{ir}/d_{ij})^\nu$	Reilly (1931)
General Unconstrained Gravity Model	$x_{ij} = k\, O_i^\alpha D_j^\beta f(d_{ij})$ special case: $\alpha = 1$, $\beta = 1$	Anderson (1955) Lowry (1964)
General Constrained Gravity Model	$x_{ij} = A_i^{\alpha-1} O_i B_j^{\beta-1} D_j f(d_{ij})$ where $A_i = \Sigma_j B_j^{\beta-1} D_j f(d_{ij})$ $B_j = \Sigma_i A_i^{\alpha-1} O_i f(d_{ij})$	Alonso (1978)
Singly Constrained Gravity Model (or One-Way Distribution Model)	A special case of the above model when $\alpha = 0$, $\beta = 1$ production-constrained $\alpha = 1$, $\beta = 0$ attraction-constrained	Wilson (1970)
Doubly Constrained Gravity Model	A special case of the above model when $\alpha = 0$, $\beta = 0$	Wilson (1970)

Notation:

F_{ij} = the force of attraction between two bodies i and j;
m_i and m_j = the respective masses of bodies i and j;
d_{ij} = the distance between bodies i and j;
p_j, p_r = the population in place j and place r, respectively;
P_{ij}, P_{ir} = the proportions of all the demand or trade attracted from i to j and r, respectively;
x_{ij} = the flow from place i to place j;
O_i = the flow originating from place i;
D_j = the flow terminating at place j;
$f(d_{ij})$ = the distance impedance function between place i and j;
A_i, B_j = the balancing factors;
g, α, β, μ, ν = the parameters;
k = a constant of proportionality.

The Lowry-type models have the theoretical foundation of economic base theory. That is, regional economic development is driven by a region's ability to produce goods for consumption outside the region. Economic activities are classified as basic sectors (which produce goods and services for consumption outside the region) and nonbasic sectors (serving inside the region); basic sectors are an engine for regional economic growth. The basic sectors generate demand for nonbasic sectors. Both basic and nonbasic sectors generate employment and support population, which in turn generate more demand for nonbasic sectors. Economic base theory assumes a causal chain from the exporting or basic industry to residential population and to nonbasic or service employment. The spatial distributions of population and service employment are simulated through gravity or spatial interaction models.

Like other models, Lowry-type models have their strengths and weaknesses. Their strengths include the following, among others:

- It has a simple and easily understandable causal structure, which can be adapted to local conditions. Different attractiveness terms can be used to represent special local characteristics.
- It is quite easily operational and transferable. Data requirements are modest.
- It has been shown that spatial allocation mechanisms mimic the results of randomness in individuals' decision processes (Williams 1977). In other words, the Lowry-type models can represent the dispersions in locational decisions that characterize real cities. Deterministic economics models lack this dispersion representation.

The Lowry-type models were criticized for their weaknesses:

- Relying heavily on the gravitational analogy, the model does not have a strong underlying theoretical foundation such as economics or behavioral theory.
- The model is basically a demand model and includes no supply side. The residential, commercial, and industrial real estate industry is entirely omitted. Rather, supply is implicitly assumed to meet the demand. The mechanisms involved in the supply/demand interactions are therefore excluded from the model framework. The assumption that supply follows demand is only valid, however, in very limited circumstances (Wilson et al. 1981). In most cases, demand continuously adjusts to supply. This lack of supply-side or pricing mechanisms prevents a true equilibrium between supply and demand (Anas 1982).
- Lack of pricing mechanisms also means that appropriate evaluation criteria are impossible to obtain. It is therefore impossible to investigate normative properties, such as welfare gains and conditions for optimality. Analyses of tax policy, mortgage rates, and other fiscal policies are also precluded (Berechman and Small 1988).
- Economic base theory underlying the Lowry model itself has its limitation. With the basic sector taken as the driving force of regional economic development, no feedback to the basic sector from the system is included.

In fact, there are other reinforcing factors in the urban system, such as housing development and local public investment.

The classical gravity model did not have a satisfactory theoretical foundation until late 1960s, when Wilson (1970) elegantly derived the gravity model using entropy-maximizing principles.

Entropy measures the uncertainty represented by a probability distribution. Given a random variable x_i ($i = 1, 2, \dots n$) with probability p_i ($i = 1, 2, \dots n$), and expectation value of some function $f(x_i)$, $E[f(x_i)]$, the entropy is defined by Shannon as

$$\Phi(p_1, p_2, \dots, p_n) = -k \sum_i p_i \ln p_i$$

Intuitively, a broad and uniform distribution represents more uncertainty than a sharply peaked one. For example, if a sports fan says that his or her team has a 50/50 chance of winning a game, this is the most uncertain situation. Entropy measures how uniform a probability distribution is. The more uniform a distribution, the higher the entropy. Maximizing entropy means that, given whatever information, it is desirable to make the distribution as uniform as possible. The entropy-maximizing process is a process of determining the most probable distribution (macrostate) which corresponds to the largest number of possible assignments (microstate). With entropy-maximizing subject to boundary constraints, a gravity model can be derived. Wilson also identified four types of gravity models: (1) unconstrained, (2) production-constrained, (3) attraction-constrained, and (4) production-attraction-constrained (see Table 9.1).

The establishment of a probabilistic or statistical approach to spatial interaction modeling by Wilson has inspired tremendous development in modeling land use, location, and transportation. The entropy-maximizing perspective and the family of models derived from it were greatly broadened when links were discovered between linear programming (LP) models (deterministic economic equilibrium) and entropy-maximizing models (Evans 1973; Senior and Wilson 1974). The doubly constrained gravity model was shown to be equivalent to a linear programming problem in the limiting case, and introduction of an entropy term in the linear programming of residential location in the Herbert-Stevens tradition generated suboptimal dispersion in the resulted nonlinear programming (NLP) models. It is important to note that these and later studies have shown equivalence, unification, and reconciliation between the entropy-based spatial interaction models and the economics-based behavior models (Anas 1975, 1981, 1982; Williams 1977).

9.2 DETERMINISTIC UTILITY, RANDOM UTILITY, AND DISCRETE CHOICE

In the idea of neoclassical microeconomics, a city is taken as a system of markets: the land market, the housing market, the transport market, the labor market, the retail market, and others. As in any other market, consumers maximize their utilities and producers maximize their profits. The price mechanisms, such as rent, transportation costs and wages, are the forces for a spatial equilibrium. The competitive equilibrium thus established represents an efficient allocation of resources.

9.2.1 DETERMINISTIC UTILITY AND OPTIMIZATION

A simple representation of utility is a deterministic function, which depends only on objective attributes of decision options but does not vary with subjective features such as perceptions, differences in preferences and tastes, or uncertainties due to incomplete knowledge and information. Given the same options and conditions, decision-makers will always make the same choices. Deterministic utility theory assumes perfect information and a competitive market. As discussed in Chapter 2, Alonso's (1964) bid–rent theory treats location choice of the household as a *trade-off between accessibility and space*. Subject to a budget constraint, households maximize their utility, which is a function of consumption of land, transportation, and composite good. A particular household has a bid–rent curve that describes a bid price the household is willing to pay at each location. In a perfect competition, each location goes to the highest bidder. In the end, buyers and sellers are equally satisfied without any motivation to change their decisions, and demand and supply for land equate to a clear land market everywhere. Under these conditions, the land market reaches equilibrium. However, establishment of equilibrium in a land market does not necessarily imply that the resulting spatial structure is optimal (Fujita 1989).

What constitutes an optimal land use really depends on what the objective is (Fujita 1989). The first operational optimal land-allocation formulation is the Herbert-Stevens model (Herbert and Stevens 1960). The model's objective is to maximize the surplus subject to a set of prespecified target utility levels for all household types. In the original form, it is a residential location model to allocate households and housing in a discrete space. Locating households choose their optimal residential location through maximizing total rents payable by all households, which is shown to be equivalent to maximizing their savings, or the surplus. It is shown that the Herbert-Stevens optimal land-use conditions are equivalent to market conditions for compensated equilibrium and thus, competitive equilibria are always efficient (Fujita 1989).

TOPAZ was the first operational nonlinear programming model of urban land use and transportation in 1970s. The TOPAZ model in its various versions is generally normative or optimizing, following the tradition of Herbert-Stevens (1960). It allocates activities to locations on the basis of maximum benefit or minimum cost. With introduction of a gravity or entropy term, TOPAZ is also predictive in allocating transport demand, in addition to optimizing land use.

Location models in the deterministic utility and optimization framework have a strong theoretical foundation in microeconomics. With this economics and behavior foundation, these models enable us to evaluate a variety of policies such as pricing mechanisms. This is a clear advantage over the Lowry-type models. These models also have some limitations.

1. The assumptions of perfect information, perfect competition, and completely rational behavior are unrealistic.
2. The assumption of determinism of utility function ignores variations in the subjective features of individual behaviors. Operationally, this results in the so-called 'lumpy' distribution, whereby allocations of households concentrate only in a few zones (Putman 1991).

3. They are very difficult to apply to a city or metropolitan area because it is impossible to aggregate individual demand or supply curves in a large population. Demand curves for the same individual vary with location (de la Barra 1989).
4. The models are usually data-hungry, and availability of the required data is problematic. It has been very difficult to estimate utilities and preference functions and the benefits of location and interaction. Use of location costs without location benefits produces unacceptable results.

9.2.2 RANDOM UTILITY THEORY AND DISCRETE CHOICE

Random utility theory holds that individuals make their decisions and choices among options to maximize their utility or satisfaction, subject to probabilistic variation (or dispersion) constraints that take into account unobserved attributes of the alternatives, differences in taste and preferences among decision-makers, and uncertainty or lack of knowledge and information (McFadden 1973; Domencich and McFadden 1975). That is, individuals behave rationally under uncertainty.

The random nature of utility comes from many sources, such as individuals, decision options, and groups. There are intra-individual and intra-option variations, whereby individuals or options do not always behave the same way (de la Barra 1989). In an aggregate case, there are always many kinds of variations within a group, no matter how homogeneous a group is. Otherwise, a group is reduced to an individual.

Two methods are generally used to generate random utility models (Williams 1977). One is the random-coefficient approach where net utilities are expressed as functions of a vector of random parameters with mean values. The other is to directly specify a functional form for each random variable in terms of a representative component that is equal for all members and a stochastic component for which a distribution is assumed. This random variable takes into account variations in tastes and preferences and those arising from unobserved factors.

Discrete choice happens everywhere every day. A common example is the choice of mode for travel between private automobile and public transit. This is a binary choice situation. Certainly, you can detail the choices further; i.e, another travel choice set can include single-occupancy vehicle (driving alone), high-occupancy vehicle (ride-sharing), bus, light-rail, heavy-rail (subway), bicycle, and walking. This is called multinomial choice, which is often represented through a multinomial logit model. As proved by McFadden (1973), the multinomial logit model can be derived from random utility maximization.

Alonso's tradition of the bid–rent theory and the discrete choice and random utility theory are two approaches with quite different assumptions. Research has shown that they produce equivalent distributions of households in space, either in perfectly competitive land markets or in markets subject to speculative land prices (Martinez 1992). This convergence is not surprising as they both have economics and behavior roots.

It is once again worth noting that the economics/behavior approach and non-economics approach reach the same solution under certain conditions. They are drastically different on the surface. Random utility and discrete choice theory deals with individual choices at a disaggregate level, while spatial interaction and entropy maximization concern the collective behavior of population at the aggregate level. The strong relationship between entropy and discrete choice models was noticed, and synthesis of entropy-maximization and utility-maximization concepts in different contexts was proposed by Anas (1975), Williams (1977), and Los (1979). Anas (1983) proved that the multinomial logit model resulting from random utility maximization was, at equal levels of aggregation, equivalent to the entropy-maximization model.

9.3   POLICY EVALUATION MEASURES

Usually, models produce an enormous amount of information, which is difficult to digest even for model practitioners, let alone decision-makers who are not familiar with models. Some form of overall measure of the policy's effects is necessary and very helpful to the decision-making process.

A distinction must be made between an *optimization* model and a *predictive* model in terms of the results. An optimization model is designed to identify the *most desirable* land use patterns for a specified objective, while a predictive model is aimed at explaining and predicting the *most likely* trends in urban structure. An optimization model can provide a Pareto-efficient urban form under certain constraints, through searching the discrete or continuum feasible-policy space; that is, it can identify the best policy from a whole continuum of possible policies. A predictive model, however, is limited to predicting the effects of one or a group of policies at a time, and in practice only a limited set of policies can be investigated. Predictive models tell the decision-maker what is likely to be the outcome of a specific set of conditions, whereas optimizing models leave the decision-maker with the decisions of how to achieve the required land use configuration and how to ensure that all individuals will conform to the optimal solution prescribed by the model. In a situation where planning control is strong and goals are explicit, an optimizing model may be the natural choice. In a situation where planning control is fairly limited and welfare goals are implicit, then a predictive model may be more appropriate. But these types of models can be complementary, with predictive models used to test the plausibility of development patterns suggested by optimization models (Webster, Bly, and Paulley 1988).

Measure of overall benefit is central to the optimization models, but the predictive models do not necessarily have such a measure. In a predictive model, an index similar to the overall benefit measure in an optimization model can be constructed to include an explicit or implicit trade-off between the various factors affecting consumer welfare, such as rents, environmental and accessibility measures, and travel time and costs. The separate expressions of these components in the overall index can help the decision-maker understand the relative significance of the policy for each component. The index will be most helpful if its unit of measurement is comparable to other activities in the decision-making process. For example, consumer benefits measured in monetary terms can be compared with policy implementation costs.

As discussed in Chapter 2, consumer surplus is an economics concept to measure consumer-welfare changes associated with changes in economic activities or public policies. Williams (1977) shows a measure of consumer surplus for logit models. When comparing two alternative policies a and b, we evaluate the difference in benefit from them:

$$\Delta S = 1/\beta \{ [\ln \Sigma_k \exp(\beta V_{ka})] - [\ln \Sigma_k \exp(\beta V_{kb})] \}$$

where ΔS is equivalent to the generalized form of the Marshallian consumer's surplus. V_{ka} is the utility function for option k evaluated for policy a, V_{kb} is the utility function for option k evaluated for policy b, and β is the parameter of the exponential function. This measure is interpreted as the aggregate expected utility accruing to the utility-maximizing actors. Similarly, Small and Rosen (1981) show how compensating variation can be derived from discrete choice models. Their formulation, similar to the above, is interpreted as the changes in indirect utility converted to dollars by the inverse of the consumer's marginal utility of income. This marginal utility of income can be obtained from the coefficient of the cost variable in discrete choice models. This type of consumer-welfare measures can be adapted to suit the specifications of the multinomial logit mode choice model in the integrated urban or travel demand modeling chain (de la Barra 1989; Johnston, Rodier, and Choy 1998). It can be segmented into various household classifications such as income. This way, distributional impacts of various land use and transportation policies can be evaluated for various income groups.

Use of the location surplus approach for evaluating land use and transportation plans is based on the maximization of a *single objective* — the location surplus function — which measures economic efficiency by taking into account the benefit arising from accessibility and attractiveness. The location surplus function is defined (Coelho and Williams 1978; Wilson et al. 1981) as

$$LS = \{ \text{interaction benefits} - \text{establishment costs} \}.$$

The benefit functions associated with a given distribution of activities usually include three terms: a term independent of the spatial configuration of the activities, a term typically containing establishment costs or benefits associated with the configurations of activities in each zone, and an interaction term expressing the mutual dependence of activities over space through the flow of people or goods between them. It has been shown that transportation cost alone is not an appropriate constituent of interaction terms in the benefit functions, as it ignores the preferences that lead to trip dispersion (Williams 1977; Wilson et al. 1981). The group surplus function is the appropriate measure of the interaction benefits.

This procedure is characterized by the cost–benefit analysis framework. The same challenges and controversies facing cost–benefit analysis are also true of the location surplus approach. Estimating costs associated with land use and transportation is not easy, and the benefits are even more difficult to estimate. Without appropriate benefit estimation, some operational optimizing models end up using only the cost component. This can, as many studies show, provide unacceptable results (Putman 1991). Most planning problems involve multiple conflicting goals and non-commensurable objectives. It can be argued that some indicators can be incorporated into the location surplus approach to reflect goals other than those associated with accessibility/attractiveness (Wilson et al. 1981). One might argue that the location surplus method can be integrated with some methods of "vector optimization" or "multi-objective programming" so that the multi-objective problem will be resolved. It is true that we cannot see serious inherent hindrances to these logical extensions. However, as the formulation becomes more and more complicated to cope with these problems, it is more and more difficult to apply it in actual planning processes.

As discussed above, an evaluation measure based on consumer welfare can be derived for various approaches, such as entropy maximization, random utility, discrete choice, and location surplus. The equivalence among these approaches has also been shown from various studies. What differs in these measures is likely to lie in what variables are included in the evaluation measure and, as a result, what policy alternatives can be evaluated in practice. Those models that incorporate price mechanisms shine because some price policies, such as taxes, user fees, and some other fiscal policies, can be adequately evaluated. Additionally, such constraints as zoning and development restrictions can be easily addressed in an optimization framework. Furthermore, if supply and demand are both considered in a model, the equilibrium can be evaluated.

All those models discussed above are silent on the equity issue. They seldom touch on distributional impacts on various socioeconomic groups. Evaluation measures of model results are aggregate in nature, telling us nothing about how the net benefits are distributed across different social groups. However, these models have different potentials for being used for equity analysis, which will be discussed later.

9.4   OPERATIONAL MODELS

Reviews and evaluation of operational urban models have shown a diversity in model development and a revival in the 1990s (Wegener 1999; Wegener 1994; Hunt, Krieger, and Miller 1999; Webster, Bly, and Paulley 1988). In the early 1980s, the International Study Group on Land-Use/Transport Interaction (ISGLUTI) started to evaluate nine models of land use and transportation in the world (Webster, Bly, and Paulley 1988). The models came from Australia, Japan, Holland, Germany, Great Britain, Sweden, and the U.S., and were applied to a wide range of cities with diverse cultures, histories, and traditions, from Amersfoort, Holland of 180,000 people to Tokyo with 28 million people. The project established a set of relevant policy tests and compared behaviors and policy evaluation results obtained by various models (Webster and Paulley 1990; Wegener, Mackett, and Simmonds 1991).

Wegener (1994) reviewed 12 operational urban models and 20 modeling centers on 4 continents. He evaluated these models using such criteria as comprehensiveness, overall structure, theoretical foundations, modeling techniques, dynamics, data requirements, calibration and verification, operationality, and actual and potential applications. He identified a few success stories of urban modeling, particularly Putman's ITLUP (Integrated Transportation and Land Use Package), also known as

DRAM/EMPAL (Putman 1983 and 1991), and Echenique's MEPLAN (Echenique et al. 1990; Hunt and Simmonds 1993). DRAM/EMPAL has become a de facto standard among metropolitan planning organizations in the U.S. (PBQD 1999). MEPLAN has seen wide applications in Europe. In an update, Wegener (1999) surveyed 15 operational urban models.

Following Wegener (1999; 1994), we take a comprehensive look at available operational urban models (Table 9.2). Of the 15 models, only MEPLAN and TRANUS model all 8 subsystems; 2 do not model land use, while 5 do not model transportation. With respect to the theoretical foundations, there is a broad consensus in using random utility or discrete choice theory to simulate the behavior of various urban subsystems. Most location submodels use random utility theory, bid–rent theory, or a combination. Almost all transportation submodels rely on random utility or entropy theory. The major difference is the degree to which the models represent the urban system in an economic sense. Ten models simulate the land market with endogenous prices, heavily relying on microeconomics theories and particularly Alonso's bid–rent theory.

It is a common practice for the models to spatially represent a city or metropolitan area with a modeling zonal system. In this aggregate framework, a spatial interaction modeling technique is often used to distribute location activities and trip origins–destinations. Activities are either implicitly located as destinations of trips through a simultaneous transportation–location process or based on accessibility and other attractiveness factors. The former approach assumes transportation determines location, or "workers decide where they want to live on their way home from work" (Wegener 1999:10). The latter approach assumes that travel is a demand derived from location decisions of a residence and workplace. An emerging modeling technique is microsimulation, which is used to simulate activities at the disaggregate level. Some researchers hold that an ideal model should be an activity-based, fully integrated market-based one (Miller, Kriger, and Hunt 1999).

In terms of dynamics, most models are quasi-dynamic; they use discrete periods, generally in 5-year intervals, to simulate urban development over time, but they are cross-sectional in each simulation period. Usually, changes in accessibility are assumed to affect the distribution of location activities in a lagging fashion. Changes in land use, on the other hand, will result in modified interactions and in turn lead to changes in travel demand within the same period. Most models use a composite modeling structure, which consists of several interconnected submodels that are processed sequentially or iteratively (also called iterative or sequential structure) within a simulation period. This structure, as Wegener (1994) points out, is especially suited for accommodating time lags or delays and various speeds of change for different urban subsystems. Within each simulation period, models use several types of equilibria (Table 9.2). Only three models use general equilibrium of location and transport in a strict economic sense. Except for one model without any equilibrium mechanism, all other models employ location equilibrium, transport equilibrium, or both.

With respect to evaluation measures, it is a common practice to compare a base case "do-nothing scenario" with an alternative policy scenario. The so-called "rule-of-a-half" benefit measure has been widely used to determine user benefits arising

TABLE 9.2
Fifteen Operational Urban Models

Model	Subsystems	Theory	Techniques	Dynamics
BOYCE (Boyce et al. 1983; Boyce 1986)	Networks Land use Employment Population Travel	Random utility for location and transport submodels	Spatial interaction for location and transport submodels	Joint equilibrium of location and transport without prices
CUFM (Landis 1994)	Land use Workplaces Housing Employment Population	Profitability for location submodels	Microsimulation and spatial disaggregation of land use	Location equilibrium only
MUS (Martinez 1992)	Networks Land use Employment Population Travel	Bid rent theory for location submodels and random utility for transport submodels	Location based on accessibility and other location factors	General equilibrium of location and transport
HUDS (Kain and Apgar 1985)	Housing Travel	Bid rent theory for location submodels	Microsimulation for location submodels	Location equilibrium only
IMREL (Anderstig & Mattsson 1991, 1998)	Networks Land use Employment Population Travel	Locational surplus for location submodels Random utility for transport submodels	Spatial interaction for location and transport submodels	Separate equilibria of location and transport
IRPUD (Wegener 1986; Wegener et al. 1991)	Networks Land use Workplaces Housing Employment Population Travel	Random utility for location and transport submodels	Microsimulation for location submodels Spatial interaction for transport submodels	Transport equilibrium only
ITLUP (Putman 1983, 1991, 1994b)	Networks Land use Employment Population Travel	Random utility for location and transport submodels	Location based on accessibility and other location factors Spatial interaction for transport submodels	Transport equilibrium only
KIM (Kim 1989)	Networks Workplaces Employment Population Travel	Bid rent theory for location submodels and random utility for transport submodels	Spatial interaction for location and transport submodels	General equilibrium of location and transport

TABLE 9.2 (CONTINUED)
Fifteen Operational Urban Models

LILT (Mackett 1990, 1991a, 1991b)	Networks Land use Workplaces Housing Employment Population Travel	Random utility for location and transport submodels	Spatial interaction for location and transport submodels	Joint equilibrium of location and transport without prices
MEPLAN (Echenique et al. 1990; Hunt & Simmonds, 1993)	Networks Land use Workplaces Housing Employment Population Travel Goods transport	Bid rent theory for location submodels and random utility for transport submodels	Spatial interaction for location and transport submodels Multiregional input-output model	Separate equilibria of location and transport
METROSIM (Anas 1992)	Networks Land use Workplaces Housing Employment Population Travel	Bid rent theory for location submodels and random utility for transport submodels	Location based on accessibility and other location factors Spatial interaction for transport submodels	General equilibrium of location and transport
POLIS (Prastacos 1986a and 1986b)	Land use Employment Population Travel	Locational surplus for location submodels	Location based on accessibility and other location factors	Location equilibrium only
RURBAN (Miyamoto & Kitazume 1989)	Land use Employment Population Travel	Bid rent theory for location submodels	Spatial interaction for location submodels	Location equilibrium only
TRANUS (de la Barra et al. 1984; de la Barra 1989)	Networks Land use Workplaces Housing Employment Population Travel Goods transport	Bid rent theory and random utility theory for location submodels and random utility for transport submodels	Spatial interaction for location and transport submodels Multiregional input-output model	Separate equilibria of location and transport
URBANSIM (Waddell 1998)	Land use Workplaces Housing Employment Population	Bid rent theory for location submodels	Location based on accessibility and other location factors and spatial disaggregation of land use	No equilibrium

Adapted from Wegener, J., *J. Am. Planning Assoc.*, 60, 1, 23, 1994. With permission.

from proposed transportation projects or policies. It is one half of the aggregate total of the products between two terms: the sum of transport demand (trips) for the two comparison scenarios multiplied by the difference in perceived costs between the two scenarios. Williams (1977:286) shows that this measure "is not only computationally inefficient, but also its marginal basis renders it inappropriate for assessing the economic values of a large number of transport systems which involve the introduction of new facilities (or movements)." Recent model development incorporates consumer-welfare measures like those discussed above. For example, ITLUP and MEPLAN contain a module for calculating consumer-surplus indicators for various groups (Wegener 1994).

One issue of debate on operational urban models has been ease of application vs. policy relevance. The Lowry-type models have been constantly criticized for missing representation of the market mechanisms and thus having an inability to evaluate pricing policies. Critics argue that Lowry-type models increasingly show serious limitations in policy analysis. Advocates of Lowry-type models counter that quality data on land price and/or value are generally unavailable or of poor quality in most metropolitan planning agencies. For the application and implementation experience of Lowry-type models, see ITLUP or DRAM/EMPAL (Putman 1994b). For the application and implementation experience of models based on economics principles, see MEPLAN and TRANUS (Hunt and Simmonds 1993). For the "best studied city in the world," see Chicago, which has been modeled in econometric housing market, land use, and transportation models (Anas 1982), in nonlinear programming equilibrium models of transportation and location (Boyce et al. 1983), and in a general equilibrium between transportation and location (Kim 1989). Actual applications of the models do suggest many difficulties in calibration and transferability for those based on the economics framework (Hunt and Simmonds 1993; Prastacos 1986b).

In summary, currently operational urban models have some common limitations (Miller, Kriger, and Hunt 1999; Wegener 1999):

- Spatial representation of urban systems is too coarse to be useful for microscale policy analysis such as neotraditional urban design, transit-oriented development, mixed land use, and transportation-demand management strategies.
- Transport submodels still use the traditional four-step travel-demand modeling approach, which is not suited for simulating behavioral responses to transportation management policies.
- Temporal representation is inadequate — too much reliance on static equilibrium and a too-long time interval.
- There is a lack of endogenous demographic processes and auto-ownership processes.
- Considerable resources are required for calibration and implementation in real-world situations.
- Most models are insensitive to social issues.

These limitations are some of the barriers for successfully integrating urban and environmental models.

9.5   INTEGRATING URBAN AND ENVIRONMENTAL MODELS FOR ENVIRONMENTAL JUSTICE ANALYSIS

Integrating urban models and environmental models can serve multifold purposes:

- Integration is crucial for improving the accuracy of environmental modeling, particularly for area- and line-source emission estimation or non-point-source pollutant generation (see Chapter 4).
- Integration is important for investigating the roles of environmental factors in location behaviors in urban systems and has the potential for improving urban modeling.
- Integration is critical for evaluating the distributional impacts of proposed policies, programs, and projects in the future.

The benefits of this integration go beyond the separate, independent fields of urban and environmental modeling. It has taken a long time for us to recognize the interaction between urban activity location and transportation, particularly the feedback effects of transportation improvements on activity location patterns (land development). Over the past decade, we have also recognized the impacts of this interactive land-use/transportation system on the environment such as air and water quality. Now it is time to emphasize the interaction between the land-use/transportation system and the environment, particularly environmental impacts of the land-use/transportation system on different population groups. We know little about the impacts of environmental risks or pollution on location behaviors of various population groups. Location and location decisions of various market participants are very important for us to understand the causation and dynamics of the distribution of various urban activities. Integrated urban models provide a powerful tool for better understanding the interconnections among these markets and the dynamics of the urban systems.

Theoretically, a land-use/transportation model based on any one or hybrid of the approaches discussed above can be linked with environmental models to investigate the role of environmental quality in population distribution and to estimate the net benefit of a policy. For example, an air quality index can be formulated based on the air pollutant concentration predicted by an air quality model. This index can be taken either as one of residential location attractiveness measures in a spatial interaction model following the entropy tradition, or as representing a kind of externality in the measure of location surplus (just like residential density used by Mattsson (1984a and b; 1987), or as an attribute in the random utility function or in the logit choice function.

Technically and practically, however, a linkage between an air quality model and any one of the approaches is not equally convenient and smooth. Those models with a "simultaneous" structure, e.g., those based on an optimization framework as described above, pose some difficulties. This is partly because of the characteristics of an air quality model. The current mainstream air quality models, including those regulatory models recommended by EPA, usually describe various pollution processes in the atmosphere. These models need an intensive data set of weather, emission, and topography. Generally, it is difficult, if not impossible, for a constraint in an optimizing framework to represent these pollution processes unless the relationship between land-use/transportation and air quality impacts is dramatically simplified. This simplification certainly is not good enough to meet EPA's regulatory requirement. Even if these pollution processes can be expressed as a constraint, that expression will be highly nonlinear. Incorporating nonlinear constraints will render the model with a nonlinear objective insoluble since there are no algorithms to handle such large problems with nonlinear objective functions and constraints (Prastacos 1986a). A way to avoid these problems is to incorporate an air quality index in the objective function or a constraint, and run an air quality model separately outside the optimization framework. This, however, will defeat the simultaneous nature of an optimization framework.

In contrast, those models with an "iterative" or "sequential" structure seem to have some advantage in this linkage. Taking a land-use/transportation policy proposal as input, an entropy-derived spatial interaction model can produce output that is necessary for running an air quality model. The output from this air quality model can be taken to calculate an air quality index or indicator, which can be used as feedback (e.g., as one of the measures that affect locational choices) in determining land-use patterns and traffic congestion. Whether this will improve the predictive power of the model is open to question since current models such as ITLUP have already achieved great predictive power without this feedback. This air quality index should be a component of the evaluation measures described above so that the net benefits of a policy can be evaluated.

As mentioned above, this air quality index can also be taken as an explicit attribute in the random utility function or in a logit choice function. The issue that may arise is how to convert the air quality index into a form with a unit consistent with those of other attributes. Besides practical application, the purpose and nature of the decision-making process have to be considered when choosing between an entropy-based model and an economics model. As emphasized by Williams (1977), when one seeks to explain variability in terms of economic rationality, it is more appropriate to use a utility-based approach rather than an information-theory approach. This is a situation where the economic implications and interpretations of a change in the vector of perceived costs are essential for a decision-making process. "If, however, the observer is not prepared to attribute the variability at the micro-level or macro-level to any particular source, or is confronted by several sources of variability, which would overtax theoretical or data capability at the level of resolution considered, the entropy maximizing method would seem to be the appropriate model building device, and in fact possibly the only approach" (Williams 1977:106).

In practice, the link between urban and environmental models has been generally one way; it has focused on how the location–transportation system affects environmental quality such as air quality, energy consumption, and, more recently, water quality. In the 1970s, impacts of transportation policies were analyzed in a couple of metropolitan areas by means of TASSIM, a transportation–air quality simulation package (Ingram and Fauth 1974; Ingram and Pellechio 1976; Kroch 1975). ITLUP and the Emission Diffusion Simulation System were developed and linked to evaluate the impacts of transportation, location and land-use policies on air quality in San

Francisco (Gross 1979; Putman 1983). Since the ISTEA of 1991 and the Clean Air Act Amendments of 1990, metropolitan planning organizations (MPOs) have used travel demand models or land-use/travel demand models to produce inputs to emission models, such as MOBILE, as part of air quality conformity determination. The link has been officially established for MPOs and state environmental departments to evaluate impacts of proposed transportation projects, programs, and plans on air pollutant emissions and potentially air quality in the future. Similar links have been used to estimate greenhouse gas emissions (Johnston, Rodier, and Shabazian 1998).

While it takes legal and federal regulatory mandates for this to happen in the land-use/transportation/air quality system, there are no similar mandates for linking land-use, transportation, and other environmental systems such as water quality, vegetation, wildlife, soil, and microclimate systems. It appears that the growth management movement at the state and local level is a driving force for the latter. For example, Maryland's Economic Growth, Resource Protection, and Planning Act of 1992 established seven new visions for growth and development of Maryland's future, emphasizing that stewardship of Chesapeake Bay and the land is a universal ethic. The Smart Growth and Neighborhood Conservation Act of 1997 gives the State programmatic and fiscal tools to assist local governments in implementing the Visions. In implementing these smart growth initiatives, Maryland Department of Planning and other state agencies recognize the impacts of the land development–transportation system on water quality and the watershed. To analyze these impacts, the link between land-use/transportation modeling and watershed modeling is initiated (Maryland Department of Planning 1998).

There has generally been a lack of environmental quality feedback on location and transportation in the integrated urban models. Without this feedback, it is impossible to assess the role of environmental factors in population distribution and the impacts of change in environmental quality on residential location due to proposed environmental policies. Other than air quality, no study has been done to include other actual or estimated environmental risk measures in location models. This feedback is just as important as the land-use/transportation feedback cycle.

To successfully integrate urban and environmental models, we need to recognize the differences between them and the potential barriers.

- Spatially, environmental models usually operate at a fine-grid level, while urban models generally employ a relatively large transportation analysis zone structure. In the GIS framework, environmental models use the raster GIS system, while urban models use lines and polygons and are suited for the vector GIS system.
- Temporally, environmental models forecast ambient concentrations at a short-time interval such as hourly, while location submodels of urban models forecast changes in location patterns of urban activities at a large time interval, such as 5 years. Although transportation submodels of urban models can forecast peak-hour traffic volume and speed, they generally operate on a daily basis.
- Although both urban and environmental models have a spatial dimension, the overall evaluation measures have an aggregate nature. The results from

both models are usually used to evaluate the aggregate impacts on the society at large, without any attention to differential distributional impacts on different segments of society.

At this time of a new millennium, we see that all these barriers are not formidable. Urban models will become increasingly disaggregate in space, time, and substantive elements. This disaggregation is possible with the help of GIS and microsimulation techniques. GIS will provide an overarching platform for database generation and management and for linking urban and environmental models. GIS makes it possible to create a huge amount of spatially disaggregate data that are essential for disaggregate urban modeling. A unified GIS platform can take advantage of the strengths of both raster and vector GIS structures. Microsimulation techniques enable simulation of individual behaviors and thus evaluation of behavioral responses to public policies (Wegener 1999). Model results will be assessed through multi-criteria evaluation models that are sensitive to equity, environmental, and efficiency criteria.

10 Equity Analysis of Air Pollution

Air pollution is one of the earliest environmental concerns and is perhaps the earliest issue for equity analysis. In previous chapters, we have discussed theories and methods for environmental justice analysis, some of which are examined in the context of air pollution. In this chapter, we begin with a brief overview of air pollutant sources, ambient air quality standards, and health impacts. Then, we review theories and methods for analyzing the relationship between air pollution distribution and population distribution, and we summarize evidence obtained in previous air pollution studies. Chapter 9 presented integrated urban models and their potential applications to environmental justice analysis. In this chapter, we discuss an application of the spatial interaction modeling approach to testing environmental inequity in the Los Angeles and Houston metropolitan areas. Finally, we examine the distributional impacts of national ambient air quality standards.

10.1 AIR QUALITY

Conventional air pollutants come from stationary sources such as fuel combustion and industrial processes, mobile sources such as cars and trucks, and natural systems. Table 10.1 shows total emissions of six criteria air pollutants from major sources nationwide in 1997. Industrial facilities and mobile sources also release toxins to air (see Table 4.1). The quantity of emission is only one factor for determining eventual environmental impacts. Weather conditions and topography are particularly important for fate and transport (dispersion) of air pollutants. Other factors include the location of emission, time and temporal patterns of emission, and height and other characteristics of stacks. All these variables contribute to the distributional patterns of ambient air quality concentrations, locally, regionally, and globally. In the U.S., ambient air quality varies widely among different regions of the country and between urban and rural areas (see Table 10.2).

Air pollutants have a variety of human health and ecological health impacts. Human health impacts range from temporary and reversible effects such as eye irritation to death (Table 10.2). Most air pollutants affect the lungs, and respiratory and cardiovascular systems. Infants, elderly, asthmatics, cardiovascular and chronic lung-disease patients are especially sensitive to air pollution. The danger of air pollution has been vividly illustrated by tragic episodes of air pollution such as those in Donora, Pennsylvania and London, England. These acute health effects have been virtually eliminated due to dramatic improvement in air quality. Air quality concentrations of CO, lead, and SO_2 decreased by more than half, and NO_2, O_3, and PM_{10} were reduced by at least 25% in the U.S. between 1978 and 1997 (U.S. EPA 1998c). Despite this progress, there are still areas with substandard air quality. It was

TABLE 10.1
National Emission Estimates by Source Category, 1997
Thousand Short Tons (% of total amount in the second row for each source)

Source Category	Carbon Monoxide (CO)	Lead (Pb)	Nitrogen Oxides (NO_x)	Volatile Organic Compounds (VOCs)	Particulate Matter (PM_{10})	Sulfur Dioxide (SO_2)
Fuel combustion	4817	0.496	10,724	861	1101	17,260
%	5.5	12.7	45.5	4.5	35.4	84.7
Industrial processes	6052	2.897	917	9836	1277	1718
%	6.9	74.0	3.9	51.2	41.0	8.4
Transportation % on-road vehicles	67,014 57.5	0.522 0.5	11,595 29.8	7660 27.2	734 8.6	1380 1.6
% non-road sources	19.2	12.8	19.3	12.6	15.0	5.2
%	76.6	13.3	49.2	39.9	23.6	6.8
Miscellaneous	9568	0	346	858	0	13
%	10.9	0.0	1.5	4.5	0.0	0.1
Total	87,451	3.915	23,582	19,214	3112	20,371
%	100.0	100.0	100.0	100.0	100.0	100.0

Source: U.S. EPA. 1998c. National Air Quality and Emissions Trends Report, 1997.

estimated that approximately 107 million people lived in counties with substandard air quality of at least one criteria pollutant in 1997 (U.S. EPA 1998c).

Ambient air quality standards have been established at both federal and state levels. The U.S. EPA has established National Ambient Air Quality Standards (NAAQS) for the following six criteria pollutants: carbon monoxide (CO), lead, nitrogen dioxide, ozone, particulate matter (PM), and sulfur dioxide (SO_2). The NAAQS for each pollutant are defined in terms of level and averaging time (see Table 10.3). There are two types of standards: primary and secondary. Primary standards protect the public health with "an adequate margin of safety" ((42 U.S.C. 7409(b)). Secondary standards protect the public welfare from any adverse "effects on soils, water, crops, vegetation, manmade materials, animals, wildlife, weather, visibility and climate, damage to and deterioration of property, and hazards to transportation, as well as effects on economic values and on personal comfort and well-being" ((42 U.S.C. 7602(h)).

For environmental justice analysis, it is helpful to examine various air pollutants using the following aspects: color, odor, generation sources, geographic variation and distribution of ambient concentrations, health and environmental impacts, and sensitive population. These characteristics have important implications for distributional impacts and equity analysis. As discussed in Chapter 2, risk perception literature has shown the importance of visual and sensible images

TABLE 10.2
Distributional Characteristics, Health Impacts, and Sensitive Subpopulations of Air Pollution

Pollutant	Leading Sources	Sites Recording Highest Ambient Level	Regions with Highest Ambient Concentrations	Areas Violating NAAQS in 1997	Health Impacts	Sensitive Population
CO	Transportation	Urban	Southern California	3 counties (population 9.1 million)	Lungs, bloodstream	Cardiovascular patients, individuals with chronic lung disease
Pb	Industrial process (metal processing)	Urban	Mid-Central	4 counties (population 2.4 million)	Kidney, liver, nervous system	Children
NO$_2$	Transportation and fuel combustion	Urban	Southern California, Mid-Atlantic	None	Pulmonary function	Children, respiratory patients
O$_3$	VOCs: industrial processes and transportation NO$_x$: transportation and fuel combustion	Suburban	Southern California, Gulf Coast, Northeast and Northcentral states	77 counties (47.9 million)	Lung, respiratory function	Active children, outdoor workers, respiratory patients (e.g., asthmatics)
PM$_{10}$	Industrial process, fuel combustion	Urban	West	7.9 million	Lung, respiratory system, premature mortality	Individuals with respiratory disease and cardiovascular disease Children Elderly Asthmatics
SO$_2$	Fuel combustion	Urban	Northeast	One county (0.1 million)	Lung, respiratory system, cardiovascular	Children Elderly Cardiovascular Chronic lung disease patients Asthmatics

Sources: U.S. EPA. 1998. National Air Quality and Emissions Trends Report. 1997.

TABLE 10.3
NAAQS in Effect in 1997

Pollutant	Primary (Health Related) Type of Average	Primary (Health Related) Standard Level[a]	Secondary (Welfare Related) Type of Average	Secondary (Welfare Related) Standard Level
CO	8-hour[b]	9 ppm (10 mg/m³)	No Secondary Standard	
	1-hour[b]	35 ppm (40 mg/m³)	No Secondary Standard	
Pb	Maximum quarterly average	1.5 µg/m³	Same as Primary Standard	
NO₂	Annual arithmetic mean	0.053 ppm (100 µg/m³)	Same as Primary Standard	
O₃	1-hour[c]	0.12 ppm (235 µg/m³)	Same as Primary Standard	
	8-hour[d]	0.08 ppm (157 µg/m³)		
PM₁₀	Annual arithmetic mean	50 µg/m³	Same as Primary Standard	
	24-hour[e]	150 µg/m³		
PM₂.₅	Annual arithmetic mean[f]	15 µg/m³		
	24-hour[g]	65 µg/m³		
SO₂	Annual arithmetic mean	0.03 ppm (80 µg/m³)	3-hour[b]	0.50 ppm (1,300 µg/m³)
	24-hour[b]	0.14 ppm (365 µg/m³)		

[a] Parenthetical value is an approximately equivalent concentration.
[b] Not to be exceeded more than once per year.
[c] Not to be exceeded more than once per year on average.
[d] 3-year average of annual 4th highest concentration.
[e] The pre-existing form is exceedance based. The revised form is the 99th percentile.
[f] Spatially averaged over designated monitors.
[g] The form is the 98th percentile.

40 CFR Part 50.

Sources: U.S. EPA. 1998c. National Air Quality and Emissions Trends Report, 1997.

in how people perceive and respond to risks. Different perceptions of and responses to risks could influence residential location choice, and, consequently, the relationship between environmental risk distribution and population distribution. Pollutant sources demonstrate who generates and benefits from pollution, and distribution of pollutant sources is the first indicator of potential impacts. The real distributional impacts depend on the distribution of ambient concentrations in relation to the distribution and activity patterns of population, particularly sensitive populations.

10.2 RELATIONSHIP BETWEEN AIR QUALITY AND POPULATION DISTRIBUTION: THEORIES, METHODS, AND EVIDENCE

10.2.1 Theories

As discussed in Chapter 2, traditional and modern theories from social sciences, such as economic, location, risk perception, and neighborhood changes, offer us insights about environmental justice in general and the relationship between air quality and population distribution in particular.

10.2.1.1 Residential Location Theory and Spatial Interaction

As discussed in Chapter 2, residential location theory tells us that households make their residential location choices based on trade-offs among space, accessibility, and environmental amenities subject to budget and time constraints (Fujita 1989). When a household buys a house or rents an apartment, the household buys or rents a bundle. This bundle includes not only the visible physical structure and its immediate surroundings but also the visible and invisible attributes associated with the location. Physical structure, such as space, comes to the immediate attention of households at different stages of the life cycle need different physical structures to satisfy their needs. Households have also other needs, social, cultural, or economic, to be satisfied. These needs or their potential supply sources are spatially defined and distributed. In fact, each and every human activity is spatially defined and, explicitly or implicitly, linked. The way human activities interact is a spatial interaction process. When deciding on a residence location, the household chooses the spatial relationship with other entities in their space of interaction. In addition to the house, they buy accessibility (i.e., travel time, mode, or distance) to workplace, shopping places, schools, recreational facilities, etc.

The household also buys, intentionally or unintentionally, environmental amenities they enjoy and environmental liabilities to which they are exposed. When choosing a residence location, the household chooses the spatial relationship with environmental amenities and liabilities. If environmental quality is a normal good, we would expect that the more affluent would buy more of it; that is, the rich would tend to locate *ex ante* where there is little air pollution, *ceteris paribus* (Liu 1996). *Ex post*, the rich, as we would expect, will tend to either fight against it or "vote with their feet." The rich generally have greater political power than the poor, and they are more likely to use their political power to protect their self-interests. If they win the battle, air quality will improve over a period of time. If they do not, they still have more choices than the poor of where they move. In either case, we would expect an inverse relationship between air quality and household income.

Therefore, we can hypothesize that the level of air pollution is a statistically significant predictor of residential location by household income, and the rich choose their residential location with the best ozone air quality while the poor live where ozone air quality is the worst. Again, when making residential location choices, households face trade-offs among various competing needs and weigh one against another according to the relative importance of those variables to them. The role of air quality in residential location choice is certainly dependent on how the household perceives and responds to it.

10.2.1.2 Risk Perception and Human Response to Air Quality

As discussed in Chapter 2, laypeople's risk perceptions are closely related to the position of a hazard within the two-factor ("dread risk" and "unknown risk") space, particularly in the "dread risk" dimension (Slovic 1987). The highest degree of dread risk is associated with catastrophic potential, fatal consequences, perceived lack of control, and perceived unequal distribution of risks and benefits, while the highest degree of unknown risk is associated with those that are new, unobservable, and/or long latent. Studies show that air pollution is situated in the middle-to-low spectrum of the two-factor space. Among 81 hazards studied in the classic analysis by Slovic and his colleagues (Slovic, Fischhoff, and Lichtenstein 1985), air pollution-related hazards [i.e., automobile (CO exhaust), automobile (airborne lead), coal burning, fossil fuels] are located in the middle-range position within the two-factor space. In another study of 37 LULUs (locally unwanted land uses) and other environmental features, heavy smog and heavy traffic are situated in the low-unfamiliarity and low-dread quadrant within the two-factor space (Slovic 1992).

According to social and cultural theories of risk, social influences, and cultural values and beliefs, socioeconomic and demographic status strongly affect human responses to risks (Short 1984; Douglas 1990; Wildavsky and Dake 1990; Rayner 1992). Those of lower socioeconomic status have a sense of less control over and more concern about their exposure to risks (Savage 1993). Human response to environmental risks is also influenced by the salience of risk in comparison with a host of competing social values. Thus, people may be concerned with some environmental risks but put them in lower priority than other social concerns.

Research has attempted to explain variations in public perception of air pollution on the basis of the nature of air pollutants, the extent of personal exposure, the socioeconomic characteristics of individuals and communities, and the extent of media coverage (Barker 1976). Public perception of air pollution was found to be influenced by tangible and observable features of air pollution (e.g., smoke, dirt), and the proportion of population who are concerned about or aware of air pollution increases with higher concentration of particulate matter. People are unaware of gaseous pollutants (e.g., SO_2, NO_2, and O_3) unless high concentrations reach olfactory thresholds or cause physical discomfort (e.g., eye irritation); events during which gaseous pollutants reach such a high concentration only occur for limited hours of the year. Survey research reveals that although recognizing the existence of serious air pollution, people deny its adverse effects on themselves (Wall 1973) or deny it being a disadvantage of living in an area (Saarinen and Cooke 1971). "Several studies indicate that air pollution has low salience, either because there are few channels for expressing dissatisfaction, or because people assign higher priorities to solving other urban problems" (Barker 1976:195). Research investigating air quality in relation to socioeconomic characteristics has produced inconclusive results.

Findings in the perception literature could offer us some insights on why human response to ozone air pollution might be different from that to environmental risks such as toxic wastes or their disposal facilities, to which the public have shown strong objections. First, the major symptoms of ozone air pollution — sore throat, cough, headache, chest discomfort, and eye irritation — are immediately observable and mostly curable. People may be concerned but not much afraid. In contrast, toxic wastes produce cancers, which generally take a long time to be diagnosed and are probably incurable. This undetectability of carcinogenicity in the latency period is responsible for a large part of the dread (Sandman 1985; Slovic, Fischhoff, and Litchtenstein 1979). Second, in terms of impact scale, tropospheric ozone pollution is more dispersed and pervasive and often has regional impacts. The population affected is often large, while the parties benefiting from the polluting activities are also large. Environmental risks associated with toxic wastes, in contrast, are quite concentrated and often have their impacts at a local scale. A small number of people surrounding a hazardous waste facility bear a substantial share of fear and risk, while society at large benefits from its service. Risks that people likely perceive as fairly distributed are more acceptable than those perceived otherwise (Sandman 1985). Third, "air pollution is chronic and insidious rather than sudden and visibly dramatic" (Barker 1976). There are certainly more dramas and stories in media reports of a toxic accident or cancer cluster than in some respiratory symptoms associated with an ozone episode. Television news coverage, the source of most people's environmental news, disproportionately focuses on acute, spectacular, and catastrophic risks rather than chronic risks (Greenberg et al. 1989). The visually dramatic stories and images in the media are easy to remember and recall; people tend to judge an event based on the "availability" of recall (Slovic, Fischhoff, and Litchtenstein 1979). Fourth, the public tolerates risks from voluntary activities much more than those resulting from involuntary hazards (Slovic 1987). Whether to locate in a downwind area is a voluntary decision under people's control, while siting a hazardous-waste disposal facility in a community imposes an involuntary risk on the residents in the community.

In short, the public perception of environmental risks may deviate from any "objective" evaluation of risks, and the perception of and response to environmental risks may vary with the socioeconomic characteristics of individuals and communities. Environmental quality is valued differently by different populations with different degrees of access to imperfect information. Various confounding factors complicate the relationship between air quality and population distribution. The complex relationship could become even more dynamic under different social and institutional forces over time.

10.2.1.3 Theories of Neighborhood Changes

Air quality could be one of the push forces that induce neighborhood changes. The push forces make the neighborhood less desirable, while the pull forces provide households with attractive opportunities outside the current neighborhood to make them want to move. Other forces include the classical invasion–succession forces, the life cycle of a neighborhood, and institutional forces such as banks, insurance companies, or universities.

10.2.2 Methods

Two approaches have generally been used to study the relationship between the socioeconomic characteristics of population and the distribution of air pollution (Asch and Seneca 1978; Liu 1996). Most studies generally took a physical measurement/correlation approach through which the researchers tried to find the association between the distribution of an air pollution exposure or risk measurement and population distribution by income and race. In the early studies, the researchers assumed a discrete space, identified existing air quality stations and their associated census tracts, and then correlated air quality readings with the socioeconomic characteristics of the monitored tracts (Freeman 1972; Asch and Seneca 1978). While an air-quality monitoring network is supposed to be designed to represent an airshed, this network does not guarantee that the population characteristics represent the whole study area. One extension is the assumption of a continuous space of air pollution, under which the isolines were constructed based on discrete monitoring sites (Earikson and Billick 1988). These discrete data were also used to interpolate values at the centroids of analysis zones, which were taken as a unit of analysis. Another variant of this approach is impact boundary delineation, based on proximity to air pollution sources (Napton and Day 1992), urban ozone-plume studies and long-term trend analysis (Liu 1996), or USEPA's nonattainment area designations (Wernette and Nieves 1992; Liu 1998). More or less, this variant assumes a relatively uniform ozone coverage in an area. The proximity method has been extensively used in most equity analyses of other types of environmental risks, partly because limitations in knowledge, data availability, and resources make it very difficult, even impossible, to delineate the actual impact boundary.

The other is an economics approach, which measures pollution damage (or benefits of avoiding pollution) and the distributional impacts of environmental policies by income and race. Gianessi, Peskin, and Wolff (1979) assessed the distribution of Clean Air Act policy benefits (or air pollution damages) and costs by ten family-income groups and two race groups on a nationwide basis. Bae (1997) used a Net Welfare Impact model to measure how air quality policies affect the welfare of individual households in the Los Angeles metropolitan area, and found low-income households and minorities enjoyed many of the benefits of air quality improvements.

While most studies describe the relationship between air quality and population distribution by income and race, they failed to incorporate other confounding variables that might contribute to such a relationship. In particular, they did not take into account various spatially interacting factors in residential location choice behaviors. Examples of such important factors are accessibility to employment location and environmental amenities. As pointed out earlier, trade-offs among these factors are an integral part of households' residential location decisions. Understanding the roles of these explanatory variables is essential for designing effective public policies for remedying any inequities. Later, we will address this methodological inadequacy through a spatial interaction modeling approach.

10.2.3 EVIDENCE

Are there inequitable distributions of air pollution? If so, who bears a disproportionate burden? National studies have generally concluded that minorities (particularly Hispanics, Asians/Pacific Islanders, and African Americans) have been at (potentially) higher risk of exposure to air pollution induced by criteria air pollutants at the national level (Gelobter 1992; Wernette and Nieves 1992; Liu 1998). Meanwhile, the rich are disproportionately represented in the substandard air quality areas. Wernette and Nieves (1992) found that Hispanics and, to a lesser degree, African Americans were substantially over-represented in the EPA-designated nonattainment areas for criteria air pollutants, but lower percentages of the poor live in these areas than percentages of either African Americans or Hispanics. Gelobter (1992) examined a population-weighted index of ambient concentration for total suspended particulates (TSP) for six income groups and two race groups from 1970 to 1984. Nonwhites had higher exposure to TSP than whites. For the country as a whole, lower-income groups had smaller exposure than upper-income groups.

The degree of disparity would be moderated under more stringent new National Ambient Air Quality Standards (NAAQS) for ozone and PM (Liu 1998). The distributional disparity for the young and elderly is not that apparent. Analysis of the national distribution of air emissions of hazardous chemicals listed in the TRI showed disparities in county-level TRI air emissions by race/ethnicity and, to a lesser extent, by household income (Perlin et al. 1995), and in ZIP-code level exposure by race and income (Brooks and Sethi 1997).

Moreover, recent studies have found important differences in the distributional disparity between urban and rural areas (Gelobter 1992; Liu 1998), and between cities and non-city areas (Liu 1998). As a deviation from the national picture, the poor bear an inequitable burden of air pollution in *city* and *rural* areas with substandard air quality. The disproportionate representation of minorities (particularly Hispanics, Asians, and African Americans) is more profound in *urban* or *city* areas with substandard air quality than at the national level (Liu 1998). This finding poses a great challenge and opportunity to city planners.

At the regional or metropolitan level, the study results are mixed and appear to vary with localities and pollutants. The findings mostly support the hypothesis made by Asch and Seneca (1978:292) that "absolute patterns of distribution at the national level could differ substantially from the relative pattern of distribution that might be typically observed within smaller (city or metropolitan) areas." Freeman (1972) found that low income and, to a larger degree, nonwhite groups were more exposed to total suspended particulates and sulfates in Kansas City, St. Louis, and Washington, D.C. Asch and Seneca (1978) found a negative association between income measures and particulates and sulfur dioxide pollution levels within the cities of Chicago, Cleveland, and Nashville. The correlation between pollution level and percentage of tract nonwhite population varied across cities and pollutants at the intra-city level in the three cities. Positive correlation was found for both particulates and SO_2 in Chicago, for SO_2 in Cleveland, and for particulates and NO_2 in Nashville, while a negative correlation was found for both particulates and NO_2 in Cleveland

and for SO_2 in Nashville. Zupan (1973) reported a significant positive correlation between ambient concentrations of sulfur dioxide and particulates and the percentages of low-income households in New York City. Earikson and Billick (1988) found that poor, mostly black, and working-class families in the inner city, particularly children, were at a higher risk of exposure to air pollution in Louisville and Detroit. For the regions of New York and Philadelphia where there has been significant regional transport of ozone and the highest ozone concentrations were often found downwind of urban source areas, Liu (1996) found that the ozone downwind areas were currently populated by a considerably larger proportion of upper-income households and whites than the source areas.

Two studies covered Houston and Los Angeles. Napton and Day (1992) selected the 5 most air-polluted areas in Texas (including Houston), and chose 40 census tracts in the pollution areas and 40 census tracts in the control areas. Through a discriminant analysis, they found that the study areas were populated by people who were younger and had higher-per-capita income than the control areas. They concluded that middle-class working people lived near their employment location. Using a regional human exposure model to account for exposure distribution by location, mobility, and activity level of the population, Brajer and Hall (1992) estimated ozone exposure concentrations in 31 districts in the south coast Basin of California and correlated population characteristics with O_3 and PM_{10} measurements. They found a statistically significant negative correlation between PM_{10} and income measures and the percentage of whites, but positive correlation between PM_{10} and the percentages of blacks and Hispanics. However, the associations between ozone exposure and these socioeconomic measures are much weaker and inconsistent. Ozone exposure was positively correlated with the percentage of families in the middle-income quintile (significant at 0.05), with the percentage of blacks (significant at 0.2), and with the percentage of Hispanics (significant at 0.1). Using the same model with a more refined unit of analysis with 126 subregions for the 1980–1982 period and 142 subregions for the 1990–1992 period, Korc (1996) found that Native Americans had the highest per capita hours of exposure followed by whites, Hispanics, Asians, and blacks, and ozone exposure disparity by race and ethnicity had diminished over time. But low-income areas may have been experiencing a higher number of per-capita hours of ozone exposure than high-income areas. Different units of analysis may contribute to the varying results in these two Los Angeles studies. It would be interesting to see whether a smaller unit of analysis affects the results.

Who generates air pollution burdens? Bingham, Anderson, and Cooley (1987) estimated distribution of the generation of five air pollutants (particulates, sulfur oxides, nitrogen oxides, hydrocarbons, and carbon monoxide) by family-income class, based on a large scale input–output model of production, consumption, family income, and residuals relationships. They found that air pollution generation was disproportionately attributable to high-income families in 1972; those with family income in the top 10% generated 4 to 6 times as much air pollution as those in the bottom 10%.

$c_{i,j}$ = transportation cost (usually travel time) between zone i and zone j;

$L_{v,i,t}$ = the amount of vacant developable land in zone i at time t;

$X_{i,t}$ = 1+ the ratio of developed-to-developable land in zone i at time t;

$L_{r,i,t}$ = the amount of residential land in zone i at time t;

$N_{i,t-1}$ = total households residing in zone i at time $t-1$;

$R_{i,t-1}$ = environmental risk index in zone i at time $t-1$; and

α_h, β_h, q_h, r_h, s_h, t_h, e_h = empirically derived parameters for type h households.

10.3.4 INDEX CONSTRUCTION AND DATA PREPARATION

Previous research tried to develop perceived (observer-based) air quality indices to link physical measures of air quality to the public appraisal of air quality. Based on which combination of air quality variables correlated best with the percentage of Los Angeles neighborhood respondents who perceived "smoggy air," Flachsbart and Phillips (1980) constructed an index consisted of prevailing visibility, O_3, and SO_2, each measured as the annual number of days when a selected standard had been equaled or exceeded. The study also found that the public generally based their perception on the quality of air that they experienced over a long period of time. These findings point the way for constructing an air quality index in DRAM.

However, data for prevailing visibility and SO_2 are so spatially sparse that intra-city spatial representation based on them is problematic. Motallebi and Allen (1991) found that the density of ozone monitors in the South Coast Air Basin of California affected estimates of residential population exposure to ozone. They concluded that the current ozone-monitoring network in the Basin was adequate for estimating population exposure at the standard levels.

In addition, air quality-monitoring data indicate that ozone has been the predominant air pollutant in both study areas (U.S. EPA 1992). For these reasons, only ozone data are used for this modeling purpose. Three forms of air quality index (ENVINDX) were constructed, namely,

- ENVINDX1 = the number of days when the pre-1997 ozone NAAQS (0.12 ppm) was exceeded during 1981–1985
- ENVINDX2 = average of annual top 3 daily maximum 1-hr ozone concentrations for 1981–1985
- ENVINDX3 = average of annual top 10 daily maximum 1-hr ozone concentrations for 1981–1985

Ozone data were obtained from the U.S. EPA's Aerometric Information Retrieval System (AIRS) database. Based on Flachsbart and Phillips (1980), the first index appears to be the best predictor of human response to air quality. The other two are used here to test the sensitivity of the models to different forms of ozone air quality index.

Discrete data in the monitoring stations were interpolated to estimate air-quality index values for all zones in the study areas (i.e., 199 zones in Houston and 772 zones in Los Angeles). There are a number of interpolation methods (see Chapter 8). The method used here is an inverse square-of-distance weighting interpolation method, which has been shown to be among the most accurate and robust methods (see Chapter 8). This interpolation procedure assumes that each point has a local influence that diminishes with distance. The ozone concentration in each zone is represented by the ozone concentration in the centroid of that zone, which is estimated according to the known concentrations at monitoring stations weighted by an inverse square-of-distance-between-them. Two interpolations were conducted to examine how sensitive the model results are to the uncertainties in the interpolation procedure. The unconstrained interpolation uses all monitoring stations in the calculation. The constrained interpolation restricts the influence to 10 km around a centroid of a zone, if there is a monitoring station in this radius. If not, the radius of influence is extended to 25 km, or further to 50 km, or even further to 75 km until there is a station.

GIS was used to identify the monitoring sites, process the data, and display the results. Table 10.4 lists some statistics for the modeling domains in the two regions. The Houston modeling domain includes Chamber County and the Houston-

TABLE 10.4
Statistics for the Modeling Domain Areas: Los Angeles and Houston

	Los Angeles	Houston
Households (in 1,000,000) in 1985	4.32	1.26
Households (in 1,000,000) in 1990	4.697	1.339
CMSA population (in 1,000,000) in 1990	14.5	3.71
CMSA median household income in 1990	$36,711	$31,488
Geographic area in sq.km (in sq. miles)	24,762 (9561)	20,282 (7831)
Residential land in sq.km (in sq. miles)	3593(1387)	1287(497)
Average ENVINDX1 (standard deviation)	287(203)	61(23)
Average ENVINDX2 (standard deviation)	0.242(0.057)	0.185(0.024)
Mobility rate in CMSA in 1985–1990	55.4%	53.6%
Average travel time to work (min.) in 1990	26.5 min. for Los Angeles-Long Beach PMSA, 25.5 min. for Anaheim-Santa Ana PMSA, 27.7 min. for Riverside-San Bernardino PMSA	26.4 for Houston PMSA
Number of analysis zones	772	199
Number of census tracts	2482	836

Note: CMSA = Consolidated Metropolitan Statistical Area, PMSA = Primary Metropolitan Statistical Area.

Data from 1990 Census, Houston-Galveston Area Council, Southern California Association of Governments, USEPA's Aerometric Information Retrieval System.

FIGURE 10.1 Analysis zones and ozone monitoring stations in the Houston modeling region.

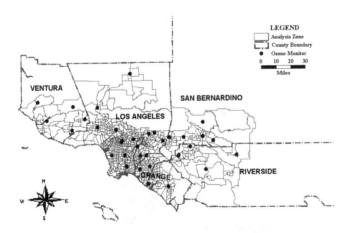

FIGURE 10.2 Analysis zones and ozone monitoring stations in the Los Angeles modeling region.

Galveston-Brazoria Consolidated Metropolitan Statistical Area (CMSA), which is made up of the counties of Brazoria, Fort Bend, Galveston, Harris, Liberty, Montgomery, and Waller (Figure 10.1). The Los Angeles modeling domain is a major part of the Los Angeles-Anaheim-Riverside CMSA and is made up of the whole counties of Los Angeles, Orange, and Venture and parts of Riverside and San Bernardino Counties (Figure 10.2).

The maps (Figures 10.3 to 10.8) show that low-income households lived mostly in the inner city and near the modeling domain boundary, while high-income households lived predominately in the suburban ring in both regions and along the coast in Los Angeles. Ozone air quality was best along the coast of both metropolitan areas. It became increasingly worse inland northwestward in Houston, and northeastward in Los Angeles. The notable exception is the inner city of Houston, where ozone air quality was among the best and the poor also lived in disproportionate numbers. These observations indicate possible different relationships between household income and ozone air quality in Houston and Los Angeles.

10.3.5 MODEL ESTIMATION

Model parameters are estimated using a gradient-search technique with a maximum likelihood criterion (Putman 1983).The gradient-search procedure is generally used to look for optimal values (maxima or minima) of a nonlinear function. It searches the optimal values by following the direction of the steepest slope at a specific point on the criterion surface. The criterion function used here is a log-likelihood function, specifically of observed and estimated numbers of households in each zone. Like the R^2 statistic in linear regression analysis, the best/worst likelihood ratio is a goodness-of-fit measure for the model. Asymptotic t-tests are used to evaluate the statistical significance of the estimated parameters.

Now let us discuss expected signs and magnitudes of the model parameters. As the distance increases from the workplace, the residential search space expands, and residential location opportunities tend to increase, but people tend to be less willing

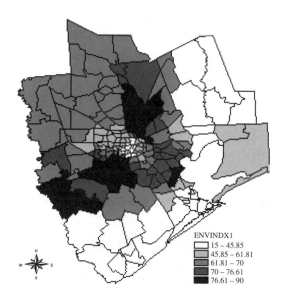

FIGURE 10.3 Number of days when the national ambient air quality standard for ozone was exceeded in the Houston region (1981–1985).

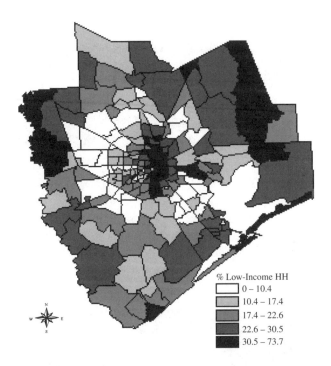

FIGURE 10.4 Low-income households distribution in the Houston region in 1985.

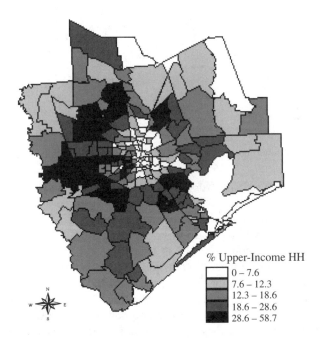

FIGURE 10.5 Upper-income households distribution in the Houston region in 1985.

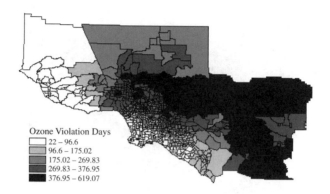

FIGURE 10.6 Number of days when the national ambient air quality standard for ozone was exceeded in the Los Angeles region (1981–1985).

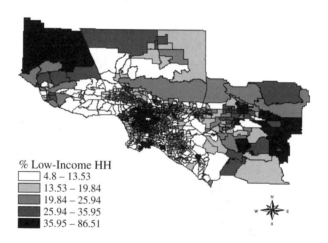

FIGURE 10.7 Percent low income households in the Los Angeles region in 1985.

to make increasingly long trips to work. It is expected that α would be positive since it is a travel-time power-term parameter associated with "a measure of residential location opportunities." β is a travel-time exponential-term parameter associated with "a measure of people's willingness to make increasingly 'long' (in time or cost) trips between work and home" and thus is expected to be negative (Putman 1983:213). Therefore, for a Tanner trip function, which is a product of a travel-time power term and a travel-time exponential term, in most cases, we would expect to observe $\alpha > 0$ and $\beta < 0$; however, it is possible to obtain other combinations (Putman 1983). If only a power term is used (instead of a power and an exponential term), α is expected to be negative. In most cases, the three land variables are expected to have positive exponent parameters; that is, the attractiveness of a zone to residential

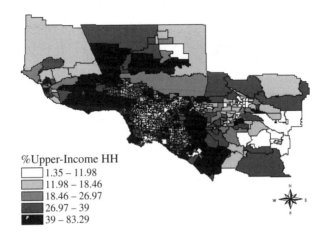

%Upper-Income HH
- 1.35 – 11.98
- 11.98 – 18.46
- 18.46 – 26.97
- 26.97 – 39
- 39 – 83.29

FIGURE 10.8 Percent upper income households in the Los Angeles region in 1985.

locators increases with the values of land variables. The magnitude of these variables might range between 0.0 and 2.0.

ENVINDX actually measures air pollution levels. The larger the ENVINDX, the worse the air quality, and the less attractive the zones for a residence. The sign of ENVINDX should be negative. According to our hypothesis, we expected that low-income households would have the largest positive parameter estimates, while upper-income households would have the most negative parameter estimates; the other types of households would be in the middle.

The models were estimated with three forms of the ozone air quality index derived from both unconstrained and constrained interpolation procedures discussed earlier. The following reports only the results from the constrained interpolation procedure, because the difference in the interpolation procedures does not change the basic results. The results from ENVINDX3 are also omitted, since they are similar to those from ENVINDX2.

10.3.6 RESULTS

10.3.6.1 Los Angeles

Table 10.5 indicates that the goodness-of-fit for the models (i.e., the best/worst likelihood ratio values) ranges from approximately 0.5 for low-income and upper-income households to approximately 0.7 for lower-middle- and upper-middle-income households. The models were reasonably well fit.

As expected, the parameter values for the vacant developable land variable are close to zero, and the other two land-use variables have their estimates in the range of 0 and 1.0 in absolute value. Also, as expected, the lagging households variables all have positive signs. Some signs for the land-use and travel variables are not as we expected. It is noticeable that α is positive for both low-income and lower-middle-income households. The positive signs imply that these two income

TABLE 10.5
Model Estimation Results for the Los Angeles Region

Variable	LIHH[a] M1	LIHH[a] M2	LMIH[b] M1	LMIH[b] M2	UMIHH[c] M1	UMIHH[c] M2	UIHH[d] M1	UIHH[d] M2
ALPHA	0.2600	0.3954	.1454	0.2093	−.0783	−.0794	−0.6138	−0.7376
	69.74	108.73	40.25	58.56	−21.67	−21.88	−184.95	−225.95
VACDEV[e]	−0.0390	−0.0467	−.0062	−0.0084	0.0111	0.0129	0.0457	0.0511
	−123.72	−147.50	−20.06	−27.12	35.80	41.52	154.39	172.85
PERDEV[f]	−0.6261	−0.7024	.0564	0.0409	.0666	.0917	−0.1061	−0.0348
	−92.19	−102.62	8.75	6.31	11.07	15.22	−20.12	−6.64
RESLND[g]	−0.3347	−0.3292	−.0409	−0.0423	.2946	.2794	0.6150	0.5851
	−276.13	−269.26	−31.65	−32.66	218.35	207.44	473.00	441.12
LAGHH[h]	1.0219	1.0267	.9330	0.9348	.7245	.7248	0.5043	0.5128
	514.33	516.79	447.95	449.08	328.69	329.01	238.87	242.00
ENVINDX	0.1840	0.6789	.0808	0.3364	−0.0317	−0.0539	−0.2555	−0.9359
	146.19	127.61	67.66	66.22	−26.41	−10.60	−228.09	−193.62
R SQ	0.5365	0.5291	0.7518	0.7451	0.7259	0.7249	0.5195	0.5089
L Ratio[i]	0.5778	0.5632	0.7519	0.7481	0.7388	0.7365	0.5202	0.5478

Note: Model M1 uses ENVINDX1; Model M2 uses ENVINDX2; see text for definitions of ENVINDX. The parameter estimates are in the first row; asymptotic t-test values are in the second row. ALPHA is a parameter for the travel-time variable. See the model formulas in Section 10.3.3.

[a] LIHH = Low Income Households,
[b] LMIH = Lower Middle Income Households,
[c] UMIH = Upper Middle Income Households,
[d] UIHH = Upper Income Households,
[e] VACDEV = vacant developable land,
[f] PERDEV = percent developable land already developed,
[g] RESLND = residential land,
[h] LAGHH = lagged total households,
[i] L Ratio = best/worst likelihood ratio.

groups tend to make longer trips to work, especially the low-income households. This seems to be contradictory to the conventional wisdom that commuters are less likely to make increasingly long trips. However, some empirical evidence shows that low-income households make longer work trips than other income groups (see Chapter 13 on the spatial mismatch hypothesis). It is also worth noting and investigating the negative signs for land variables in the low-income and low-middle-income household models. It is plausible that there is a negative relationship between low- and low-middle-income households and vacant developable land in the base year (Putman 1983). It is also likely that these negative signs result from multicollinearity among these land variables. In this nonlinear model estimation, the statistics for evaluating multicollinearity are not readily available, and further investigation is needed.

The signs, magnitude, and significance of ENVINDX estimation results are mostly consistent with our expectations, and thus support our hypotheses. All parameter estimates for ENVINDX are statistically significant as their asymptotic t-test values range from 26 to 228 in absolute value, substantially larger than 2. As expected, the parameter estimates for ENVINDX decrease with household income. The parameter estimates for ENVINDX are positive and largest for low-income households, and smallest and most negative for upper-income households. For the models using ENVINDX1, the parameter estimates for ENVINDX decrease to about 0 for the lower-middle- and upper-middle-income households. For the models using ENVINDX2, the inverse relationship between households by income and air quality is even stronger; the parameter estimates for the lower- and lower-middle-income households become more positive, while the parameter estimates for the upper-middle- and upper-income households turn more negative.

In short, the model estimation for the Los Angeles region provides evidence for supporting all hypotheses; that is, the level of ozone air pollution is a statistically significant predictor of household distribution by income. Low-income households live where air pollution is the worst, while upper-income households choose the residential location with the best ozone air quality.

10.3.6.2 Houston

The model parameter estimates and goodness-of-fit measures are shown in Table 10.6. The best/worst likelihood ratio values indicate good fit for the models, with the best fit for the middle-income and lower-middle-income households.

All signs for the land-use and travel variables with a couple of exceptions are as we expected. Like the Los Angeles models, the low-income household models have unexpected signs for travel-time parameter estimates and residential land parameter estimate. For the other household income categories, the models have expected parameter estimates for α and β. The low-income household models have negative α and positive β and $|\alpha| \gg |\beta|$. This trip function implies that low-income households have a slight preference for longer rather than shorter trips to work. This finding is consistent with some empirical evidence that shows that low-income households make longer trips to work than other income groups (see Chapter 13 on the spatial mismatch hypothesis).

TABLE 10.6
Model Estimation Results for the Houston Region

Variable	LIHH[a]		LMIH[b]		MIHH[c]		UMIH[d]		UIHH[e]	
	M1	M2	M1	M2	M1	M2	M1	M2	M1	M2
ALPHA	-1.1939	-1.2080	0.5459	0.5378	1.3418	1.4850	2.9989	3.1468	2.6507	2.8519
	-216.61	-219.28	82.97	81.78	185.14	202.85	352.99	366.39	306.47	324.90
BETA	0.3423	0.3435	-0.0493	-0.0539	-0.2706	-0.3117	-0.6721	-0.7087	-0.6107	-0.6586
	250.28	251.20	-31.20	-34.14	-157.30	-179.44	-341.04	-355.80	-302.23	-321.40
VACDEV[f]	0.1865	0.1883	0.0779	0.0844	0.0629	0.0563	-0.0484	-0.0565	-0.3413	-0.3578
	178.38	180.10	72.98	79.04	58.68	52.52	-44.13	-51.45	-324.37	-339.14
PERDEV[g]	2.0562	2.0710	0.7963	0.8649	0.1288	0.1831	-1.3597	-1.3954	-3.1126	-3.2109
	213.67	215.18	82.14	89.29	13.46	8.69	-140.79	-144.33	-326.31	-336.08
RESLND[h]	-0.2029	-0.1992	-0.1890	-0.1817	-0.0933	-0.0827	0.0895	0.1014	0.4888	0.5036
	-76.19	-74.74	-67.69	-65.42	-32.98	-29.19	30.88	34.87	162.90	167.28
LAGHH[i]	0.9934	0.9901	1.1496	1.1378	1.0839	1.0766	1.0296	1.0211	0.9117	0.9051
	389.90	388.50	434.14	429.65	413.64	410.14	392.32	388.32	333.34	330.17
ENVINDX	0.0050	-0.0155	0.1225	0.3603	0.1467	0.3714	0.1301	0.4577	0.1629	0.7759
	0.92	-0.70	21.83	15.82	26.01	16.24	23.27	19.94	27.99	24.07
R SQ	0.7826	0.7830	0.9218	0.9218	0.9245	0.9247	0.8567	0.8577	0.5594	0.5578
L Ratio[j]	0.8180	0.8181	0.9357	0.9354	0.9416	0.9412	0.8719	0.8724	0.6358	0.6363

Note: Model M1 uses ENVINDX1; Model M2 uses ENVINDX2: see text for definitions of ENVINDX. The parameter estimates are in the first row; asymptotic t-test values are in the second row. ALPHA and BETA are parameters for the travel-time variable. See Section 10.3.3 for model formulation.

[a] LIHH = Low Income Households,
[c] LMIH = Lower Middle Income Households,
[d] MIHH = Middle Income Households,
[e] UMIH = Upper Middle Income Households,
[f] UIHH = Upper Income Households,
[g] VACDEV = vacant developable land,
[h] PERDEV = percent developable land already developed,
[i] RESLND = residential land, LAGHH = lagged total households,
[j] L Ratio = best/worst likelihood ratio.

The signs and magnitude of parameter estimates for ENVINDX are mostly contrary to our expectations with a couple of exceptions. Except for the low-income households in Model 2, parameter estimates for ENVINDX are all positive. The upper-income household category has the largest parameter estimate of ENVINDX, while the low-income household category has the smallest estimate. Basically, parameter estimates for ENVINDX increase with household income. Except for the low-income household category, ENVINDX parameter estimates have their asymptotic t-test values larger than 2, and thus are statistically significant.

In short, air quality is a statistically significant predictor for the distribution of households by income in the Houston metropolitan region. Contrary to our hypothesis, the rich chose their residential location with the worst ozone air quality while the poor lived where ozone air quality was the best.

10.3.7 DISCUSSIONS AND CONCLUSIONS

The model estimations for Houston and Los Angeles provide consistent support for the first hypothesis but inconsistent results for the second hypothesis; that is, the level of air pollution is a statistically significant predictor of household distribution by income. Los Angeles data support the hypothesis that the rich choose their residential location with the best ozone air quality while the poor live where ozone air quality is the worst. Contrary to our expectation, the results from the Houston case reject this hypothesis. As discussed earlier, households make their residential location choices based on trade-offs among a number of attributes. The findings from this study indicate that the role of air quality (particularly ozone) in residential location choice could vary with localities, and trade-offs among and spatial interaction of major location factors may lead to different relationships between air quality and population distribution in different metropolitan areas.

These findings show a stronger relationship than three previous studies (Napton and Day 1992; Brajer and Hall 1992; Korc 1996), although they are not strictly comparable. There are some differences in study area coverage; Houston is one of five areas covered by Napton and Day (1992), and this study covers a larger area than Brajer and Hall (1992) and Korc (1996). There are also some differences in research methodology. In particular, this research takes into account interaction among land use, transportation, residential and employment location, and ozone air quality. Another difference is that this study uses ambient ozone concentration and a more refined spatial representation of 772 zones, while the Brajer and Hall study uses exposure measurement and a 31-district representation, and Korc (1996) used 126 subregions for 1980–1982 and 142 subregions for 1990–1992.

The model estimations using data from two interpolations and three forms of air quality index show that the basic findings are robust in terms of the relative magnitude and goodness-of-fit measures. Uncertainties associated with choice of interpolation procedures and an air quality index do not change the basic results in the two study areas. For this spatial interaction modeling approach to be successfully applied to other metropolitan areas, a spatially well-represented network of air quality monitoring stations and careful choice of an environmental quality or risk index are needed. Ideally, an indicator or index should be used that best represents human response to environmental risks or quality in a study area. This may require data from actual surveys of local residents. When there are no such data and no adequate resources to carry out surveys in the study area, we have to rely on the findings of previous studies for other areas. In this case, using multiple indices is a way to test the sensitivity of the model to the uncertainties of environmental index choice.

The strikingly different findings for Los Angeles and Houston raise a series of questions for further research. Why does the relationship between air quality and population distribution by income in Los Angeles differ from that in Houston? Could the difference in actual or perceived risks be associated with ozone air pollution and spatial differentiation in the two areas lead to the different relationships? We do not know exactly, but, as Table 10.4 shows, the two areas did vary greatly in spatial differentiation of ozone pollution distribution and in the duration and degree of detection of ozone in 1981–1985. Apparently, the magnitude of ozone pollution was higher and more easily detectable in the Los Angeles area than in the Houston area. As noted earlier, previous studies have usually concluded that direct sensory perception of air pollution (particularly gaseous air pollutants) is related to the visual and olfactory stimuli of air pollutants (Barker 1976). Previous research also indicated a sort of threshold that people could tolerate above which they began to report "smoggy" air (Flachsbart and Phillips 1980).

As argued earlier, human response to ozone air pollution might be different from that to environmental risks such as toxic wastes and their disposal facilities (to which the public have shown strong objections) because of their different risk characteristics. Tropospheric ozone pollution arises from diverse sources, produces dispersed regional impacts, may lead to chronic, easily detectable and curable symptoms, and generates unsensational publicity. Another important feature is the long-range transport that leads to detachment between source and impact areas. The downwind areas are not as visible as the source areas of urban cores and industrial complexes. These unique characteristics may result in discounted importance of ozone pollution in a residential location decision. The Houston findings are in agreement with those from New York and Philadelphia that the rich live in the downwind areas, where the worst episodes occur (Liu 1996).

Could other residential location factors in Houston overshadow the role of ozone air quality? Some pull–push, institutional, social, and cultural forces that affect residential location choice might not be well represented in the model. Examples are peculiar environmental amenities, other environmental externalities, major institutional resources, social symbols of neighborhoods, and social and cultural values of ethnic neighborhoods to minorities. The distribution of these variables that people might really consider in their residential location choice might be different from that of ozone in Houston. Air pollution is only one of many environmental hazards that people use to make their residential location choices. The relationship between household distribution and the other environmental stresses in Houston needs further investigation.

Could the urban form and mobility pose some restriction for Houston but not for Los Angeles so that people able to afford avoiding bad air were able to do so in Los Angeles but not necessarily in Houston? Data show that there was not much difference between Houston and Los Angeles in terms of overall mobility rate, travel time and distance, and place of work at the county and MSA levels (Table 10.4).

Houston had a higher mobility rate for the central city residents than Los Angeles. Los Angeles had a higher degree of employment dispersion distribution than Houston. These data do not seem to suggest evidence for a mobility constraint for Houston.

Could the difference in the social and demographic composition in the two areas matter? They may differ in racial composition (e.g., African vs. Hispanic Americans) and residence settings (e.g., urban vs. rural). The Los Angeles region has a larger proportion of low-income households who are urban and Hispanic than the Houston area. They are also more likely to be new immigrants. As demonstrated in previous research, cultural values and beliefs are an important determinant in human perception of risks and hazards. These cultural differences might lead to differences in residential location patterns for households of the same income type. This hypothesis could be formally tested with a survey of the two areas.

To address environmental justice concerns, policy-makers and planners need to take a comprehensive look at the underlying forces that shape the sociodemographic landscape of a metropolitan area. As demonstrated in the two cases, the role of environmental variables in the residential location choice process might not be unidirectional. It might be compounded by how people perceive and respond to the environmental variables and the relative importance of other variables. Finding an association between environmental risk distribution and population distribution is only the beginning. Understanding why this association comes into being can help design effective public policies that can correct and prevent any inequity in the long run. A spatial-interaction modeling approach helps planners understand various interacting forces and trade-offs.

Largely because of the Clean Air Act Amendments of 1990 and the Intermodal Surface Transportation Efficiency Act of 1991, the land use/transportation interaction has been rising on the planners' horizon and they are now concerned about the air quality and transportation congestion implications of the spatial interaction process. While planners' efforts in this area will advance the goals of environmental protection and economic efficiency, they may have different equity implications, which deserve more attention. This is a unique opportunity through which planners can make their contributions. As demonstrated in this study, existing urban models can help make it happen; they can explicitly address the distributional impacts of development and conservation policies, and forecast the dynamics of the various interacting forces and their equity implications. By predicting potential conflicts from basic goals in the future and taking conflict-resolving action today, planners can help make a better tomorrow.

10.4 EQUITY ANALYSIS OF NATIONAL AMBIENT AIR QUALITY STANDARDS

10.4.1 PROBLEM DEFINITION

On November 27, 1996, EPA proposed new National Ambient Air Quality Standards (NAAQS) for particulate matter (PM) and ozone (O_3) (U.S. EPA 1996). EPA proposed two new primary $PM_{2.5}$ standards: 15 $\mu g/m^3$, annual arithmetic mean, and 50 $\mu g/m^3$, 24-h average. Retaining the current annual primary PM_{10} standard of 50 $\mu g/m^3$, EPA also proposed to revise the current 24-h primary PM_{10} standard of 150 µg/m³ by replacing the one-expected-exceedance form with a 98th percentile form, averaged over 3 years at each monitoring site within an area. The proposed ozone standard replaces the current 1-h standard of 0.12 ppm and one-expected-exceedance annually over 3 years, with an 8-h standard of 0.08 ppm, the average third highest concentration over 3 years determining the status of compliance. After an extensive comment period, EPA announced final new standards, which are slightly different from those originally proposed.

According to EPA Administrator Carol M. Browner, the proposed standards "would provide new protection to nearly 133 million Americans, including 40 million children;" but who are these people who will benefit from the new standards, and who currently live in the areas with substandard air quality? While the scientific basis of revising the standards has been debated, the distributional impacts of the proposed standards have received little attention. EPA conducted a draft Regulatory Impact Analysis (RIA) for alternative proposal standards for ozone and particular matter (EPA 1996c; 1996d). These analyses assessed the costs, economic impacts, and benefits associated with the implementation of the proposed NAAQS. Like most benefit–cost analyses, the RIA contributed little to our understanding of the distributional impacts of the proposed standards. In the following, we examine the characteristics of the population living in the projected nonattainment areas under the proposed standards, and compare them with those in the current nonattainment areas.

10.4.2 METHODS

Based on the 1993–1995 air quality data from the Aerometric Information Retrieval System (AIRS), EPA identified counties that did not meet the existing NAAQS for PM_{10} and O_3, and projected counties that would not meet EPA's originally proposed standards. According to initial estimates, implementation of the proposed standards would more than triple the number of nonattainment areas, from 106 counties to 335 counties for ozone and from 41 counties to 167 counties for particulate matter. Because of the lack of $PM_{2.5}$ monitoring data, EPA derived $PM_{2.5}$ results based on monitored PM_{10} data; hence these findings are subject to significant uncertainty. The current AIRS is also inadequate for capturing ozone pollution in rural areas because the AIRS monitoring sites are usually located in urban and suburban areas. Using limited 1-year data from networks designed to characterize rural air quality, Chameides, Saylor, and Cowling (1997) found that the proposed ozone standard would bring many more rural counties into nonattainment than projected from the AIRS data. Moreover, by simply being part of Consolidated Metropolitan Statistical Areas (CMSAs), some counties without monitors would appear to be in nonattainment. These uncertainties should be kept in mind when interpreting the following results that are based on the EPA's projections.

The 1990 census data at the county level were used to characterize current and projected nonattainment areas. Data were aggregated for the nonattainment areas as a whole and, alone, for *additional* new nonattainment areas (hereafter, the term additional is used to emphasize those areas that are currently in compliance and would be out of compliance under the new proposed standards). In addition, the nonattainment areas were grouped into urban and rural counties and compared. Urban counties are those in the census-defined Metropolitan Statistical Areas (MSA). Differences between city and non-city nonattainment areas were also investigated, based on the city-level census data. City nonattainment areas are cities with at least 100,000 people in the nonattainment areas. The remaining areas are the non-city nonattainment areas, which include suburban and rural counties. The analysis could be resolved to a much smaller geographic scale, as in case studies of Houston and Los Angeles. However, this is not practical for a national-level study that uses the AIRS database, because it has inherent limitations in spatial representation at an intra-city geographic level.

The population variables analyzed include census-defined race and ethnicity measurements, income measurements (household income distribution aggregated into five groups and percentage of persons living in poverty), and sensitive subpopulations (percentage of children, 5 and under and 6 to 11; adolescents, 12 to 17; the elderly, 65 and over; resident workers employed in the construction sector as a substitute measure for outdoor workers).

The concept of location quotient was used as a measure indicative of the spatial concentration of a socioeconomic or demographic variable relative to the country as a whole. It is defined as the ratio of the population segment or employment sector share for a study area to that for the country. If a location quotient has a value less than one, the population segment is under-represented in the study area relative to the country; if the quotient is greater than one, it is over-represented. GIS was used to spatially identify current and projected nonattainment areas, identify the cities with at least 100,000 residents in the nonattainment areas, process data, and display the spatial distribution of location quotients.

10.4.3 Results and Discussion

Implementation of EPA's proposed standards would significantly alter the demographic and socioeconomic characteristics of resident populations in substandard air quality areas. Under the proposed standards, nearly half of the country's population would live in the projected ozone nonattainment areas, comprising about 8% of the country's land area, compared with currently less than one third who reside in about 4% of the country's land area. The proportion of U.S. population living in the PM nonattainment areas would increase from 12 to 30%.

10.4.3.1 Nonattainment Areas as a Whole

The projected nonattainment areas exhibit demographic and socioeconomic characteristics that differ considerably from existing ones (Figure 10.9). Projected ozone and $PM_{2.5}$ nonattainment areas would be over-represented by Hispanics, Asians/Pacific Islanders (hereafter referred to in aggregate as "Asians"), and African Americans and under-represented by American Indians and whites. Although the overall pattern would remain the same as the existing standards, the distributional disparity would decline sharply for Hispanics and Asians, and the proportion of whites would slightly increase; however, African-American representation would increase considerably from a current under-representation to an over-representation in the projected $PM_{2.5}$ nonattainment areas.

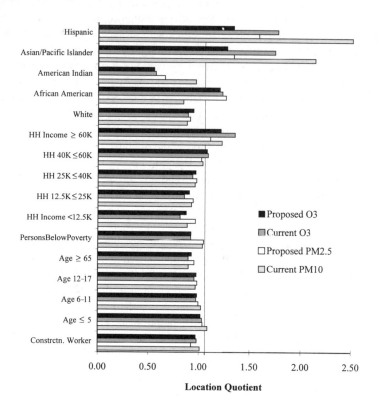

FIGURE 10.9 Distributional inequities in nonattainment areas for current and proposed ozone and PM standards. Aggregate location quotients over all nonattainment areas for current and proposed ozone and particulate matter standards reveal distributional inequities. A location quotient > 1 indicates overrepresentation of a population group (e.g., Hispanics) in nonattainment areas; a value < 1 indicates underrepresentation.

These changes result from *additional* nonattainment counties that have significantly different demographic and socioeconomic characteristics than existing ones. Percentages of Asians and Hispanics in existing ozone nonattainment areas, 5.1 and 15.6%, respectively, are more than twice as high as those in the *additional* areas, 1.6 and 5.9%, respectively. Whites are under-represented in current ozone nonattainment areas (71.8%), but are slightly over-represented in *additional* areas (81.5%). The *additional* PM nonattainment counties have a high over-representation of African Americans; a nearly even representation of Hispanics and whites; and an under-representation of American Indians and Asians. In contrast, the existing areas have an under-representation of African Americans and whites; a high over-representation of Hispanics and Asians; and an even representation of American Indians.

As are existing nonattainment areas, the projected ozone and PM nonattainment areas would be over-represented by the rich and under-represented by the poor. However, with implementation of proposed standards, the comparative disparity would decrease because of additional nonattainment counties. These *additional* areas, on average, are poorer than existing ones but are still slightly richer than the country as a whole.

Analysis of the population age structure indicates that projected nonattainment areas for O_3 and $PM_{2.5}$ contain a smaller proportion of the elderly and a slightly larger proportion of children, 5 and under, than the country. This disparity would also decline because of additional nonattainment counties.

At the aggregate level, Hispanics, followed by Asians, would benefit most from air quality improvement in the current nonattainment areas. African Americans, followed by whites, would benefit most in the *additional* nonattainment areas. The relatively rich would benefit most from air quality improvement in current and *additional* nonattainment areas.

10.4.3.2 Spatial Distribution and Regional Differences

The disparities show substantial spatial variations and regional differences. Minorities are concentrated regionally and in the largest cities (Figures 10.10 and 10.11). For example, Hispanics reside widely in the Southwest and West; African Americans are regionally concentrated in the Southeast; large numbers of Asians reside in California; and American Indians are found throughout the West. Although whites are and would be under-represented in the nonattainment areas at the aggregate level, they are over-represented in 58 and 67% of the nonattainment counties for current and proposed ozone standards, respectively. In contrast, Asians, African Americans, and Hispanics are over-represented in 29–32% of current ozone nonattainment counties; over-representation would decline to less than 15% of the projected nonattainment counties for Hispanics and Asians and increase slightly to 32% for African Americans. A similar pattern occurs in PM nonattainment counties, except for a large increase in over-representation of African Americans, from 7 to 26% of the nonattainment counties.

To check the influences of regional variations on reported outcomes, states were also used as reference areas for calculating location quotients. Although over-representation occurs in 64% of current ozone nonattainment counties for whites, it ranges from 38 to 43% for Hispanics, African Americans, American Indians, and Asians. Under the proposed 8-h ozone standards, the proportion of over-representation will increase to 70% for whites and decrease to 31 to 40% for minorities. Aggregation of nonattainment counties at the state level gives a different picture: Asians are over-represented in 77% of 26 states with ozone nonattainment counties, followed by African Americans and Hispanics (62%), whites (42%), and Native Americans (23%). The median location quotient for those over-represented states (an indication of the degree of over-representation at the state level) is highest for African Americans (1.36), followed by Hispanics (1.24), Asians (1.19), Native Americans (1.07), and whites (1.03). A similar pattern emerges for the projected O_3 nonattainment areas.

Current 1-Hour Ozone Nonattainment Areas
  Under-representation
  Over-representation
Projected 8-Hour Nonattainment Areas (Additional)
  Under-representation
  Over-representation

FIGURE 10.10 Distribution of population groups in ozone nonattainment areas. The spatial representation (by county) of population according to race and ethnicity, income, and age in nonattainment areas for current and proposed ozone standards highlights the distribution of impacts on selected population groups. Over-representation is flagged by location quotients > 1; under-representation is flagged by values < 1. (a) African American; (b) American Indian. Data from 1990 Census. Nonattainment areas are based on 1993–1995 EPA data.

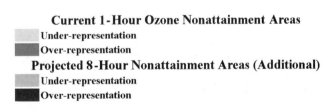

Current 1-Hour Ozone Nonattainment Areas
Under-representation
Over-representation
Projected 8-Hour Nonattainment Areas (Additional)
Under-representation
Over-representation

FIGURE 10.10 Distribution of population groups in ozone nonattainment areas. The spatial representation (by county) of population according to race and ethnicity, income, and age in nonattainment areas for current and proposed ozone standards highlights the distribution of impacts on selected population groups. Over-representation is flagged by location quotients > 1; under-representation is flagged by values < 1. (c) Asian/Pacific Islander; (d) Hispanic. Data from 1990 Census. Nonattainment areas are based on 1993–1995 EPA data.

Current 1-Hour Ozone Nonattainment Areas
  Under-representation
  Over-representation
Projected 8-Hour Nonattainment Areas (Additional)
  Under-representation
  Over-representation

FIGURE 10.10 Distribution of population groups in ozone nonattainment areas. The spatial representation (by county) of population according to race and ethnicity, income, and age in nonattainment areas for current and proposed ozone standards highlights the distribution of impacts on selected population groups. Over-representation is flagged by location quotients > 1; under-representation is flagged by values < 1. (e) White; (f) Children age 5 and under. Data from 1990 Census. Nonattainment areas are based on 1993–1995 EPA data.

Current 1-Hour Ozone Nonattainment Areas

■ Under-representation
■ Over-representation

Projected 8-Hour Nonattainment Areas (Additional)

■ Under-representation
■ Over-representation

FIGURE 10.10 Distribution of population groups in ozone nonattainment areas. The spatial representation (by county) of population according to race and ethnicity, income, and age in nonattainment areas for current and proposed ozone standards highlights the distribution of impacts on selected population groups. Over-representation is flagged by location quotients > 1; under-representation is flagged by values < 1. (g) Persons age 65 and over; (h) Persons below poverty. Data from 1990 Census. Nonattainment areas are based on 1993–1995 EPA data.

Current PM$_{10}$ Nonattainment Areas

▢ Under representation
▨ Over representation

Projected PM$_{2.5}$ Nonattainment Areas (Additional)

▨ Under representation
■ Over representation

FIGURE 10.11 Distribution of population groups in particulate matter nonattainment areas. The spatial representation (by county) of population according to race and ethnicity, income, and age in nonattainment areas for current and proposed particular matter standards highlights the distribution of impacts on selected population groups. (a) African American; (b) Asian/Pacific Islander. Data from 1990 Census. Nonattainment areas are based on 1993–1995 EPA data.

Current PM₁₀ Nonattainment Areas
Under representation
Over representation
Projected PM₂.₅ Nonattainment Areas (Additional)
Under representation
Over representation

FIGURE 10.11 Distribution of population groups in particulate matter nonattainment areas. The spatial representation (by county) of population according to race and ethnicity, income, and age in nonattainment areas for current and proposed particular matter standards highlights the distribution of impacts on selected population groups. (c) Hispanic; (d) Children age 5 and under. Data from 1990 Census. Nonattainment areas are based on 1993–1995 EPA data.

10.4.3.3 City vs. Non-City Nonattainment Areas

In O_3 nonattainment areas, cities with at least 100,000 people account for 47% of the population and 5.5% of land area. In the PM_{10} nonattainment areas, they account for 52% of the population and 3.5% of land area. Under the proposed standards, nonattainment city share of the population would slightly fall to 41 and 49%, respectively, for O_3 and $PM_{2.5}$. Significant differences occur between the *city* and *non-city* nonattainment areas (Figure 10.12a–d).

Overall, *city* nonattainment areas have and would continue to have much larger shares of minority populations than *non-city* nonattainment areas. The proposed standards would initiate a decline in over-representation of Asians and Hispanics in *city* areas. On the other hand, African-American over-representation would increase slightly. Current over-representation of Asians and Hispanics in *non-city* areas would decline to a level of under-representation. African Americans are and would still be highly under-represented in the *non-city* areas. Whites are currently over-represented in *non-city* O_3 nonattainment areas, and would be, to a greater degree, over-represented in *non-city* nonattainment areas for both O_3 and $PM_{2.5}$.

Generally, *city* nonattainment areas are now and, if proposed standards are implemented, would continue to be poorer than the country as a whole, while *non-city* nonattainment areas are and would be richer than the country. Proposed standards would bring in poorer *city* and *non-city* areas into nonattainment.

Non-city areas, where elderly representation is and would be close to the national average, have and would continue to have a larger share of seniors than *city* counterparts, where they are and would be under-represented. The difference between the city and non-city areas would decrease slightly. There is and would be little difference in the representation of children, 5 and under, in city and non-city areas. *Non-city* areas have and would continue to have a larger share of residents employed in the construction industry than *city* areas.

Because non-city nonattainment areas include suburban and rural areas, rural nonattainment areas have overall distributional patterns that are similar to non-city areas. There are, however, some differences.

- American Indians are and would be highly over-represented in the rural nonattainment areas for PM, in contrast to an under-representation for non-city areas.
- Rural nonattainment areas are and would be poorer than non-city areas and, concerning PM impacts, poorer than the country now and in the future. The proposed standards would bring additional rural counties that are poorer than the current ones into nonattainment. Moreover, the disproportionate representation of the economically disadvantaged could be higher if the AIRS covered more rural counties.

Because of their predominance in nonattainment areas, urban counties as a group (i.e., those in the census MSAs) show the same distributional impact pattern as the nonattainment areas described above (see Table 10.7).

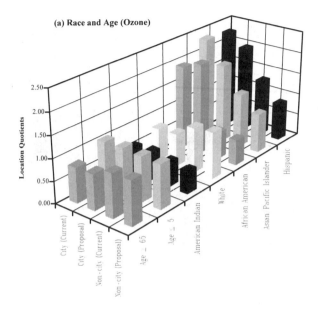

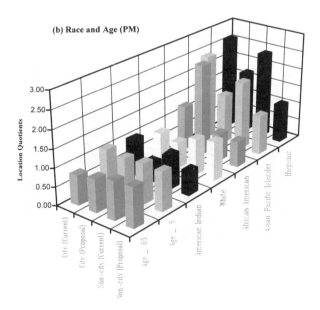

FIGURE 10.12 Distributional inequities in cities vs. non-city areas. The representation of race and ethnicity and income in city and non-city nonattainment areas for the current and proposed ozone and PM standards demonstrates distributional inequities in cities. City non-attainment areas are cities with at least 100,000 residents in nonattainment areas, and all other areas are classified as non-city nonattainment areas.

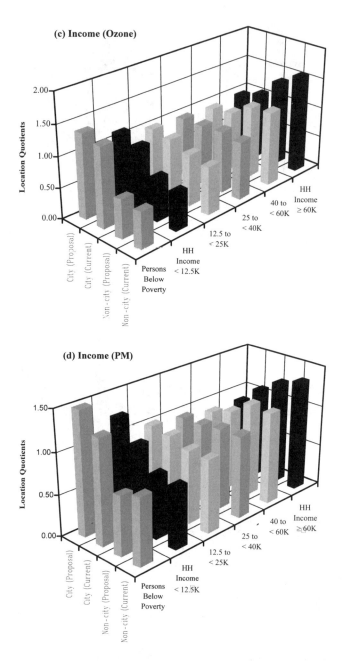

FIGURE 10.12 Distributional inequities in cities vs. non-city areas. The representation of race and ethnicity and income in city and non-city nonattainment areas for the current and proposed ozone and PM standards demonstrates distributional inequities in cities. City non-attainment areas are cities with at least 100,000 residents in nonattainment areas, and all other areas are classified as non-city nonattainment areas.

10.4.3.4 Major Findings

On the national scale, Hispanics, Asians, and, to a much lesser degree, African Americans are very disproportionately represented in the current nonattainment areas for O_3 and PM, and are, therefore, at a potentially higher risk of exposure. This disparity would generally be moderated under the proposed standards because of additional nonattainment areas that have population characteristics closer to average. The most striking examples of disproportionate representation of Hispanics, Asians, African Americans, and the poor currently occur and would continue to occur under the proposed standards in large *city* nonattainment areas. In addition, the poor would also be over-represented in rural nonattainment areas.

Disparity, or lack thereof, in nonattainment areas mirror overall spatial population distributions in the country. They generally reflect characteristic differences between resident populations in urban and rural areas and between those in city and non-city areas. Minorities, except American Indians, are concentrated in urban areas, especially in large cities. As was found in several urban plume studies, the highest O_3 concentrations often occur downwind of large urban and industrial centers. In concert with previous findings from a New York and Philadelphia study, assuming that cities are contaminant source areas and non-city nonattainment areas are downwind areas, the above results appear to indicate a nationwide trend in nonattainment areas: Wealthy households and whites, who disproportionately populate downwind areas are, and would continue to be, at a potentially higher risk of exposure to O_3 than minorities and the poor, who disproportionately inhabit source areas.

TABLE 10.7
Who Would be Overrepresented in the Projected Nonattainment Areas under the Proposed Ozone and Particulate Matter Standards?

Population Variables	Nonattainment Areas	Nonattainment Areas		Nonattainment Areas	
		City	Non-city	Urban	Rural
African American	Yes	Yes	No	Yes	No
American Indian	No	No	No	No	No for O_3 Yes for $PM_{2.5}$
Asian/Pacific Islander	Yes	Yes	No	Yes	No
Hispanic	Yes	Yes	No for O_3 Yes for $PM_{2.5}$	Yes	No for O_3 Yes for $PM_{2.5}$
White	No	No	Yes	No	Yes
The poor (HHI[a] < 12.5K)	No	Yes	No	No	Yes
The young (Age ≤ 5)	Yes	Yes	No for O_3 Yes for $PM_{2.5}$	Yes	No for O_3 Yes for $PM_{2.5}$
The elderly (Age ≥ 65)	No	No	No for O_3 Yes for $PM_{2.5}$	No	Yes

[a] HHI = Household income in thousands of dollars

10.4.3.5 Implications for Environmental Policy

This study demonstrates that different uniform national air quality policies have distinctive distributional impacts on affected populations. To ensure equal protection of all people regardless of race, ethnicity, and socioeconomic status, policy-makers must assess potential distributional impacts associated with implementing each policy alternative. Such an exercise was not undertaken during the standards-making process for O_3 and PM NAAQS. The debate over new standards was dominated by consideration of the underpinning scientific basis of the proposed standards and the benefit–cost analysis.

Much of the present inequity, clearly evident with respect to the nationwide distribution of O_3 and PM, may be remedied by achieving compliance with existing and new standards. If new, tightened standards help make nationwide compliance a reality, they can be considered socially beneficial even if total benefits are not commensurate with total costs.

Tightened O_3 standards would address long-range regional transport of O_3 from Midwest pollution sources and benefit the Northeast corridor areas that contain a cluster of major cities with a disproportionate number of minorities and economically disadvantaged populations. EPA's new standards implementation package includes a trading plan for emissions from utilities in the 37 states east of the Rocky Mountains, aimed at achieving the most cost-effective pollution reduction. A critical issue that deserves more attention in designing trading schemes is their pollutant redistribution impact. Implementation of trading schemes could create hot spots of air pollution that disproportionately impact minority and low-income communities. The distribution impacts of EPA's trading plan should be carefully examined to ensure that emission trading will not cause disproportionate, adverse impacts.

Although states shoulder the primary responsibility for developing implementation strategies that achieve required standards, actual hands-on responsibility for implementing strategies largely lies with metropolitan and local government officials. As found in this study and other investigations, there are substantial variations in the spatial distribution of air pollution burdens that must be addressed. The relative pollutant-impact distribution pattern typically observed in city or metropolitan areas may differ substantially from absolute patterns of distribution at the national level. Therefore, policy-makers at the state, metropolitan, and local levels should carefully examine equity issues in their domains of responsibility. This study indicates a likelihood that the inequity is concentrated in large cities, where other kinds of environmental inequities and social injustice often exist; causes of these inequities may be related. Cities should be priority areas of concern in the environmental justice policy arena. Concerted efforts are needed and should address the range of inequities that can occur. Municipal officials need to examine fundamental social, economic, and political forces that may cause these inequities — discrimination barriers in the housing market, discriminatory practices in siting polluting industries, uneven enforcement of land-use regulations and environmental laws, unbalanced provision of municipal services, and social prejudice.

Although the inequitable distribution of air pollution impacts was first documented more than 20 years ago, it has received less attention than other environmental injustices. Long-term trends in development and implementation of uniform national air quality policies have led to dramatic, nationwide air quality improvements and, to some extent, have benefited socially and economically disadvantaged groups of our society. However, inequities remain.

This situation suggests that uniform national air quality policies alone are inadequate for addressing equity issues; specific equity-targeted policies are needed. Some recent trends suggest this may be possible, for example, policy responses and proposed legislation focusing on the inequities of hazardous and toxic waste management, which is the most publicized environmental justice issue. Lack of adequate attention to air pollution distribution inequities in some measure reflects the unique nature of air pollution and consequent difficulties in addressing these inequities in a balanced manner. Air pollutants, particularly O_3, arise from diverse sources; produce widespread, dispersed regional impacts; pose relatively low-to-moderate perceived risks; and generate relatively unsensational publicity and greater public tolerance compared with other environmental justice issues. This observation, in conjunction with the preceding analysis, strongly suggests that regional cooperation is especially important for achieving air quality and environmental equity goals. To effectively address air pollution-related inequities, policies and strategies that consider the uniqueness of air pollution distribution impacts are needed, especially on regional and metropolitan scales.

11 Environmental Justice Analysis of Hazardous Waste Facilities, Superfund Sites, and Toxic Release Facilities

This chapter deals with three types of waste facilities: hazardous waste facilities, Superfund sites, and toxic release facilities. For each one, we briefly discuss basic concepts about these wastes and waste facilities. Next, we review major environmental justice studies on each type of facility, with particular attention to the debate in the literature. Finally, we discuss some methodological issues and the potential for improvement.

11.1 EQUITY ANALYSIS OF HAZARDOUS WASTE FACILITIES

11.1.1 HAZARDOUS WASTES

A waste is hazardous if it has one or more of the following characteristics (U.S. EPA 1997b):

- Ignitability. Ignitable wastes can cause fire. Waste oils are examples.
- Corrosivity. Corrosive wastes, such as batteries, are acids or bases that can corrode metal, i.e., storage tanks.
- Reactivity. Reactive wastes such as explosives are unstable and can cause explosions, toxic fumes, gases, or vapors when mixed with water.
- Toxicity. Toxic wastes such as certain heavy metals are harmful or fatal when ingested or absorbed. Toxicity is defined through a laboratory procedure called the Toxicity Characteristic Leaching Procedure (TCLP).

By definition, EPA determines that three categories of specific wastes are hazardous and publishes the list:

- Source-specific wastes from specific industries, such as petroleum refining or pesticide manufacturing.
- Nonspecific source wastes from common manufacturing and industrial processes.
- Commercial chemical products in an unused form, such as some pesticides and some pharmaceutical products.

Hazardous wastes are solid wastes that meet any of the following criteria. Solid waste is discarded material, including garbage, refuse, and sludge (solids, semisolids, liquids, or contained gaseous materials). U.S. EPA (1997b:7) defines hazardous wastes as "those that:

- Possess one or more of the four characteristics of hazardous waste.
- Are included on an EPA list of hazardous waste.
- Are a mixture of nonhazardous and hazardous waste listed solely for a characteristic (e.g., dirty water mixed with spent solvents).
- Derive from the treatment, storage, or disposal of a hazardous waste (e.g., incineration ash or emission control dust).
- Are soil, ground water, or sediment (environmental media) contaminated with hazardous waste.
- Are either manufactured objects, plant or animal matter, or natural geological material (debris) containing hazardous waste that are intended for disposal (e.g., concrete, bricks, industrial equipment, rocks, and grass)."

The Resource Conservation and Recovery Act (RCRA) of 1976 and its subsequent amendments in 1980 and 1984 set forth a framework for managing hazardous wastes (under Subtitle C) and solid wastes (under Subtitle D). RCRA regulations adopt a "cradle to grave" approach to manage hazardous waste from its generation until its ultimate disposal. The two key components of this approach are the tracking system that monitors hazardous waste at every point in the waste cycle and the permitting system that manages facilities that receive hazardous wastes for treatment, storage, or disposal, or TSDFs. Treatment facilities use various processes (such as incineration or combustion) to alter the character or composition of hazardous wastes. As a result of treatment, some wastes are recovered and reused, while others are dramatically reduced in terms of quantity. Storage facilities temporarily hold hazardous wastes until their treatment or disposal. Disposal facilities contain hazardous wastes permanently. A landfill, the most common disposal facility, disposes of hazardous wastes in carefully constructed units that are designed to protect groundwater and surface-water resources.

TSDFs must obtain a RCRA permit in order to operate. A RCRA permit establishes the waste management activities that a facility can conduct and the conditions under which it can conduct them. The permit outlines facility design and operation, lays out safety standards, specifies facility-specific requirements, and describes activities that the facility must perform, such as monitoring and reporting. Exemptions from obtaining a RCRA permit include businesses that generate hazardous waste and transport it off site without storing it for long periods of time, businesses that transport hazardous waste, and businesses that store hazardous waste for short periods of time without treatment.

11.1.2 EQUITY ANALYSIS OF HAZARDOUS WASTE FACILITIES

As discussed in Chapter 1, it was the issue of siting a hazardous waste facility that first sparked national attention to environmental justice. The 1982 Warren event received the attention of the U.S. Congress, which requested the United States General Accounting Office (GAO) to investigate "the correlation between the location of hazardous waste landfills and the racial and economic status of the surrounding communities" (GAO 1983:1). The GAO studied offsite landfills in the 8 southeastern states that comprise the EPA's Region IV. For the four offsite hazardous waste landfills identified in the region, the study concluded that blacks were the majority of the population in three of the four host communities and at least 26% of the population had income below the poverty level. This was the first major study of regional scope that found inequitable distribution of hazardous waste facilities by race and income.

The methodology used in the GAO study included onsite and telephone interview, EPA and state file review, and census data analysis. The geographic unit was census-designated areas for three host communities, and township for the Warren County host community (labeled as "Area A" in the report). Census maps were used to identify the facility sites. Data and maps also included adjacent census-designated areas or townships that have borders within about 4 miles. However, the report did not show any data for the aggregated area including adjacent census-designated areas or townships. The report's conclusion was based solely on the census areas or townships where the facilities were located. Examinations of the original location maps in the report and the maps using 1990 boundaries show that all four facilities were near borders of census areas or townships, and could have impacts on adjacent census areas or townships. Been (1994) revisited this study and found that the data in the GAO report did not match the data from the census publications. She concluded that the GAO boundaries did not correspond to the Census Bureau's geographic units. Using the county subdivisions that were closest to the GAO's areas, she found that all four host communities were disproportionately populated by blacks at the time of the siting (with 1970 as the baseline for three sites and the 1980 for one site).

11.1.2.1 Cross-Sectional National Studies

The second study triggered by the Warren County event was "Toxic Wastes and Race in the United States: A National Report on the Racial and Socio-Economic Characteristics of Communities with Hazardous Waste Sites," commissioned by the United Church of Christ Commission of Racial Justice in 1987. This was "the first national report to comprehensively document the presence of hazardous wastes in racial and ethnic communities throughout the United States" (UCC 1987:ix).

The study chose the potential distributional impacts from commercial or offsite rather than onsite hazardous waste facilities on the basis that these facilities' location decisions were more likely affected by factors other than proximity to hazardous waste generation activities. The study identified 415 operating commercial hazardous waste facilities as of May 1986, using the EPA's Hazardous Waste Data Management System (HWDMS) and Environmental Information Ltd.'s 1986 directory *Industrial and Hazardous Waste Firms*. Residential 5-digit ZIP code areas were used to define "communities." The study recognized the different magnitudes of environmental risks posed by these facilities in residential ZIP code areas and established four groups of 5-digit ZIP code areas having:

- No operating commercial hazardous waste TSDFs
- One operating commercial hazardous waste TSDF that is not a landfill
- One operating commercial hazardous waste landfill facility that is not one of the five largest
- One of the five largest commercial hazardous waste landfills or more than one operating commercial hazardous waste TSDF

The size of landfills was defined on the basis of landfill capacities.

Five statistical tests (see Table 11.1) were used to test the following hypotheses: "(1) The mean minority percentage of the population was a more significant discriminator than the other variables for differentiating communities with greater numbers of commercial hazardous waste facilities and the largest landfills. (2) The mean minority percentage of the population was significantly greater in communities with facilities than in those without" (UCC 1987:11).

This study found that the mean minority percentage of the population in ZIP code areas with one operating commercial hazardous waste facility was approximately twice as large as that in ZIP code areas without a facility (24 vs. 12%). ZIP code areas with two or more facilities or one of the five largest landfills had an average minority percentage that was more than three times that in ZIP code areas without a facility. Predominantly black and Hispanic communities hosted three out of the five largest commercial hazardous waste landfills in the U.S.: Emelle, Alabama (79% black); Scotlandville, Louisiana (93% black); and Kettleman City, California (78% Hispanic). They accounted for 40% of the nation's total commercial landfill capacity. After controlling for regional differences and urbanization, the minority percentage of the population was a more significant discriminator than the other variables in differentiating the level of commercial hazardous waste activity. The UCC report concluded that "[R]ace proved to be the most significant among variables tested in association with the location of commercial hazardous waste facilities. This represented a consistent national pattern" (UCC 1987:xiii).

Critics argue that the UCC study suffers from several methodological limitations. As discussed in Chapter 6, use of ZIP codes as a geographic unit of analysis has been attacked on several grounds. In particular, ZIP code areas are overly aggregated and too large and, as a result, the findings are vulnerable to ecological fallacies (Anderton et al. 1994). In addition, the study failed to control for urban and rural differences. The geographic nature and size of rural geographic units such as ZIP codes and census tracts are substantially different from urban ones. These differences are likely to confound the results. To account for the urban/rural differences, Anderton (1996) called for controlled comparisons and multivariate analyses. The UCC study's use of statistical methods is also criticized. Acknowledging the generally sound research design, Greenberg (1993) argued that the study downplayed the matched-pair test, which he considered as a particularly important tool. The matched-pair tests controlled for local variations in market conditions and socioeconomic status by comparing host ZIP codes with the parts of their surrounding counties without commercial facilities. The matched-pair test results showed that mean family income was a more significant variable than percent minority. Mean family income

TABLE 11.1
Comparing Major Methodological Issues and Findings of Three Cross-Sectional National Studies

	UCC 1980	UMass 1980	Been 1990
Data Year			
Environmental Risks	415 Commercial TSDFs	446 Commercial TSDFs	608 Commercial TSDFs
Unit of Analysis	5-digit ZIP code	Census tracts	Census tracts
Universe	Residential 5-digit ZIP code areas in the contiguous U.S. (35,406 ZIP codes or 96% of the total in the nation)	SMSAs with at least one TSDF facility in the contiguous U.S. (32,003 census tracts or 68% of all tracts in the nation)	Continental U.S. (about 60,600 census tracts)
Number of Host Areas	369	408	Approximately 600
Control areas	35,037 non-host residential 5-digit ZIP codes	31,595 non-host census tracts in SMSAs with at least one TSDF	Approximately 60,000 non-host census tracts
Variables			
Race	Minority defined as Hispanics and non-Hispanic non-white (blacks; Asian and Pacific Islanders; American Indian, Eskimo and Aleu; other)	Blacks or African Americans, Hispanics	Blacks or African Americans, Hispanics; Minority defined as all races other than white and all Hispanics
Income	Mean household income	Percentage of families at or below poverty line Non-farm family of four Percentage of households receiving public assistance income	Median family income Percentage of people living in poverty

continued

TABLE 11.1 (CONTINUED)
Comparing Major Methodological Issues and Findings of Three Cross-Sectional National Studies

Control variables	Mean value of owner-occupied homes	Mean value of housing stock	Median housing value
	Pounds of hazardous waste generated per person	Percentage employed in manufacturing and industry	Percent workers in manufacturing
	Number of uncontrolled toxic waste sites per 1000 persons	Percentage males in the civilian labor force who are employed	Percent people not receiving high school diploma
			Percent employed in professional occupations
			Mean population density
Statistical Methods	Discriminant analysis	T test, Wilcoxon rank sum test, and logistic regression	t test, logit regression
	Difference of means test		
	Matched-pairs test		
	Non-parametric versions of the difference of means and matched-pairs tests		
Inequity by Race/ethnicity?	Yes	No	Yes for Hispanics and Minority
			No/yes for African Americans
Inequity by Income?	Yes	Yes for bivariate analysis	Yes for bivariate analysis
		No for multivariate analysis	No for multivariate analysis
Is Race/ethnicity more significant than income?	Yes	No	Yes for multivariate analysis
			No for bivariate analysis

Date from: UCC 1987; Anderton et al. 1994; Been 1995; Mohai 1995.

was statistically significant in 8 of 10 EPA regions and 10 of 43 states, but percent minority was statistically significant in only 5 of 10 EPA regions and 5 of 43 states.

A study conducted at the University of Massachusetts reached very different conclusions than the UCC study (Anderton et al. 1994). They concluded that "no consistent national level association exists between the location of commercial hazardous waste TSDFs and the percentage of either minority or disadvantaged populations" (Anderton et al. 1994:232). The UMass study used census tracts as its geographic unit of analysis. The UMass study also focused on commercial TSDFs, but it included only those in SMSAs tracted in 1980 that opened for business before 1990 and were still in operation. The TSDF data were extracted from the Environment Institute's 1992 Environmental Services Directory (ESD), the earlier version of which was used in the UCC study. In contrast to the UCC study, the UMass study did not take into account the magnitude of potential environmental risks associated with commercial TSDFs.

The UMass study conducted a series of analyses. The first analysis tested the difference between census tracts with TSDFs and those without TSDFs but within SMSAs that had at least one facility. The second analysis compared TSDF tracts with surrounding areas that included any tract that had at least 50% of its area within a 2.5-mi radius from the center of a TSDF tract. The third analysis combined TSDF tracts with their surrounding areas and compared the aggregated area with the remaining tracts of the SMSAs. The fourth analysis was a series of logistic regressions (presence of a TSDF as a function of census tract characteristics) by EPA Regions. This analysis was done to control for the multivariate effects on the relationship between the location of TSDFs and various variables.

These analyses provided two different pictures. The first and fourth analyses found no significant association between TSDs and the variables of percentage black and percentage Hispanic. However, the second and third analyses demonstrated that the surrounding areas were populated by a significantly larger proportion of blacks than the TSDF tracts, and the aggregated areas including TSDF tracts and surrounding areas had significantly larger proportions of blacks, Hispanics, families below poverty, and households receiving public assistance than the remainder of the SMSAs. These results agreed with the ZIP code-based study by the UCC. The authors dismissed these findings on the grounds that there was no evidence to believe that the larger unit of analysis is more appropriate than census tracts and too large a geographic unit may lead to "aggregation errors" or "ecological fallacy" by obscuring differences within these areas. Instead, the authors concluded that manufacturing employment was the most significant predictor for the location of TSDFs.

This study sparked a heated debate. Critics challenged the UMass study on several grounds. One challenge was the motivation behind the UMass study as critics pointed out that the UMass study was funded by WMX Technologies, Inc., the largest commercial handler of solid and toxic wastes in the world (Goldman 1996). Other challenges touched on several methodological issues such as selection of control population, choice of geographic units of analysis, and selection of variables (Goldman and Fitton 1994; Mohai 1995; Goldman 1996).

Although the UMass authors attributed the contradictory findings solely to the choice of units of analysis, critics claimed that the control populations were the primary reason (Mohai 1995; Goldman 1996). The UCC study's experiment group consisted of residential ZIP code areas with at least one commercial hazardous waste facility (369 ZIP code areas), and its control group included all residential ZIP code areas that did not have a facility. The UMass study's experiment group consisted of 408 census tracts with at least one commercial TSDF, and its control group was made up of 31,595 census tracts without a facility, which were located within SMSAs with at least one commercial TSDF. The UMass study universe was limited to census tracts in SMSAs with at least one commercial TSDF in the contiguous U.S., which consisted of 32,003 census tracts (68% of the total 47,311 census tracts in the nation in 1980). It excluded from analysis all tracts outside SMSAs (about 3,000 in 1980) and those tracts inside the SMSAs that did not have a commercial TSDF.

Estimations show that the mean minority percentages in the two studies were very close for the experiment group (around 25%), but differed dramatically for the control group (Mohai 1995; Goldman 1996). The minority percentage in the UMass study's control group was more than twice as large as that of the UCC study (12%) (see Table 11.2). Critics believed that the differences in comparison populations accounted for the major differences in findings in the two studies.

The UMass researchers' rationale for choosing the comparison group was two-fold. First, siting and plausible siting candidates are constrained and the existing constraints should be reflected in evaluating environmental inequities (Anderson, Anderton, and Oakes 1994). The UMass researchers argued that the facility-siting process can be simplified as a two-step process. Facility locators first look at various large market regions, and then decide on specific locations within a specific market region based on a number of factors, including political, technical, legal, economic, and other constraints. Second, lumping together metropolitan and rural areas would introduce bias since there are dramatic differences in the socioeconomic and demographic composition between urban and rural areas (Oakes et al. 1996). Been (1997) argued that using the presence and absence of a TSDF within a metropolitan area or rural county to eliminate certain areas from the potential siting universe is inappropriate and "extremely rough" to represent the siting processes.

TABLE 11.2
Empirical Results of Cross-Sectional National Studies

Study	Base Year	Sample	Cases	Black %	Black Host/Non-host Ratio	Hispanic %	Hispanic Host/Non-host Ratio	Minority %	Minority Host/Non-host Ratio
UCC	1980	Host ZIPs	369					25.2	2.05
		Non-Host	35,037					12.3	
UMass	1980	Host	408	14.5	0.95	9.4	1.2		
		Non-Host	31,595	15.2		7.7			
Been	1990	Host	600	14.4	1.07	10.3	1.32	27.2	1.13
		Non-Host	60,000	13.5		7.8		24.2	
UCC II	1993	Host						30.8	2.14
		Non-Host						14.4	

Furthermore, the two studies address two different research questions because of the different control populations. "In effect, the UCC study addresses the question of where hazardous waste facilities are most likely to be located, regardless of whether these areas are urban or rural. The UMass study, on the other hand, addresses the question of where within metropolitan areas currently containing a facility such facilities are likely to be located" (Mohai 1995:648). Moreover, the UMass study's choice of comparison population may have made the questionable assumption that excluded census tracts are not suitable for siting commercial TSDFs. Critics argued that there was no justification for this exclusion, and alternative sites for commercial TSDFs were much broader (Goldman and Fitton 1994; Mohai 1995). They were quick to point out that some of the well-known TSDFs were located in rural areas such as Emmelle, Alabama and Warren County, North Carolina, which hosted two of the five largest commercial hazardous waste landfills in the country mentioned above. The rural nature may be an attractive siting factor for hazardous waste facilities. For example, one of siting criteria for the State of North Carolina for selecting a landfill site in the well-known Warren County case was that the landfill should be in an area "isolated from highly populated areas" (GAO 1983:A9). Obviously, it could be argued that the UMass study excluded some feasible sites while attempting to eliminate some unfeasible sites.

Clearly, not all places are potential candidates for the placement of a commercial hazardous waste facility. You cannot possibly consider the Mall area in Washington, D.C. or the Inner Harbor area in Downtown Baltimore as a potential site. There have been local zoning and land-use regulations since early in the twentieth century, which establish the constraints for land uses that may pose a potential "nuisance" to the neighbors. There have also been technical constraints for the placement of hazardous waste facilities. All of these make some areas unsuitable for further consideration. Therefore, it is reasonable to assume that potential sites are not the whole country, but the UMass elimination method is problematic. This leads to an important question: How can we devise such a list of potential alternative sites for hazardous waste facilities? A GIS-based suitability analysis can offer some help (see Chapter 8).

What effects does the UMass exclusion have on the findings? Been (1995) examined the impacts of excluding these SMSAs and rural tracts. By dropping 18,000 non-host tracts from the analysis for the 1990 data that were included for the 1980 data in the UMass study, Been (1995) found that the mean percentage of African Americans in the non-host tracts increased from 13.46 to 15.66%. This resulted in a higher mean percentage of African Americans in the non-host tracts than for the host tracts, although not statistically significant. The most dramatic change was the increased mean percentage of Hispanics from 7.83 to 9.15%, which meant it was no longer statistically significantly different from the host tracts (10.34%). The concern that limiting the control population as was done in the UMass study would increase the comparison benchmark appears to be borne out. As a result of geographic coverage limitation, the minority percentage in the control population would be approximately 3 percentage points higher than without this limitation. However, even without dropping these cases, the control groups (non-host tracts) have a much higher percentage of minorities than in the UCC study. The UCC study and its recent update reported the mean percentage of minority for non-host 5-digit

ZIP code areas as 12.3 and 14.4%, respectively, for 1980 and 1993 (UCC 1987; Goldman and Fitton 1994), compared with 24.2% for the non-host census tracts for 1990 (Been 1995).

Obviously, this is a difference of at least 10 percentage points, and only 3 percentage points could be attributed to the geographic coverage limitation in the UMass study. This demonstrates that the difference in the geographic coverage of the control (or comparison) groups alone does not explain the whole story. In other words, limiting the study to the SMSAs with at least one facility in the UMass study is only one reason for the dramatic difference in research findings. The differences in the units of analysis play some role. It is more reasonable to say that both the units of analysis and the control populations played significant roles in reaching the striking difference in findings.

Although the UMass authors argued that ZIP code areas were too large, critics claimed that census tracts may be too small for representing the impact areas of commercial TSDFs. As discussed in Chapter 6, neither census tracts nor ZIP code areas are ideal units of analysis by random sampling, although census tracts may have a greater chance of being the right size for an impact area between 0.8 and 28 square miles. None of the previous studies has ever examined the size distribution of host areas for commercial TSDFs used in their analysis, whether it is census tract, ZIP code, or MCD. Nor have these studies determined where these TSDFs sites are located in their units of analysis and whether the border effect could render their units of analysis less representative of the true impact area. It is not clear to us whether choice of different units of analysis will bias the results one way or the other for the case of commercial TSDFs.

Regardless of these differences, a census tract-level study (Been 1995) confirms the ZIP-code-based UCC study that there was an inequitable burden of commercial TSDFs on minorities as a whole (see Tables 11.1 and 11.2). The mean percentage of minorities in the host tracts was significantly higher than for the non-host tracts in 1990, although the difference was not as large as was found in the UCC study. The UMass study did not include a variable measuring the minority as a whole and thus is not directly comparable with the UCC and Been studies. The UCC study did not have a break-out of the minority. The UMass and Been studies included blacks or African Americans and Hispanics but offered different pictures. The UMass study found no evidence of any inequity for these two groups in both bivariate and multivariate analyses. However, the Been study showed consistent, inequitable impacts on Hispanics but an inconsistent relationship between African Americans and the location of commercial hazardous waste facilities. A bivariate analysis and a multivariate analysis without the population density variable did not show any distributional disparity for African Americans, but a multivariate analysis with the population density variable indicated otherwise.

The bivariate analyses in the three studies found inequitable distribution by income and class, although using different measures. All multivariate analyses show a reduced role of income in the location of commercial hazardous waste facilities, while having some differences in the results. In the UCC study, mean household income remained statistically significant for the country as a whole and for three out of ten EPA regions. In the UMass study, the percentage of families living below the poverty line was statistically significant but in the wrong direction. In the Been study, median family income and the percentage of persons living below the poverty line were either no longer statistically significant or in the wrong direction.

The Been study did extensive work in improving data quality. Although its data sources are the same as those used in the UCC and UMass studies, it did more work on data quality control. It established a more complete universe of commercial TSDFs by cross-referencing two databases: ESD and EPA's RCRIS.

While these studies examined operational facilities, one study focused on facilities that ceased hazardous waste operations during the 1989–1995 period (Atlas 1998). It explored what motivated these facilities to cease operation: Did political activism in the host communities affect the facilities' closure decision? Did race and income matter in such decisions? The study hypothesized that facility closures were related to the following community characteristics: race, income, education, occupation, length of residence, population levels, government employees, drinking water wells, and children. Using EPA's databases, the study identified a total of 595 commercial treatment, disposal, or recycling facilities of hazardous waste management that operated at some time between 1989 and 1995. The geographic units of analysis are the concentric rings with 0.5- and 1-mi radii surrounding the facilities. Census Bureau GIS software was used to derive socioeconomic variables for the rings based on census tract data. The procedure assumes that socioeconomic characteristics are evenly distributed in a census tract. It calculates the proportions of people residing in each of the two rings based on blocks and uses them as weights to estimate community characteristics for the tracts that partially fall in the rings. Using a logit model structure, the study found no evidence that political activism characteristics of the host communities affected the facilities' closure. The models explained very little of the variation in facilities' statuses. The models did not include facility production and operational variables, which may have affected the facilities' decisions. It is not clear whether incomplete model specification might affect the model estimation results. The longitude and latitude data used in the study are also suspect because there are numerous errors in the database.

11.1.2.2 Regional Studies

With all these limitations to nationwide studies, several studies focused on one county, one metropolitan area, or one state, and made some improvements in some methodological issues. Mohai and Bryant (1992) targeted three counties surrounding the city of Detroit, in order to "examine the relative strength of the relationship of race and income on the distribution of commercial hazardous waste facilities in the Detroit area." They conducted a survey with "a stratified two-stage area probability sampling design" in the 3 counties and an oversample within 1.5 mi of 16 existing and proposed facilities. They obtained race and income data for 793 respondents, 289 of which were within 1.5 mi of existing and proposed facilities. They also measured the distance between these 289 respondents and one of the 16 facilities. The data indicate that for a minority resident in the three-county area, the chance of living within a mile of a hazardous waste facility was about four times as large as that for a white resident. Two multiple linear regressions were used to examine whether race and income each had an independent relationship with the distance to a facility. They found that "(t)he relationship between race and the location of commercial hazardous waste facilities in the Detroit area is independent of income in each of the analyses. And ... it is the race which is the best predictor" (Mohai and Bryant 1992:174). This study overcame some of the limitations associated with census-based geographic units by using the circle approach and a sample survey. However, the regressions failed to take into account some independent variables other than race and income and resulted in a great deal of unexplained variance (adjusted R^2 values of 0.04 and 0.06). As discussed in Chapter 7, incomplete specifications may bias the regression results. In addition, the linear model assumes a linear relationship between exposure and distance from the source: exposure at 1 mi is exactly 10 times that at 0.1 mi. This assumption is hardly plausible and may misrepresent the true relationship (Pollock and Vittas 1995).

Boer et al. (1997) studied the location of hazardous waste TSDFs in Los Angeles County, California. It was not limited to commercial TSDFs as in the previous studies. The small area covered in the study allowed the researchers to make two major improvements: identify the geographic locations of TSDFs more accurately and introduce land use/zoning variables (i.e., percentage of land zoned for residential use, percentage of land zoned for industrial use). Census tract was their geographic unit of analysis. Using both univariate analysis and multivariate logit model, the authors confirmed some of the claims made on both sides of the debate:

1. Race and ethnicity were significantly associated with TSDF location, as suggested by environmental justice advocates;
2. There was a significant association between TSDF location and manufacturing employment and industrial land use, as suggested by critics of environmental justice;
3. Income had first a positive then a negative effect on the probability of a TSDF location.

The authors concluded that "communities most affected by TSDFs in the Los Angeles area are working-class communities of color located near industrial areas" (Boer et al. 1997:793).

11.1.3 METHODOLOGICAL ISSUES

Several data issues complicate a national study of TSDFs. First, the true universe of commercial TSDFs is difficult to identify because each database has different coverage. Previous studies relied mostly on two databases: Environmental Information Ltd.'s Environmental Service Directory (ESD) and EPA's RCRIS database. The ESD tends to understate the universe of commercial TSDFs, by as much as 17% (Been 1995). It also includes some less risky facilities that are not subject to RCRA regulations. The RCRIS database also tends to bias the universe of commercial TSDFs, but for different reasons. The RCRIS database does not have a field to flag "commercial" status, and only has a field indicating whether the facility receives offsite waste. Although this offsite receipt indicator can serve as a substitute for commercial status, it is sometimes missing from the database. The RCRIS can miss true commercial TSDFs by as much as 18% (Been 1995). This may result in an underestimation of the universe of commercial TSDFs. Meanwhile, the RCRIS can also overstate the universe by including facilities that have been closed or in the process of closing. A stratified sample survey of firms in the 1992 RCRIS found that nearly 47% of facilities surveyed were no longer in business, could not be located from reported data, or were incorrectly recorded (Oakes, Anderton, and Anderson 1996). A telephone survey found that about 80 out of 612 commercial TSDF facilities identified in the 1994 RCRIS had closed or were in the process of closing, or no longer had working phone numbers; another forty were not commercial, or did not currently accept hazardous waste for treatment, storage, or disposal, or had never opened.

The true universe of commercial TSDFs is difficult to identify because it changes year by year. EPA's Biennial Reporting System (BRS) contains information about facilities from the Hazardous Waste Reports that must be filed every 2 years under RCRA. The facilities in BRS include Large Quantity Generators of waste and TSDFs for RCRA hazardous wastes on site in units subject to RCRA permitting requirements. BRS data have been collected since 1989. BRS reported 400 treatment and disposal facilities in 1989, 415 in 1991, 371 in 1993, and 333 in 1995 (Atlas 1998). Furthermore, there has been a substantial amount of entries and exits among the facilities. Only 179 facilities operated in each BRS year from 1989 to 1995, comprising 30% of the 595 facilities that operated at any time during the same period. Approximately 29% of the facilities did not operate for 2 consecutive BRS years. At least 25% of the facilities in one BRS year were absent in the next BRS year. One major cause for these changes was the changing definition of a RCRA hazardous waste. In 1990, EPA changed the TCLP and added 25 more chemicals to the original 18 chemicals for which allowable concentration levels had been established. This change resulted in more wastes being classified as hazardous. EPA also defined other types of wastes as hazardous in 1992 and 1995 (EPA 1995c). In 1991, EPA also defined other types of processes as hazardous waste management (EPA 1991b). As a result of these changes, some originally non-hazardous waste facilities became hazardous waste facilities and were required to obtain a permit. Some facilities may have ceased accepting the newly defined hazardous wastes or closed.

The uncertainty and variability in the universe of commercial TSDFs may affect research findings. Been (1995) found that inclusion of those facilities that were not subject to RCRA regulations skewed the results away from a finding that facilities were sited disproportionately in communities of color. How temporal variability in the universe of commercial TSDFs changes the results is not clear.

The true universe of commercial TSDFs at the time of siting is even more illusive to define, complicating any attempt to study the socioeconomic characteristics of host neighborhoods at the time of siting. EPA's RCRIS database reports the date of the facility's existence. This date can be when the facility began its hazardous waste operations, or when construction on the facility began, or when operation is expected to begin (EPA 1996g). Many of the dates in the database are when the facilities first became subject to RCRA regulations, which may be long after the siting date (Atlas 1998). The hazardous waste management permit system was first established in

1980. In addition, legal definitions of hazardous wastes have changed over time. Therefore, the dates reported in RCRIS database could be biased toward the recent years. Indeed, while Been (1997) reported 29 commercial hazardous waste facilities that opened in or after 1990, other siting sources documented far less. In fact, siting experts and industries have been very frustrated with the siting impasse since the early 1980s. In 1981, EPA predicted that 50 to 125 large facilities would be needed to avert a capacity crisis. McCoy and Associates, perhaps the best source of siting information, reported that not one single facility was sited from 1983 to 1986. After 1986, a few new facilities came on line. Only one new hazardous waste land disposal facility (in Last Chance, Colorado) and fewer than ten new hazardous waste treatment and incinerator units were reported (Gerrard 1994). "Although no one seems to know the exact number of successful sitings that have taken place, it is quite certainly far short of the 50 to 125 large facilities" predicted by EPA (Szasz 1994: 114–115).

The negative impacts of this data problem are at least twofold: First, it is biased toward recent years, which could be long after the actual siting date. This makes any conclusion from an analysis of siting disparity based on such data unreliable. Second, it could result in a lumping together of facilities that were originally sited for hazardous or non-hazardous waste management. To the extent that siting processes and decisions may be different for hazardous and non-hazardous waste facilities, the analysis would be like comparing apples and oranges. The ultimate impacts of these biases on research findings need further investigation.

Although most studies focused on the current association between hazardous waste facilities and host-community characteristics, few studies examined whether inequity or lack thereof was also true when the facilities were sited. Cross-sectional studies answer the question of whether there is an association between location of environmentally risky facilities or LULUs and society's disadvantaged groups at the time of data point. Longitudinal studies explore the question of how the association has changed over time, particularly since the facility siting time. Both types of studies were important for design of effective public policies for remedying any environmental injustice. The first type of studies tells us whether there is inequitable distribution of environmental risks that need policy intervention, but it tells us little about how any inequity comes into being and how government should intervene. The second type of studies could answer the question of whether any siting bias contributed to the inequity. If yes, siting policies may be justified to ensure a fair share of environmental burdens across society. We will discuss dynamics analysis in Chapter 12.

11.2 EQUITY ANALYSIS OF CERCLIS AND SUPERFUND SITES

11.2.1 CERCLIS AND SUPERFUND SITES

The Comprehensive Environmental Response, Compensation, and Liability Act of 1980 (CERCLA), as amended by the 1986 Superfund Amendments and Reauthorization Act (SARA), regulates inactive and abandoned hazardous waste sites. CERCLA authorizes EPA to identify contaminated hazardous waste sites, and EPA maintains an inventory through the Comprehensive Environmental Response, Compensation, and Liability Information System (CERCLIS). The most dangerous sites that pose a "substantial health threat" to human, are placed on the National Priority List (NPL) for cleanup under the Superfund program. These sites are commonly known as Superfund sites. To be on the NPL, a site has to undergo a discovery process and a screening and prioritization process. During the discovery process, EPA is notified of a potential dangerous site, starts its investigation, and records the site in the CERCLIS. Then, EPA conducts a preliminary assessment on the site's potential risk. After the preliminary assessment and site investigation, the sites are screened using the Hazard Ranking System (HRS). Sites with scores greater than 28.5, an arbitrary threshold, are placed on the NPL. Alternatively, States and Territories can designate one top-priority site regardless of HRS score. Once placed on the NPL, a site generally proceeds through the remedial program. The Superfund cleanup process consists of the following steps:

- Preliminary Assessment/Site Inspection (PA/SI)
- HRS Scoring
- NPL Site Listing Process
- Remedial Investigation/Feasibility Study (RI/FS)
- Record of Decision (ROD)
- Remedial Design/Remedial Action (RD/RA)
- Construction Completion
- Operation and Maintenance (O&M)
- NPL Site Deletions

HRS is a numerical screening system that uses the information from the preliminary assessment and site inspection to assess the relative potential risk of sites (EPA 1992c). It considers three categories of factors and four pathways. The three factor categories are

- Likelihood that a site has released or has the potential to release hazardous substances into the environment
- Characteristics of the waste (e.g., toxicity and waste quantity)
- People or sensitive environments (targets) affected by the release

The following four exposure pathways are scored and combined using a root-mean-square equation:
- Groundwater migration (drinking water)
- Surface water migration (drinking water, human food chain, sensitive environments)
- Soil exposure (resident population, nearby population, sensitive environments)
- Air migration (population, sensitive environments)

The ROD is an important milestone in the Superfund site cleanup process. It is a public document that explains which cleanup alternatives will be used to clean up a Superfund site. It is created from information generated during the RI/FS.

11.2.2 Hypotheses and Empirical Evidence

CERCLIS and Superfund sites raise environmental justice concerns that are different from other noxious facilities. CERCLIS and Superfund sites reflect historical practices of private and public sectors in dealing with facilities with hazardous potentials. If these sites were once manufacturing plants, they reflect the then siting outcome. If these sites were once dumping grounds, they demonstrate the historical practice of waste management. In either case, Superfund sites are the results of past practice and are identified as posing a threat to human health or the environment. The surrounding communities were exposed to these actual risks at these sites until Superfund cleanup. Once cleaned up, Superfund sites no longer pose any unacceptable risk. In this regard, Superfund sites are different from TSDFs, which are regulated and might not expose the surrounding communities to actual hazardous wastes. Therefore, Superfund sites reflect actual environmental risks (of the pre-cleanup periods) more accurately than TSDFs. The spatial distribution of Superfund sites indicates the distribution of risk burdens on different population groups. Any disproportionate distribution of Superfund sites constitutes environmental inequity. One hypothesis is that an inequitable distribution of Superfund sites results from race or class biases in historical siting and dumping practices. To the extent that CERCLIS sites pose any potential risks, this hypothesis also applies to the CERCLIS sites.

Another hypothesis is that the NPL designation and cleanup processes reflect the political power of different population groups. In particular, since minority and poor communities tend to be politically powerless and disenfranchised, they have little ability to exercise their influence on the NPL designation and Superfund cleanup processes. For the NPL designation process, inequity may be suggested if NPL sites have a smaller proportion of minorities and the poor (Anderton, Oakes, and Egan 1997). Unique to Superfund sites are the clean-up processes that involve government, host communities, and responsible parties. As hypothesized, any disparity in the pace of Superfund cleanup reflects unequal enforcement of federal laws and regulations. Such inequity may be indicated if minority and poor host communities are less likely to have a ROD (Zimmerman 1993), or a longer remedial time.

Studies have examined both the distributional patterns of CERCLIS and Superfund sites and potential biases in NPL designation and cleanup progresses (see Table 11.3). Four national studies analyzed the spatial patterns of CERCLIS or NPL sites using different units of analysis such as ZIP codes (UCC 1987), county (Hird 1993), Census Places or MCDs (Zimmerman 1993), and census tracts (Anderton, Oakes, and Egan 1997). They did not find income inequity, but offered mixed evidence about distributional disparity by race.

The second part of the UCC report (1987) focused on the distribution of CERCLIS sites. It was descriptive, with its primary purpose being to document the presence of uncontrolled toxic waste sites in racial and ethnic communities. The study found that 3 out of 5 five African- and Hispanic-Americans (57.1 and 56.6%, respectively) and approximately half of all Asian-Pacific Islanders and American Indians (52.8 and 46.4%, respectively) lived in communities with uncontrolled toxic waste sites. Overall, more than half of the nation's population (54%) resided in such

TABLE 11.3
CERCLIS and Superfund Studies

	UCC 1985	Hird (1993) 1989	Zimmerman (1993) 1990	Anderton, Oakes, and Egan (1997) 1995
Data Year				
Facilities	18,164 CERCLIS sites	788 NPL sites	825 NPL sites	15,427 CERCLIS sites, of which 1392 are NPL sites
Unit of analysis	5-digit ZIP Code	County	Census Places/MCD	Census tracts
Universe	Continental U.S.	3139 counties	Continental U.S.	61,258 census tracts
Number of host areas	7975	Over 500 (estimated)	622	9,093 CERCLIS tracts including 1088 NPL tracts
Control areas	U.S. States Metropolitan Areas	Over 80% of counties without any NPL sites	The U.S. 4 Census Regions	About 59,000 non-CERCLIS tracts; 47,000 non-CERCLIS tracts in MAs or rural counties with at least one CERCLIS site; 8000 non-NPL CERCLIS sites.
Variables				
Race	Minority Black Hispanic Asian/Pacific Islander American Indian	Nonwhite	Blacks Hispanics	Blacks Hispanics Native Americans
Income		Percentage of residents below the poverty line	Percentage of persons below the poverty level Per capita income Household income	Percentage of families below the 1989 poverty line Percentage of households receiving public assistance income

continued

TABLE 11.3 (CONTINUED)
CERCLIS and Superfund Studies

Control variables		Amount of hazardous waste generated at the state level, % manufacturing, % college educated, % owner-occupied housing, % unemployed, Median housing value, Population density HRS score, State priority, Federal site, Year final on NPL, Congressional subcommittees	HRS score, Controversy, Congressional voting average, # NPL sites-same county, -same city, Area, population, median house value, Population density, % owner occupancy, % change in population, % 12 or more years schooling	Mean value of owner-occupied housing, % employed in manufacturing and industry, % males in the civilian labor force who are employed, % Persons with 1+ year of college, Total Persons, Density
Statistical methods	Population-weighted average	Tobit model, Ordered probit model	Arithmetic mean, Population-weighted average, Probit model	t-test, Poison regression, Cox proportional hazards regression
Inequity by race/ethnicity?	Yes	Yes for distribution of NPL sites, No for the pace of NPL site cleanup	Yes for blacks and Hispanics in distribution of NPL sites (weighted averages), No based on unweighted averages, Yes for blacks for ROD decision	No for distribution of CERCLIS and NPL based on bivariate analysis, Yes but small for the incidence of CERCLIS sites based on multivariate analysis, Yes but small for blacks in the likelihood and pace of NPL designation
Inequity by income?		No.	No.	No for distribution of CERCLIS and NPL, Yes in the likelihood and pace of NPL designation
Is race/ethnicity more significant than income?		Yes.	Yes.	No.

Sources: UCC (1987); Hird (1993); Zimmerman (1993); Anderton, Oakes, and Egan 1997.

communities. Also, at the aggregate level, 56.3% of the minority population lived in communities with uncontrolled toxic waste sites, compared with 53.6% for whites. The study concluded that race was an important factor in describing the distribution of uncontrolled toxic waste sites. These numbers represent population-weighted averages. The strengths and weaknesses of this statistic were discussed in Chapter 7. Unlike its analysis of the commercial TSDFs, the UCC study of CERCLIS sites did not employ statistical methods to test the statistical significance of differences.

Hird (1993) used a Tobit model to examine the distribution of NPL sites by county as a function of the amount of hazardous waste generated, potential political mobilization, and socioeconomic characteristics, after controlling for the urban/rural differences and the pre-Superfund residential growth. A consistent model result is that the economically advantaged counties were more likely to have more NPL sites, contrary to most expectations. In addition, NPL sites were likely to be located in manufacturing counties with a higher percentage of nonwhite or college-educated residents. A separate model for urban counties alone results in the insignificance of race and manufacturing variables.

Zimmerman (1993) examined the spatial distribution of Superfund sites by focusing on over 800 NPL sites out of 1,090 non-military and non-DOE NPL sites in the continental U.S. This set excluded those sites in rural areas whose community populations were below 2,500 in 1980. The geographic units of analysis are Census Places, or MCDs where places do not exist. Comparison populations were census-defined geographic regions and the nation. Two types of statistics indicated different results. The arithmetic means of socioeconomic variables across the host communities were comparable to those of the regions where the communities are located. On the other hand, the percentage of blacks and Hispanics in the NPL-host communities aggregated as a whole was larger than the national figures (18.7 vs. 12.1% for blacks, 13.7 vs. 9% for Hispanics, respectively). This disparity was attributed to a few large urban areas where minority populations were overrepresented. No disparity was found for the poverty population.

Hird's and Zimmerman's studies provide analytical insights on Superfund programs, but they both suffer from geographic units of analysis that are too large. As indicated in Chapter 3, the county is so large a geographic unit that any site-based equity analysis using it as a unit of analysis runs the risk of committing an ecological fallacy. As discussed in Chapter 6, Census Places vary widely in terms of population and area sizes. In this case (Zimmerman 1993), Census Places or MCDs with NPL sites have a wide range of both population and area size: a median 1990 population of 17,929 and mean population of 87,945 with a standard deviation of 277,811; a median area size of 15.2 square mi and mean area of 39.4 with a standard deviation of 94.2 square mi. Clearly, the data have skewed distributions, and the arithmetic mean statistic is skewed because of extreme values.

Anderton, Oakes, and Egan (1997) addressed the equity concerns about the spatial distribution of 1,5427 CERCLIS sites and 1,392 NPL sites (a subset of CERCLIS sites) at the census-tract level as of July 1995. Two comparison groups were employed: all other (non-CERCLIS or non-NPL) census tracts in the country and all other tracts in metropolitan area or non-metropolitan counties where there was at least one existing site. The results showed that CERCLIS sites were located in census tracts that were typically less black (11.6% for host tracts vs. 13.7% for all other tracts vs. 14.1% for all other tracts in metropolitan or rural counties with at least one CERCLIS site), less Hispanic (6.97 vs. 7.99 or 8.29%, respectively), but more Native American (1.2 vs. 0.81 or 0.75%, respectively). The CERCLIS or NPL host neighborhoods had a smaller percentage of college-educated residents, a lower average value of owner-occupied housing, and a lower population density, which are contrary to previous findings for NPLs at the County or Places/MCD level. Similar to the findings of Hird (1993), the CERCLIS sites had a higher percentage of residents employed in industrial sectors. In short, neither racial nor income biases were found in the discovery stage of CERCLIS sites. Similar patterns were found for the NPL-host tracts as compared with all other tracts in the country.

Their multivariate analyses produced mixed results with respect to racial bias (Anderton, Oakes, and Egan 1997). Poisson regression models were used to examine the relationship between the number of CERCLIS and NPL sites in a neighborhood and neighborhood characteristics. After controlling for residential density and metropolitan area designation, the model shows that the average percentage of Native Americans, and to a much lesser degree, blacks and Hispanics was positively associated with the number of CERCLIS sites in a neighborhood. In contrast, the average percentage of poor families and to a much lesser degree, blacks, was negatively associated with the number of NPL sites in neighborhoods with CERCLIS sites. None of these effects, however, are substantive. The single substantively large effect is from the metropolitan area indicator: being in a metropolitan area increased the number of CERCLIS sites by nearly 38%. Overall, both models have little predictive power (0.02 to 0.06 for pseudo R^2), which raises questions about potential model misspecifications (see Chapter 7).

These three studies also examined the second hypothesis in the context of the Superfund cleanup pace, and two of them found some evidence of racial bias. Hird (1993) constructed an ordered probit model to represent the three progressive stages (RI/FS, ROD, and RA) as a function of the site's HRS score, state and congressional political influence, socioeconomic characteristics, and residents' potential political mobilization. Control variables included the year when the site was designated as final on the NPL, whether federal money was the principal cleanup fund, and if it was a federal facility. The model results showed that the most important variables in explaining the cleanup progress were the HRS scores, a federal fund, a federal facility, and the year for the site designation. The pace of cleanup had no relationship to the county's socioeconomic characteristics, including race and income.

However, Zimmerman (1993) found some evidence of racial bias. She used a probit model to examine the relationship between ROD status and socioeconomic characteristics of host communities. The higher the proportion of blacks in the host communities, the less likely the site had a ROD. The opposite was true for Hispanics. The probit model did not control for the length of time during which a site had been on the NPL. Descriptive statistics indicate that NPL sites designated earlier were more likely to have RODs and more likely to have lower proportions of blacks in the host communities. This implies that early NPL designation process may have some bias.

Similarly, Anderton, Oakes, and Egan (1997) found plausible but substantively small racial bias in the NPL designation and remediation processes. They addressed whether there was any evidence that poor or minority neighborhoods were less likely to have NPL designation from among CERCLIS sites in a timely fashion. A proportional hazard regression analysis suggested some plausible biases in the likelihood and pace of a NPL designation. Other things being equal, a higher percentage of blacks or poor families decreases the likelihood and pace of NPL designation. These multivariate analysis results are different from those from bivariate analyses. Comparison of NPL-host tracts and non-NPL CERCLIS showed that NPL neighborhoods had significantly less blacks and Hispanics and fewer families below the poverty line, were less densely populated, but more well-educated and had a higher average housing value than non-NPL CERCLIS neighborhoods. These bivariate analyses did not provide evidence for hypothesized bias in the distribution of NPL sites because of the prioritization process.

A few studies focus on states or metropolitan areas. Stretesky and Hogan (1998) investigated the relationship between 53 Superfund sites and socioeconomic characteristics of host communities in the State of Florida. Their geographic units of analysis were based on census tracts. Two groups were used to represent the host communities: census tracts with at least one NPL site (totaling 49); census tracts with at least one NPL site and those adjacent tracts (totaling 276 tracts). Comparison groups were all other tracts, 2356 and 2129, respectively, for the two groups. Bivariate analysis indicated that Superfund tracts had a higher percentage of blacks (22 vs. 15%), Hispanics (16.9 vs. 9.3%), and the poor (16.5 vs. 12.3%) than non-Superfund tracts. After controlling for urban indicator, population density, median housing value, and median rent, percentage blacks and Hispanics were still statistically significant for predicting the presence or absence of Superfund sites. Income variables, however, were no longer significant. Longitudinal analysis was used to examine the racial and ethnicity changes in the Superfund tracts over the years 1970, 1980, and 1990. The percentage of blacks and Hispanics in the Superfund tracts increased between 1970 and 1990. A logistic regression for 1980 indicates that race and ethnicity were much weaker predictors of the presence of Superfund sites in 1980 than in 1990. The authors concluded that environmental injustice does exist in Florida and its likely cause is indirect, rather than direct, forms of discrimination. Like Anderton, Oakes, and Egan (1997), both models have so little predictive power (0.1 for R^2) that one wonders about potential model misspecifications.

11.2.3 Methodological Issues

Like TSDF studies, several methodological issues confront Superfund studies. Previous research has failed to deal with the issues of impact areas and border effect, which are discussed in Chapter 6. Although it is encouraging to see that recent Superfund studies use smaller geographic units of analysis than early studies, these efforts have seldom taken into account the potential impact boundary of Superfund sites and the relative location of Superfund sites within existing census geography. EPA has considerably enhanced the GIS database, including the Superfund site boundary. This database is useful for delineating more accurate geographic units of analysis for environmental justice research.

Like TSDFs, CERCLIS and Superfund sites are a moving target. Any study with a specific data year is a snapshot. More challenging than other facilities is that CERCLIS and Superfund sites are "discovered" and then "deleted" after cleanup. This means it is possible that some potential hazardous sites are still unidentified and thus missing from the existing database. To the extent that this unknown set of sites represents different spatial patterns from the existing one, this measurement error could distort the truth about the distributional impacts of CERCLIS and Superfund sites. Although we have no way of knowing the exact direction for this potential bias, we can take a look at how discovery in the past has affected research findings over the years.

Table 11.4 records the number of CERCLIS and Superfund sites each year since 1980. The number of CERCLIS sites steadily increased from 8689 in 1980 to 39,099 in 1994. The 1995 data reflect the removal of over 24,000 sites from the Superfund inventory as part of EPA's Brownfields initiative to help promote economic redevelopment of these properties. The number of Superfund sites also steadily increased from 160 in 1982 to 1374 in 1995. The pace of Superfund cleanup has picked up since mid-1990s. Once sites are cleaned up, they are deleted from the NPL. Meanwhile, new sites are proposed to the NPL. As of the end of fiscal year 1997, there were 1405 total NPL sites. As of December 8, 1999, 10,589 CERCLIS sites were active, and another 31,467 were archived for the country as a whole. Possessions had an additional 188 active sites and 422 archived sites. No research has examined how this dynamic process affects our understanding of the relationship between these sites and socioeconomic characteristics of host communities. It is worth some attention in future research.

11.3 EQUITY ANALYSIS OF TOXICS RELEASE FACILITIES

11.3.1 Toxics Releases Inventory

The Emergency Planning and Community Right-to-Know Act of 1986 mandated establishment of the Toxics Release Inventory (TRI). The law, also known as Title III of the Superfund Amendments, has two purposes: to encourage planning for response to chemical accidents and provide the public with information about possible chemical hazards in their communities. It requires certain manufacturers to report to EPA and the States the quantities of over 300 toxic chemicals that they release directly to air, water, or land, or that they transport to offsite facilities. EPA compiles these reports into an annual inventory—the TRI— and makes it available to the public in a computerized database.

A facility is required to report if it

- Has ten or more full-time employees, and
- Manufactures or processes over 25,000 pounds of the approximately 600 designated chemicals or 28 chemical categories specified in the

TABLE 11.4
CERCLIS and NPL Sites

Year	CERCLIS sites	NPL sites
1980	8,689	na
1981	13,893	na
1982	14,697	160
1983	16,023	551
1984	18,378	547
1985	22,238	864
1986	24,940	906
1987	27,274	967
1988	29,809	1,196
1989	31,650	1,254
1990	33,371	1,236
1991	35,108	1,245
1992	36,869	1,275
1993	38,169	1,321
1994	39,099	1,360
1995	15,622	1,374
1996	12,781	1,210
1997	9,245	1,194

Notes: CERCLIS = Comprehensive Environmental Response, Compensation, and Liability Information System. NPL = National Priorities List. The 1995 data reflect the removal of over 24,000 sites from the Superfund inventory as part of EPA's Brownfields initiative to help promote economic redevelopment of these properties.

Sources: U.S. Environmental Protection Agency. 2000. *Superfund Cleanup Figures.* Office of Emergency and Remedial Response. http://www.epa.gov/superfund/whatissf/mgmtrpt.htm. Accessed 2/14/2000.

U.S. Environmental Protection Agency. 2000. Inventory of CERCLIS and Archive (NFRAP) Sites by State as of December 8, 1999. Office of Emergency and Remedial Response. http://www.epa.gov/superfund/sites/topics/archinv.htm. Accessed 2/14/2000.

Council on Environmental Quality. 1997. Environmental Quality. Washington, D.C.

regulations, or uses more than 10,000 pounds of any designated chemical or category, and

• Engages in certain manufacturing operations in the industry groups specified in the U.S. Government Standard Industrial Classification Codes (SIC) 20 through 39, or

• Is a Federal facility in any SIC code.

Standard Industrial Classification (SIC) primary codes 20 to 39 include among others, chemicals, petroleum refining, primary metals, fabricated metals, paper, plastics, and transportation equipment. In May 1997, EPA added seven new industry sectors that will report to the TRI for the first time in July 1999 for reporting year 1998 (see Table 11.5).

TABLE 11.5
TRI Program Expansion

Type of Expansion	Reporting Year	Description
Chemical expansion	1993	Addition of certain Resource Conservation and Recovery Act (RCRA) (58 FR 63500) chemicals and hydrocholorofluorocarbons (HCFCs) (58FR 63496)
	1995	Addition of 286 chemicals and chemical categories
	2000	Adding seven chemicals and two chemical compound categories
Facility expansion	1994	Inclusion of any Federal facility in any SIC
	1998	Addition of seven industry sectors: • Metal mining (SIC code 10, except for SIC codes 1011, 1081, and 1094) • Coal mining (SIC code 12, except for 1241 and extraction activities) • Electrical utilities that combust coal and/or oil (SIC codes 4911, 4931, and 4939) • Resource Conservation and Recovery Act (RCRA) Subtitle C hazardous waste treatment and disposal facilities (SIC code 4953) • Chemicals and allied products wholesale distributors (SIC code 5169) • Petroleum bulk plants and terminals (SIC code 5171) • Solvent recovery services (SIC code 7389)
Threshold reduction	1990	Threshold for manufacturing and processing was reduced to 25,000 pounds
	2000	Lowering reporting thresholds for 18 chemicals and chemical categories that meet the EPCRA section 313 criteria for persistence and bioaccumulation. Finalizing two thresholds based on the chemicals' potential to persist and bioaccumulate in the environment: 100 pounds for PBT chemicals and to 10 pounds for that subset of PBT chemicals that are highly persistent and highly bioaccumulative.
Program expansion	1991	Inclusion of reporting pollution prevention activities: amounts of chemicals that are recycled, used for energy recovery, and treated on-site.

Source: U.S. EPA (1999b)

In 1989, EPA released the first year of data, for 1987 (U.S. EPA 1989b). Since then, the TRI program has expanded considerably (see Table 11.5). The 1987 TRI covers manufacturing facilities that produced, imported, or processed 75,000 or more pounds of any of the 328 TRI chemicals, or otherwise used 10,000 pounds or more of a TRI chemical. The first threshold has been reduced to 25,000 pounds. The number of TRI chemicals has been increased to over 600. Federal facilities have been added. Seven industry groups have been added to the 1998 TRI. EPA has proposed lowering the EPCRA section 313 reporting thresholds for certain persistent, bioaccumulative toxic (PBT) chemicals and to add certain other PBT chemicals to the section 313 list of toxic chemicals.

The TRI data are available online and in a variety of computer and hard-copy formats. Online access includes the National Library of Medicine's TOX-NET system, the Right-to-Know Network (RTK NET), and EPA's Environfacts system. TRI program information is available on the EPA's TRI Web site at http://www.epa.gov/opptintr/tri.

The TRI database includes information about:

- What chemicals were released during the preceding year
- How much of each chemical went into the air, water, and land
- How much of the chemicals were transported away from the reporting facility for disposal, treatment, recycling, or energy recovery
- How chemical wastes were treated at the reporting facility
- The efficiency of waste treatment
- Pollution prevention and chemical recycling activities

The TRI database has several caveats:

1. TRI reports only the amount of release and transfer, representing neither exposure nor risks posed by TRI chemicals. TRI chemicals transferred offsite may undergo treatment before final disposal, which may result in a lesser amount and toxicity than the original transfers. Furthermore, the toxicity of TRI chemicals ranges widely.
2. TRI does not cover all manufacturing activities. Small manufacturing facilities are missed in the database, including those with fewer than ten full-time employees and those manufacturing, processing, or otherwise using the listed chemicals below threshold amounts.
3. TRI does not cover all sources of toxic releases. Besides the manufacturing industry, non-manufacturing industrial processes, use and disposal of consumer products, agricultural uses of chemicals, and mobile sources also generate toxic wastes but are not covered in TRI.
4. TRI data are estimates rather than actual measurements. Facilities are not required to measure or verify their data.
5. Companies can claim the chemical identity as trade secret, using a generic chemical name.
6. Companies have a learning curve in TRI reporting and gradually improve the quality of the reported data.

Taken together, there is a tendency for TRI to underestimate real toxics releases. Certainly, TRI represents a small part of all emissions. It is important to emphasize that by nature, TRI only records toxics.

11.3.2 NATIONAL STUDIES AND EVIDENCE

Environmental justice studies using the TRI have been increasing rapidly over the past few years. The rich database from the TRI, although its quality varies and is uncertain, shows potential for environmental justice analyses being much richer than identifying the spatial distribution of facility location. National studies have been conducted with geographic units of analysis at the county level (Perlin et al. 1995) and at the ZIP-code level (Brooks and Sethi 1997; Ringquist 1997).

Unlike the mixed results in the Superfund studies, TRI studies ususally found racial inequities in the spatial distribution of TRI facilities. Perlin et al. (1995) showed disparities in county-level TRI air toxics emissions by race/ethnicity and, to a lesser extent, by household income. Ringquist (1997) found racial bias — more than income bias — in the distribution and density of TRI facilities and in the concentration of TRI pollutants at the ZIP-code level. Brooks and Sethi (1997) reported a significant and positive relationship between the proportion of blacks and potential exposure.

Clearly, from our discussion in previous chapters, these studies have inadequate geographic units of analysis. They use the proximity assumption and emission measures, a crude proxy that fails to account for actual environmental impacts (see Chapter 4). Although these studies have various limitations, they offer us some analytical insights and potential for improvement. Ringquist (1997) showed that different statistical models were useful for different concerns: probit models for examining the presence or absence of TRI facilities; negative binomial event count models for explaining the number of TRI facilities; truncated negative binomial event count models for explaining facility clustering; weighted least-square regressions for describing the total weight of TRI pollutants; and marginal effects analysis for ranking the relative importance of independent variables. Overall, control variables such as manufacturing employment, private wells, median housing age, and urbanization were the most powerful predictors for the presence, number, and emission magnitude of TRI facilities. Race/ethnicity variables were secondary, and class attributes were the least powerful.

Perlin et al. (1995) used an emission index approach. The population emission index (PEI) is a population group-specific, average air emission value:

$$\text{PEI}_s = \sum_c R_c n_{cs}/N_{CS}$$

where R_c is the emission of TRI chemicals released in county c; n_{cs} is the number of people in population group S in county c; C is the set of all counties of interest, and N_{CS} is the total population in population group S in the set of all counties C. This index gives more weight to counties with a large population base of a particular population group and/or a large emission rate.

The population emission ratio (PER) for a population group can be calculated by comparing the PEI for that group (e.g., the low-income group) to the PEI for the reference group (e.g., the middle-income group). Disparity is suggested if the PER is considerably larger than 1. A county has a high PER if:

1. The county has large emissions relative to the reference emission value and a large population group c relative to the reference group, or
2. The county has low emissions relative to the reference emission value and a low population group c relative to the reference group.

The reference emission value may be the median emission value for whites for all counties in the study area. In other words, disparity will occur if minorities are overrepresented in high emission counties or underrepresented in low emission counties. Results show that less than 10% of the 3137 counties in the United States satisfy the first condition for any racial/ethnic group. However, these counties, most of which are urban, have more than 43% of the national population of blacks, Asians, other races, and Hispanics.

These groups are more likely to reside in a county with higher TRI air releases than whites. Meanwhile, higher TRI air emission counties tend to be higher in annual household income than counties with median household income for the nation. This finding that both minorities and the rich are positively associated with pollution is consistent with those from county-level studies (Gelobter 1992; Hird 1993; Liu 1998). This finding is valid if and only if the underlying assumption is met that everyone in a county experience roughly the same level of impact from the environmental problem of interest. This holds for relatively ubiquitous environmental problems such as ozone but is problematic for environmental problems whose impacts are localized, such as some air toxics.

The emission index approach is an important first step in moving toward exposure-based methodology for environmental justice analysis, although emission is among the poorest measures of exposure (see Chapter 4).

An improvement over the emission measure is to take into account the toxicity of TRI chemicals. Brooks and Sethi (1997) constructed an index that incorporates toxicity of TRI chemicals and distance from the emission source. The so-called index of exposure for ZIP code i was defined as

$$X_i = \sum_{k \in N(i)} E_k \, (s - d_{i,k})/s$$

where $E_k = \sum_j E_{kj}/T_j$, the toxicity-weighted emissions for ZIP code k; E_{kj} is the emission of chemical j from ZIP code k; T_j is the Threshold Limit Value (TLV) of chemical j (which is the amount of airborne concentration in mg/m^3 of a chemical to which a worker may be exposed in a normal 8-h workday and a 40-h workweek without adverse health effects); $N(i)$ is the set of neighboring ZIP codes whose centroids lie within a distance s of the centroid of ZIP code i; $d_{i,k}$ is the distance between the respective centroids of ZIP codes i and k. Since sulfuric acid has a TLV of one, this index can be interpreted as "weighted pounds of sulfuric acid equivalent."

The index is really a toxicity-weighted, distance-weighted emission measure that incorporates the influence of toxicity and neighboring ZIP-code areas and does not represent exposure as defined in Chapter 4. Although it is still a poor substitute for exposure, it is better than the emission amount measure.

Subpopulation-weighted means showed that racial and ethnic groups and the poor tended to be at high risk of exposure to air toxics. Meanwhile, they benefited the most from reduction of toxics release during the 1988 to 1992 period. Results also showed a U-shaped relationship between income and the potential exposure index, whereby the poorest had the highest exposure, followed by some high income groups.

After controlling for various socioeconomic and political factors in a log-linear regression, a significant and positive relationship was found between the proportion of blacks and potential exposure. Income measures (proportion of persons below poverty and median household income) have U-shaped relationships of the opposite kind with the potential index. The effect of poverty on potential exposure, while U-shaped, is positive for the variable value above 36% and negative otherwise. The threshold was deemed too high to be plausible. The effect of median household income on potential exposure is negative for the variable value above $67,000 and positive otherwise. Opposite to the descriptive statistics, this inverse U-shaped relationship indicates that both the highest income groups and the lowest income groups experience lower potential exposure. This inverse U-shaped relationship was also confirmed in the case of toxic releases in Los Angeles (Sadd et al. 1999a). Representing collective action participation, voter turnout has a negative effect on exposure, with 1% larger voter turnout leading to 3% lower exposure. Socioeconomic characteristics and political participation also explain the changes in potential exposure between 1990 and 1992. Again, a positive relationship was found between proportion of blacks and change in potential exposure, while a negative relationship existed between voter turnout and change in potential exposure.

11.3.3 REGIONAL STUDIES AND METHODOLOGICAL IMPROVEMENTS

Like their TSDF counterparts, regional studies focus on smaller geographic areas such as one county, one metropolitan area, one state, or one census region. Small areas make it feasible to collect more detailed or accurate data and allow some innovative improvements in some methodological issues.

Pollock and Vittas (1995) investigated the functional forms of exposure with respect to distance in the context of TRI in Florida. Three functional forms were explored: linear, square root, and natural log. They argue that the linear function is less plausible for representing the proximity-exposure relationship than others. It makes sense to assume that those living nearby bear the largest brunt of impacts from a noxious facility. As discussed in Chapter 4, evidence shows that economic impacts decline nonlinearly from the facility. However, we do not have an *a priori* theoretical basis on how steep this decline is. The square root function represents a mild decline slope, but the logarithmic function has a steep one. Pollock and Vittas (1995) chose the logarithmic function of proximity as a measure of exposure. A linear model was run to regress this exposure on seven background variables at the block-group level. High exposure areas are more densely populated, more urbanized, and having higher proportions of workforce in manufacturing and wholesaling. They tend to have lower rent and housing values. Residual analyses indicate that low-income African Americans were at the highest risk of exposure to TRI facilities in Florida. While this study improves over previous proximity-based research, it fails to incorporate the TRI information on the types and magnitude of toxic releases. As a result, at the same distance to two release sources with dramatically different emission amount and chemical toxicity, the same potential exposure is assumed.

Using the toxicity-weighted emissions based on the TLVs, Bowen et al. (1995) conducted a statewide assessment of spatial equity at the county level for the State of Ohio and an intra-metropolitan assessment of tract-level data for Cuyahoga County. The statewide analyses of the 1987 to 1990 TRI data found a high correlation between proportion of minority and level of toxic release at the county level. However, the tract-level analyses did not show such relationship but reported income inequity. Recognizing the boundary effect, the authors classified census tracts into three groups: "clean" tracts without TRI facilities in them or in adjacent tracts; "potentially exposed" tracts with TRI facilities in adjacent tracts only; "dirty" tracts with one or more TRI facilities. Moran's I statistics were used to test for spatial autocorrelation in the tract-level partial correlation analysis. First, spatial differences were employed to correct the spatial autocorrelation, if found. Zero-order correlation analysis indicates negative association between population density and emissions and toxicity-weighted emissions, while partial correlation analysis shows mostly negative but insignificant relationships. Concurring with Anderton et al. (1994), Bowen et al. (1995) argued that urban/rural differences confounded the analyses because of coincident concentrations of industry, minority populations, and toxic releases in urban counties. They concluded that a county-level analysis is not appropriate for equity analysis. In addition, the author hypothesized a "U"-shaped relationship, whereby "older toxic-waste sites are located in inner-city urbanized areas with high incidences of minority and poor housing, and the newer waste sites are located near higher income suburbs" (Bowen et al. 1995:658).

The urban/rural dichotomy or the urban analysis approach provokes a debate (Mohai 1995; Downey 1998). Critics of the urban-focused equity studies contend that these studies fail to account for why minorities and the poor are concentrated in the urban cores in the first place (Downey 1998). This narrow focus also fails to compare the population at large. Based on this argument, a statewide model is favored over a citywide model. Using the ZIP-code level data, Downey (1998) estimated linear models of TRI emissions on race and income for the State of Michigan, urban areas, and the Detroit Metropolitan Areas. The results are quite similar to those of Bowen et al. (1995). The statewide model shows race to be a better predictor than income of TRI emissions, while urban models show otherwise. But interpretation of the results is dramatically different. According to the environmental justice definition based on the institutional models (see Chapter 1), Downey (1998) argues that a racial disparity in the distribution of environmental hazard is evidence of environmental racism, no matter whether other variables are controlled for. Therefore, the findings that race is significant in the statewide analysis and in the bivariate urban analysis are consistent with the claim of environmental racism based on the institutional model. Downey (1998) also argues that the race-vs.-income debate is moot because race and income are highly correlated and it is very difficult to separate them. It should be noted that only race and income are taken as independent variables in the multiple regressions. Like Mohai and Bryant (1992), the model specification is incomplete and has certainly missed relevant variables. As discussed in Chapter 7, this model misspecification would result in biased estimates of model coefficients.

11.3.4 Methodological Issues

As indicated earlier, the TRI database provides a unique opportunity to move beyond the proximity-based approach to equity analysis. As reviewed above, a couple of studies have incorporated toxicity data in their analyses. The TLV used in these studies, however, is not intended for use in an environmental setting (Jia, Guardo, and Mackay 1996). TLV (developed by the American Conference of Governmental Industrial Hygienists) and Permissible Exposure Limits (developed by the Occupational Safety and Health Administration) take into account economic cost and technical feasibility. Their use for risk assessment is considered to be scientifically questionable. EPA has been moving in the direction of incorporating toxicity information in its projects. In the Sector Facility Indexing Project (SFIP), EPA proposed to combine TRI data with toxicity risk factors and make the information available to the public via the Internet. The simple toxicity weighting, however, has its limitations. It fails to consider any exposure data such as acute or ecological effects. From emission to exposure, we need further information about the physical, chemical, and biological properties of chemicals that affect persistence and pollutant fate, as well as media properties that also affect transport and fate of chemicals. We also need to know population characteristics. While this consideration may call for a full risk assessment approach, some simplifying assessment methods can be called upon for environmental justice analyses.

Various methods have been designed to incorporate chemical emissions, toxicity, persistence, bioaccumulation, and exposure into a hazard index. Toxicity equivalency potentials methods are promising for application in equity analysis of toxic releases. These methods provide a common metric to evaluate different adverse effects. Four methods — toxicity-based scoring, sustainable process index, concentration/toxicity equivalency, and human toxicity potential — have been proposed and evaluated for assessing human health impacts of chemical emissions (Hertwich, Pease, and Mckone 1998).

The toxicity-based scoring method is the simplest and uses only the toxicity of a chemical, like the toxicity-weighted emission measure discussed above. Occupational TLVs are used as a toxicity potency factor (Horvath et al. 1995). This method requires pretty strict assumptions for aggregating different chemicals. Aggregation of different chemicals is justified for the same media, similar effects, and when the overall environmental persistence is similar for all chemicals (Jia, Guardo, and Mackay 1996). Therefore, it is desirable to have separate indices, for example, for air, water, and soil and for carcinogenicity and ozone depletion. It does not account for the variation in persistence. The sustainable process index method incorporates persistence as well as toxicity (Krotscheck and Narodoslawsky 1996). Therefore, long-lived chemicals are given more weight than short-lived chemicals. The toxic potency factor uses ambient environmental quality standards. However, this method does not include intermedia transport processes such as evaporation, deposition, and rain-out. These processes affect the fate of chemicals and ultimately exposure. To correct this deficiency, the concentration/toxicity equivalency method uses a multimedia model to account for intermedia transport processes (Jia, Guardo, and Mackay 1996). So-called environmental mobility fractions are used to represent the fraction of the original mass of a chemical in the receiving media that is transported to another environmental media. Separate indices are required for each environmental media. A decision rule is needed to compare releases in different environmental media (Hertwich, Pease, and Mckone 1998). The human toxicity potential method uses a multimedia environmental fate model and a multipathway exposure model to estimate the health risks (Guinée and Heijungs 1993). As the most comprehensive, this method accounts for toxicity, persistence, pollutant fate, and exposure pathways. For a toxic potency factor, it uses allowable daily intake values or values derived from animal data. It integrates across all media and includes direct and indirect exposure pathways (Hertwich, Pease, and Mckone 1998).

Application of these four methods to the same chemicals indicates that the relative toxicity values generated vary by three orders of magnitude. In particular, the toxicity-based scoring method generates results that are substantially different from those estimated by the other three methods. Hertwich, Pease, and Mckone (1998) recommend use of the human toxicity potential method for assessing toxic emissions. The drawback for the human toxicity potential and the concentration/ toxicity equivalency methods is that they require considerable data. However, the authors claim that they have collected sufficient data for 214 of 423 TRI chemicals released into air or water.

The Pratt Index, similar to the human toxicity potential method discussed above, was used for a preliminary assessment of the relationship between health impacts of TRI air toxic releases and poverty in the City of Minneapolis (McMaster, Leitner, and Sheppard 1997). This is the most sophisticated method of toxic impact assessment that has been used so far in environmental justice analyses. In the Pratt Index, fugacity models are used to estimate potential chemical concentrations in a specific environmental compartment (i.e., air, water, soils, sediment) after a standardized amount is release into the air (Pratt et al. 1993). Then, the potential concentrations are converted to potential exposure. Human toxicity indicators such as Reference Concentration (RfC) are used to weight the potential exposure. This index provided a much different view of potential toxic risk in the city than the emissions. Overlay of the block-group level concentrated poverty data with the index showed inequitable distribution of potential risk by income, although no statistical analysis was conducted.

Despite being clearly better than total emissions for characterizing a source's potential risk, toxicity equivalency potentials methods generally lack a spatial dimension. They treat the source as a point and leave the analyst to decide the impact area for inclusion in environmental justice study. It is clearly most desirable to integrate toxicity equivalency potentials methods, dispersion models, and GIS into one framework. The integration of the latter two was discussed in Chapter 8.

To make the above framework really work, we need to improve our existing database. Although EPA's Locational Data Improvement Project has been making strides in improving the accuracy of EPA's locational data, the accuracy of TRI's facility location data has been a major concern. The poor quality of TRI facility location data in the early years has been documented. A New York study shows that location data of TRI data were inaccurate for a considerable number of facilities (New York State Parks Management and Research Institute 1993). The study found that 53% of the facilities were within 1 km of the correct location, 35% within 1 to 5 km, 7% within 5 to 10 km, and 5% within more than 10 km. A Minneapolis/St. Paul study revealed a smaller degree of inaccuracy: 41% within 100 m of the correct location, 32% within 0.1 to 0.5 km, 11% within 0.5 to 10 km, 12% within more than 10 km, and another 4% with missing and transposed values (Werner 1997).

11.4 SUMMARY

In this chapter, we review studies on three controversial types of facilities: TSDFs, Superfund and CERCLIS sites, and TRI facilities. Since the landmark UCC study of 1987, we have seen an explosion of studies on these facilities. Within the past decade, we have made remarkable strides in knowledge and databases of these areas. However, the debate continues, and there are still a lot of challenges and opportunities. Cross-sectional national studies have shown clear evidence of racial disparity in the spatial distribution of TRI facilities, but provided mixed results in the distributional patterns of TSDFs and Superfund sites. Evidence indicates lack of income inequity in the distribution of Superfund sites but offers mixed findings about TSDFs and TRI facilities. The mixed results can be interpreted differently. Although the UMass study concluded that there was no significant racial disparity, critics still found that the UMass results pointed to racial inequity based on the fact that the host communities and surrounding areas have higher proportions of minorities (see Table 11.2).

The conflicting evidence is certainly related to a series of methodological issues such as unit of analysis, comparison population, statistical methods, and data quality. With few exceptions, almost all studies are proximity and census geography based. These methods have many limitations, as discussed in Chapters 4 and 6. We need to use more accurate measures and methods of environmental, health, economic, social, and psychological impacts, as shown in Chapter 4. We also need to account for the siting processes of various LULUs. Data have been becoming increasingly available, but data quality needs to catch up.

12 Dynamics Analysis of Locally Unwanted Land Uses

There are two fundamental and sometimes contentious questions about environmental equity or justice. First of all, do socioeconomically disadvantaged groups of the society bear a disproportionately high burden of environmental risks and hazards? This question is the focus of most environmental justice studies, which can be called statics analysis. In previous chapters, we have reviewed the methods used and the findings obtained from these studies. Most studies have found that the poor and minorities now bear a disproportionate burden of potential or actual exposure to environmental hazards from air pollution to toxic wastes, while a few offer conflicting evidence.

Why are some environmental risks distributed disproportionately in the neighborhoods of minorities and the poor? Another debate is focused on the dynamics and causation of this inequity and on possible policies to remedy the problem. The environmental justice movement attributes this inequity to racism and discrimination in the siting decision processes of environmentally risky facilities and locally unwanted land uses (LULUs) and in the process of enforcing and/or complying with environmental laws and regulations. On the other side of the debate, business and pro-business interest groups contend that siting processes have not intentionally discriminated against the poor and minorities. They argue that the communities surrounding their facilities were populated by middle- or upper-middle-income households and whites when the facilities were sited. This is the so-called issue of "which come first" — the poor and minorities or the LULU. Some environmental justice movement advocates argue that this issue is irrelevant (Bullard 1994).

Understanding the dynamics and causation of environmental equity is crucial for developing effective public policies that can address environmental inequity for the long term (Been 1994; Hamilton 1995). If the current inequitable distribution of LULUs results from siting processes that are motivated by racial prejudice and discrimination, it is a violation of the United States Constitution and the Civil Rights Act of 1964, and the government can take legal action under these laws. Governments can also design public policies to encourage fair share of LULU burdens. For example, New York City adopted a new City Charter in 1989, which mandated design of a set of criteria of fair share "among communities of burdens and benefits associated with the city facilities" (Weisberg 1993). The City Planning Commission later adopted the Fair Share Criteria, which received a national award from the American Planning Association. The long-term effectiveness of such policies is dependent on the dynamics of post-siting processes. If market forces induced by the

LULUs make the host communities home to more poor people and people of color, the effectiveness of such siting regulation will be difficult to sustain in the long run.

While there is mounting evidence for the *current* association between LULUs and the poor and minorities, there is a paucity of data for or against the *then* association. Even more complicated and largely unknown are the effects of LULUs on neighborhood characteristics. Moreover, the method for analyzing environmental equity dynamics and causation is inadequate. In this chapter, we make a methodological critique of existing dynamics studies. Next, we present a conceptual framework for equity dynamics analysis, which incorporates technical methods presented in previous chapters and theories and hypotheses of neighborhood changes presented in Chapter 2. Using this framework, we re-visit the Houston case, testing the market dynamics hypothesis and examining alternative hypotheses. Finally, we conclude the chapter with a brief summary.

12.1 METHODOLOGICAL ISSUES IN DYNAMICS ANALYSIS

Only a few studies examine the *then* association, and even less the post-siting dynamic processes. Hamilton (1993) examined the relationship between the planned capacity changes for hazardous waste facilities during the period from 1987 to 1992 and the political power, race, and income of the facilities' host neighborhoods at the ZIP code level. He found that those neighborhoods targeted for capacity expansion had a higher proportion of nonwhite population (25%) than those without such expansion plans (18%). He also used the data to test three hypotheses of why exposure to environmental risks might vary by race: pure discrimination, differences in willingness to pay, and propensity to engage in collective action. He found that collective action, measured in terms of actual voter turnout in a county, offered the best explanation for selecting neighborhoods for capacity expansions. The percentage of nonwhite population was, however, not statistically significant.

As the first to examine the post-siting dynamics, Been (1994) extended two widely publicized studies — the GAO study and the Bullard study. In a regional study of EPA's Region IV, encompassing eight southeastern states as reviewed in Chapter 11, the GAO (1983) concluded that the host communities were disproportionately black and poor. In a study on location of solid waste facilities in Houston, Bullard (1983) found that solid waste facilities were disproportionately sited in predominantly black neighborhoods and near black schools. He claimed that institutionalized discrimination contributed to this inequity. These two studies, along with most other studies, did not examine the socioeconomic characteristics at the time the facilities were sited. Filling this gap, Been (1994) analyzed the socioeconomic status of the host neighborhoods (defined at the census tract level) at the time of siting of those facilities in these two studies. She confirmed the disparity of burdens that these facilities imposed on minorities and the poor at the time of siting.

Moreover, a market dynamics hypothesis was proposed for explaining the current correlation and the change process from *then* to *now* (Been 1994). That is, the current association between locations of LULUs and locations of minorities and the poor might result from market dynamics. The externalities associated with LULUs could make the host neighborhood undesirable to upper-income households and make housing more available to low-income households by reducing property values. As a result, market forces would eventually turn the host neighborhood into a poor one. Racial discrimination in the housing market could make the neighborhood a community of color. To test this hypothesis, Been (1994) analyzed the post-siting changes of some socioeconomic characteristics of the host neighborhoods, using 1970, 1980, and 1990 census data. This analysis produced inconsistent results for the two cases. It was found that, after siting, the levels of poverty and percentages of blacks in the host neighborhoods in Houston increased and property values went down. Accordingly, she concluded that "...such analysis provides considerable support for the theory that market dynamics contribute to the disproportionate burden LULUs impose upon people of color and the poor" (Been 1994:1405). In the extension of the GAO study, however, no such support was found.

Oakes, Anderton, and Anderson (1996) conducted the first national tract-level longitudinal study of communities with commercial TSDFs. The data include 476 commercial TSDFs identified as operating in 1992 and tract-level variables from the 1970, 1980, and 1990 censuses. One complex data issue facing a longitudinal study is the changes in tract boundaries and areas covered under the census tract program. As discussed in Chapter 6, a census tract can change its boundary in the form of splitting, merging, or shifting because population grows or declines over time or because of municipal boundary changes due to annexations. In addition, census programs tracted more and more areas in the country until they covered the whole country in the 1990 census. The 1970 Census of Population and Housing Summary Tape File (STF) 3A "identifies 34,586 tracts, 1980 STF 3A identifies 43,300 tracts, and 1990 STF 3A identifies 66,258 tracts" (Oakes, Anderton, and Anderson 1996: 130). Based on estimates from the Bureau of Census, approximately 18% of all tracts changed their boundaries significantly between 1970 and 1980, and approximately 21% did so between 1980 and 1990 (Been 1997).

Two comparison approaches were used in longitudinal analysis. First, an aggregated comparison over time was done for all TSDF tracts taken together to all non-TSDF tracts taken together. This approach ignores the different areas covered under the three censuses and assumes different universes for different censuses. For example, TSDFs were broken into three groups based on their operations beginning during 1960 to 1969, 1970 to 1979, and 1980 to 1989, respectively. They compared the characteristics of communities hosting the 1970 to 79 TSDFs with those without TSDFs before the siting (i.e., 1970), to determine whether there was a siting bias. They also evaluated the changes after the siting (i.e., 1980 and 1990), to look at the possible impacts of TSDFs on community characteristics.

The second is a consistent area approach. To address the incompatibility issues discussed earlier, Oakes, Anderton, and Anderson (1996) used a stratified random sampling and aggregation approach. They drew a 4-group stratified random sample of non-TSDF tracts within 1, 2, 3, and greater than 3 mi from the nearest TSDF tracts in the 1990 census, and sampled 150 tracts from each stratum. These sampled tracts and all TSDF tracts were reconciled to corresponding tracts in previous censuses, and aggregation was made to generate consistent geographic areas over the three censuses. This aggregation involved aggregating additional tracts for split, merged, or changed tracts until the boundary was the same across three censuses.

The study found "no support for claims of stark national patterns of systematic bias in the siting of commercial TSDFs" (Oakes, Anderton, and Anderson 1996:146). The communities hosting commercial TSDFs are "best characterized as areas with largely white and disproportionately industrial working-class residential areas" (Oakes, Anderton, and Anderson 1996:147). The longitudinal analyses found no evidence that the TSDFs had a significant impact on the host community character-istics in terms of the percentage of minority composition. Instead, the "changes in TSDF communities are similar to those in other more industrial areas" (Oakes, Anderton, and Anderson 1996:147). These changes reflect general population trends.

Been (1997) used a different sampling method. She randomly drew five 1% samples of all the tracts in the censuses of 1970 and 1980, respectively. The com-parison group is also different from the UMass researchers. She used all non-host tracts (or the samples) as the comparison group.

The limited tract coverage in the 1970 and 1980 censuses made the analyses biased toward urban areas. Been (1997) acknowledged that this bias could understate the relationship between race, ethnicity, and facility location. Because facilities were sited in untracted areas, 33 facilities (13% of the total) had to be dropped for 1970, and 23 facilities (11% of the total) had to be dropped for 1980.

Been (1997) found no substantial evidence that facilities were originally sited in communities that were African American or poor, but did find evidence that these facilities were sited in communities that were disproportionately Hispanic at the time of siting. Further, working class and lower-middle-income neighborhoods were found to be disproportionately represented in host communities at the time of siting. The study found little evidence for the market dynamics hypothesis that host neigh-borhoods became poorer and increasingly populated by racial and ethnic minorities after siting of a TSDF.

This study's methodology was built upon its previous one (Been 1995). A major extension is measuring socioeconomic characteristics of host communities in 1970 and 1980 and making them comparable with the 1990 data in terms of geographic coverage. Non-host tracts were sampled for 1970 and 1980 to reconcile tract bound-ary changes over time.

By examining the host neighborhood characteristics at the time a LULU was sited, we are able to determine whether LULUs were initially disproportionately sited in the neighborhoods of the poor and minorities. This would establish a benchmark on which to compare *then* with *now*, that is, to see what has happened to the host neighborhoods since the LULUs were sited. This comparison, however, does not establish a causal linkage between the presence of LULUs and the change in neighborhood characteristics. The following reasons, among others, are important.

First, as we know, correlation is not causation. The coexistence of a LULU and neighborhood changes does not necessarily mean one causes the other. A LULU is neither a necessary nor sufficient condition for neighborhood changes. We discussed this point in theories of neighborhood change in Chapter 2. We can easily find neighborhoods that have changed from white to black without the presence of LULUs. Similarly, there are neighborhoods that are neither deteriorating nor chang-ing from white to black because of the presence of LULUs.

Second, the forces motivating neighborhood changes may have been set in motion at national, metropolitan, or local levels before those LULUs were sited. The forces may include changes in real income, changes in population and households, and changes in government interventions (Grigsby et al. 1987). These forces and others have mobilized neighborhood changes in cities and may have overshadowed the influences of LULUs, if any, throughout the last few decades. In order to understand the relationship between LULUs and neighborhood change, we have to understand the forces that drive the changes.

Third, in order to know whether the changes of neighborhood characteristics during the post-siting period have anything to do with the siting of LULUs, we have to look at how the neighborhood changed before siting of a LULU and how much the changes in the LULU neighborhoods differ from those in the non-LULU neighborhoods. Without examining the pre-conditions and controlling other confounding factors, one cannot pinpoint the presence of "treatment" effects. If changes before and after siting are not significantly different, then we are not in a position to argue convincingly that LULUs have contributed to the post-siting neighborhood changes. More likely, the forces inducing the pre-siting changes are still in effect after siting. If post-siting declines of the host neighborhoods are significantly faster than the pre-siting ones, we may say that LULUs could be a potential contributor to the changes. Still, we are not sure of a causal relationship until we control the effects of other variables on neighborhood changes. Examining the data only at the time of LULU siting was simply insufficient to make any inferences about potential contribution of LULUs. Any conclusions linking the LULUs and neighborhood changes without controlling other variables or examining the pre-siting conditions could be, therefore, misleading.

Fourth, besides the time dimension, there is a spatial dimension of neighborhood change. A city is a spatially interacted and integrated system. Spatial processes have been one of the major driving forces shaping the urban landscape of demographic, social, and economic differentiation. An analysis of change in the LULU-hosting neighborhoods must take into account spatial influences, or at least compare them with the changes in the neighborhood without LULUs.

Finally, theories of neighborhood change offer us alternative explanations for how and why neighborhoods change over time and space. They would enable us to better understand neighborhood dynamics and policy implications for addressing neighborhood changes. In Chapter 2, we discussed theories of neighborhood changes and examined what role(s) LULUs could play in these theories and how Been's hypothesis fits into them. We will discuss alternative hypotheses in the context of the Houston case.

12.2 FRAMEWORK FOR DYNAMICS ANALYSIS

As a way to integrate previous findings and guide future research, I propose an analytical framework for the analysis of environmental equity dynamics and causation (see Figure 12.1). As argued earlier, an analysis of post-siting changes alone is insufficient to test the causation hypothesis. To test whether LULUs have made a significant contribution to the current association between population distribution

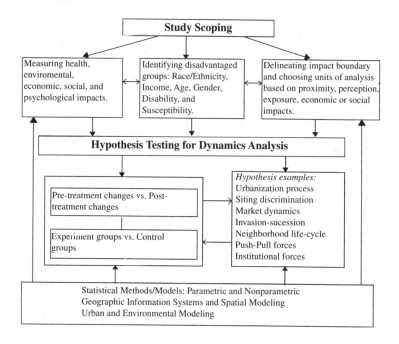

FIGURE 12.1 Analytical framework for dynamics analysis.

and LULUs, we need to conduct an analysis of two series of dynamics. First, we examine the dynamics of LULU neighborhoods in both pre-siting and post-siting periods, and test the difference between pre-siting and post-siting changes. This analysis will enable us to understand whether there is any significant deviation of the post-siting dynamics of LULU neighborhoods from the past. No significant deviation may suggest that the LULUs might have little effect on the current association between population distribution and LULUs. Significant deviation, however, does not necessarily mean that the LULUs cause the deviation. What it suggests is that the LULUs might contribute to the deviating situations in the post-siting period, *provided that all other variables are controlled.* If we can control variables other than the presence of LULUs, we can identify the presence of a treatment effect due to LULUs by comparing LULU and non-LULU neighborhoods. This necessitates the second test.

Second, we examine the dynamics of LULU and non-LULU neighborhoods in both pre-siting and post-siting periods, and test the difference between LULU and non-LULU neighborhoods. In this test, we select a group of control neighborhoods. By testing the difference between LULU and control neighborhoods, we are able to identify whether it is the LULUs that make the LULU-neighborhoods differ significantly from the non-LULU neighborhoods. If a significant difference is found, we may conclude that LULUs contribute to the current association between population distribution and LULUs. An insignificant difference may suggest that the LULUs have little effect on the current association between population distribution and LULUs. In other words, LULU and non-LULU neighborhoods may have undergone the same or similar social changes, which may have nothing or little to do with siting of LULUs. We may also want to examine possible forces that drive the neighborhood dynamics.

Third, we may want to test some possible hypotheses that are based on the theories of neighborhood changes such as invasion–succession, neighborhood life cycle, push–pull, and institutional forces. They provide competing hypotheses for explaining why there is a current association between population distribution and environmental risk distribution.

Particular attention should be paid to the methodological issues discussed in Chapters 3 through 9. It is worth repeating the importance of statistical methods, models, and GIS in dynamics analysis. The validity of an environmental equity analysis strongly depends upon the right choice of statistical methods/models and the correct interpretation of the results. We should pay particular attention to the underlying assumptions of different statistical methods/models and examine how the actual research issue meets these assumptions (see Chapter 7). The nature of the space dimension complicates use of ordinary statistical methods/models. The space dimension is an inherent feature of the relationship between population distribution and environmental risk distribution. Ironically, previous environmental equity studies usually ignore the space dimension and the spatial relationship. GIS is a powerful tool to deal with spatial data. As discussed in Chapter 8, GIS is particularly useful for delineating the impact boundary, choosing appropriate units of analysis, integrating data for population and risk modeling, and presenting the modeling results.

Specifically, we may have the following formulation for testing a significant difference between the pre-siting and post-siting changes and between the LULU and non-LULU neighborhoods. Let PRE_i be a variable for the ith observation for the before period and $POST_i$ be a variable for ith observation for the after period. And let $Z_i = POST_i - PRE_i$. The model is

$$Z_i = \theta + e_i$$

where θ is the treatment effect and e_i is a random variable. The null hypothesis is that there is no treatment effect. The "treatment" here is siting of solid waste facilities. The null hypothesis is thus $\theta = 0$. The alternative hypothesis is that there is a "positive" treatment effect, that is, $|\theta| > 0$.

For the percentage of black population, we would expect the post-siting change to be larger than the pre-siting change if there is a treatment effect. For the relative median family income, we would expect its decline to be larger (that is, negatively smaller) in the post-siting period than in the pre-siting period. Relative median family income is the ratio of median family income in a census tract to median family income for the county. This concise measure conveys both space and time, allowing us to compare different census tracts and their changes over censuses.

$POST_i$ and PRE_i can be measured in two ways: (1) absolute change, which is the *absolute* change in the relative family income or the proportion of black population during the before or after period; and (2) relative change, which is the *percentage* change in the relative family income or the proportion of black population during the before or after period. The before and after periods are each 10 years. These paired replicate data represent "pre-treatment" and "post-treatment" social changes.

Next, I apply this framework to the Houston case discussed earlier. The null hypothesis is that solid waste facilities in Houston do not make the host communities home to more blacks and poor people.

12.3 REVISITING THE HOUSTON CASE: HYPOTHESIS TESTING

12.3.1 DATA

Data preparation is similar to Been's (1994) and differs only in the addition and analysis of pre-siting data. Specifically, census tract was used as a study unit of a neighborhood. When a solid waste facility is located on the border of a census tract, the bordering tracts are combined as a unit of analysis. As a result, there are nine observations for the solid waste facilities (mini-incinerators and landfills) in Houston, which is part of Harris County, Texas (Figure 12.2). Of these nine observations, two facilities opened in 1970, one in 1971, and four in 1972. The 1970 census data were taken as a baseline for before and after for these seven observations. The difference between 1970 and 1960 was taken as the pre-siting change, while the difference between 1980 and 1970 was taken as the post-siting change. Another site was opened in 1978, and the 1980 census data were used as its baseline. Finally, two adjacent facilities were sited in mid-1950s, but comparing

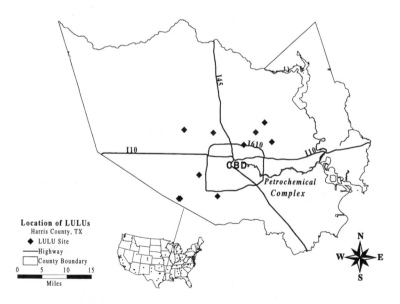

FIGURE 12.2 Location of solid waste facilities in Houston.

1940 and 1950 census data with later census data is problematic due to the large change in the census boundary. Instead, the 1970 census data were taken as the baseline. With a small number of minorities in the neighborhood in 1960, this approximation would probably introduce little bias for the purpose of comparing neighborhood changes. If any, the bias would be in the conservative direction. A test using the data without this observation was also conducted to see if this observation has any effect on the test results.

There were some changes in census tract boundaries in the past five censuses. In particular, the 1960 census data have quite different census tract boundaries than the 1970 census. Only one of the eight observations using the 1960 census data has exactly matching boundaries between 1960 and 1970. The remaining seven were approximated. This could introduce some bias. Furthermore, there is possible spatial dependence among four observations clustered in the northeast Houston Central Business District (CBD). To examine whether these problems affect the results, three tests of different combinations of census data were conducted. One test is to combine the three spatially adjacent observations into one. Another is a test without these three spatially adjacent observations. A third is to use a dataset with more comparability between 1960 and 1970. That is, instead of combining adjacent LULU tracts, I combined those tracts that are comparable among censuses.

The nine observations with hypothesized influence by solid waste facilities are a test group. A control group is selected to see whether the population characteristic changes in this test group differ from those for the rest of the population. The criteria for selecting a control observation can be socioeconomic status such as black percentage of population and household (family) income. In this case, the first criterion is a similar black composition of neighborhood population and a similar change during the pre-siting decade. Second, these control neighborhoods are located in the same sections of Houston as their respective counterparts of LULU neighborhoods. Therefore, they are more likely to be subject to the same forces driving neighborhood change. It is hoped this can control variables other than the presence of LULUs at a similar level for each matched pair. Overall, the test group would demonstrate a significant difference from the control group, if there is a significant "treatment" effect associated with LULUs. Nine observations were selected to form a control group based on the above criteria.

12.3.2 Tests

Statistical tests are needed because there are various uncertainties and errors in the observations. The control group is based on a sample, and thus has sampling errors. The original census data have both sampling and non-sampling errors for the family income variables and non-sampling errors for population by race variables. The sampling errors and the random portion of non-sampling errors could be captured in the standard errors.

Two test methods were applied: two-sample t-test and two-sample Wilcoxon rank sum test (Hollander and Wolfe 1973). Two-sample t-test has more stringent assumptions than two-sample Wilcoxon rank sum test. The former requires the assumption of normal distribution for the population while the latter does not. On the other hand, parametric tests are generally more efficient than nonparametric methods. They will more likely make a correct rejection of the null hypothesis. Both tests were used for the data.

12.3.3 RESULTS

Table 12.1 shows the test results for black composition changes and for relative median family income changes. These tests indicate no significant difference between the test group and the control group. That is, the LULU neighborhoods are not significantly different from those without LULUs in terms of the pre-siting vs. post-siting changes in neighborhood characteristics such as the percentage of blacks and relative median family income, thus showing that the presence of LULUs did not make a significant difference in affecting these changes.

If the solid waste facilities are not the major cause for host communities becoming home for more poor people and minorities in Houston, we want to know what the possible driving forces for such changes might be. The theories of neighborhood changes, as discussed in Chapter 2, offer some possible hypotheses. In the following, we will discuss three such hypotheses: invasion–succession, life cycle of the housing stock, and other environmental risks.

TABLE 12.1
Two-Sample Test for the PRE-POST Difference between LULU Tracts and Non-LULU Tracts in Houston

Black Population Percentage

	Absolute Change				Relative Change		
	T-Test	Wilcoxon Rank Sum Test			T-Test	Wilcoxon Rank Sum Test	
Mean	0.85, −3.47	W	90	Mean	199.9, 1092.3	W	81
Std Dev	21.9, 25.3	n	9	Std Dev	855.0, 3222	n	9
T	0.39			T	−0.80		
p	0.35	p	0.365	p	0.22	p	0.365
δ^2 test	equal			δ^2 test	unequal		

Relative Median Family Income

	Absolute Change				Relative Change		
	T-Test	Wilcoxon Rank Sum Test			T-Test	Wilcoxon Rank Sum Test	
Mean	−1.08, −0.084	W	87	Mean	−14.93, −10.94	W	83
Std Dev	0.134, 0.167	n	9	Std Dev	13.36, 19.95	n	9
T	0.34			T	0.50		
p	0.37	p	0.46	p	0.31	p	0.43
δ^2 test	equal			δ^2 test	equal		

Note: Std dev = Standard deviation, T = t-test value, p = Probability, δ^2 = Variance, W = Wilcoxon Rank Sum Test value.

12.4 DISCUSSION OF ALTERNATIVE HYPOTHESES

12.4.1 INVASION-SUCCESSION HYPOTHESIS

In Houston, minorities, particularly blacks, have been highly segregated from whites. The first wave of freed blacks after the Civil War was concentrated in so-called "Freedmen's town," part of which is now the Fourth Ward (Bullard 1987). By 1950, three major neighborhoods had about two thirds of Houston's black population. The 1950 census indicates that 66.9% of the total nonwhite population lived in the census tracts where nonwhites were the majority (McEntire 1960). In 1980, 78% of the black population lived in census tracts where the majority of the population was black (Shelton et al. 1989). In contrast, the concentration rate was 37% for Hispanics. In 1990, blacks were the majority of the population in 98 census tracts (out of 571 populated tracts) in Harris County, accounting for 53% of the county's black population. In contrast, Hispanics were the majority in 14 census tracts, accounting for about 12% of the country's Hispanic population.

Census data of 1960, 1970, 1980, and 1990 were used to examine the changes in black percentages of census tract population in the LULU tracts, the tracts near these LULU tracts, and those tracts with a substantial presence of black population. The 1960s and 1970s witnessed dramatic increases in the percentage of blacks in some tracts, while the 1980s saw much smaller changes for most. Visual examination of the changes does not reveal any deviation of the LULU tracts from the non-LULU tracts.

As noted earlier in theories of neighborhood changes, population growth and diffusion may lead to spatial dispersion of population and invasion–succession. The rapid black population growth particularly after World War II contributed to the expansion of the black enclaves into adjacent territories. In fact, Houston's black population nearly doubled in the decades from 1930 to 1950. It only took 1 decade (from 1950 to 1960) for the black population to increase by 71.5%. The growth rates decreased to 47.2 and 39.1%, respectively, for 1960s and 1970s.

The geographic expansion of the black enclaves as a result of population growth in Houston was documented in the early literature (Bullock 1957: 61-62): "… the [Third Ward] is rapidly expanding to include adjacent tracts. The expansion is a result of increased population pressure from the Fourth Ward area, but is occasioned by the successful infiltration of Negro families into the white areas adjacent to Third Ward." This outward expansion started in the 1950s and continued for the following 3 decades. There were two expansion paths: from the Fifth Ward northeast and from the Third Ward south (Bullard 1987). It also seems clear that transportation lines have some constraints on geographic expansion. As demonstrated in previous studies (O'Neill 1981), the geometric boundaries defined by transportation lines such as highways have geographically constrained neighborhood transitions.

Geographic expansion may explain why some LULU neighborhoods have experienced an increase in the percentage of blacks. Clearly, three northeast LULU neighborhoods and one south LULU neighborhood fall in the two outward expansion paths. In 1960, only one of them was a predominantly black neighborhood. In 1970, blacks became the majority in another neighborhood. In 1980, all but one were

predominantly black neighborhoods. All of them have increases in the black percentage post-siting of the LULU, while two of the four have post-siting increases
smaller than pre-siting increases. Another LULU neighborhood (Tract 434) in the
southwest, not adjacent to the south black enclave but near it, had a small increase
in the percentage of blacks (from 4.1 to 7.6%) in the 1970s but a dramatic increase
(from 7.6 to 33.0%) in the 1980s. A similar pattern exists for those tracts between
this LULU tract and the south black enclave. Clearly, the expansion did not knock
on their door until the 1980s. Overall, the expansion is wavelike. The inner tracts
started earlier, maybe 1950s, while the outside tracts followed. The more outward
the black enclaves expand, the wider the expansion paths. Eventually, the black
districts in south and northeast Houston had a fan-like shape (Figure 12.3).

12.4.2   Life-Cycle Hypothesis

According to the life-cycle model discussed in Chapter 2, the age of housing stock
and density are two of the most important factors that determine changes of neighborhood characteristics over time. Like other metropolitan areas, Houston has older
housing stock in the center city than in its suburban areas. Second to the central city
is the eastern near-suburban area. There are sectoral and a centric patterns in the
age of the housing stock in the Houston-Galveston metropolitan area.

     A closer look at the black neighborhoods reveals a significant difference in
housing stock from the rest of the neighborhoods in Harris County. Let us compare
the 98 black-dominated census tracts with the rest of the county using the 1990
census data. On an average, the percentage of the structures built before 1949 is
22.6% for these black neighborhoods, significantly higher than the 13.4% for the

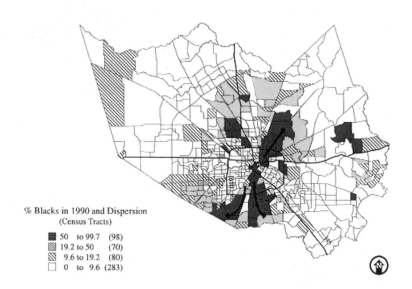

FIGURE 12.3 Distribution of the African-American population in Houston in 1990 and
spatial dispersion patterns.

TABLE 12.2
Percentage of Structures Built by Year in Different Neighborhoods in Harris County

Neighborhoods		Pre-1950	1950s	1960s	1970s	1980s	Median	Units
Black Tracts	Mean	22.6	28.3	24.8	16.4	7.8	1960	1468
(n = 98)	Median	16.6	28.8	21.8	11.4	4.2	1958	1249
Other	Mean	13.4	15.8	19.2	28.3	23.3	1968	2187
(n = 471)	Median	2.8	11.5	16.5	26.8	13.6	1970	1882
Minority Tracts	Mean	23.3	23.5	22.4	21.0	9.9	1961	1770
(n = 169)	Median	16.6	21.4	19.6	15.4	5.4	1960	1435
Other	Mean	11.5	15.6	19.2	28.5	25.2	1969	2187
(n = 400)	Median	2.5	10.0	16.4	27.7	11.3	1971	1920
LULU	Mean	6.7	21.6	24.4	32.2	15.1	1965	1729
Neighborhoods	Median	5.8	16.7	26.0	27.4	11.6	1965	1121
(n = 9)								
Other	Mean	15.3	17.6	19.7	26.1	21.3	1966	2084
(n = 535)	Median	4.2	14.1	16.6	23.4	10.6	1967	1762

Data source: Bureau of the Census, 1992b. Census of Population CD ROM. The first row in each category is the mean, and the second is the median. Black (or minority) tracts = tracts where blacks (or minority) people are no less than 50% of population.

rest of the county (see Table 12.2). Similarly, the minority-dominated neighborhoods as a whole have older housing stock than the others.

Let us look at the LULU tracts in particular. Compared with the rest of Harris, LULU neighborhoods have a smaller percentage of the oldest and the youngest houses and a larger percentage of housing units built in the 1950s, 1960s, and 1970s. Home ownership shows little difference from the rest of county. Compared with the black-dominated tracts as a whole, the LULU neighborhoods appear to have younger housing stock. This does not suggest any evidence for the hypothesis that LULU neighborhoods as a whole become home to more poor people and people of color because of the aging of housing stock.

As discussed earlier in the life-cycle model, neighborhoods will decline as they age. The old age of housing stock in the central city or in the black- or minority-dominated neighborhoods means that these neighborhoods are likely to decline. A city report describes a pattern of housing deterioration in Houston's core area (HCPC 1980). The area surrounding CBD was mostly in decline. In particular, the worst declining area was concentrated east and northeast of the CBD. Factors listed, which contributed to the housing deterioration, include "a high incidence of non-resident ownership, resident financial inability to provide routine maintenance, and the general age of the housing stock" (HCPC 1980:V-26). Three LULU neighborhoods studied here fall in the study area of HCPC. From the deterioration pattern and area, it is hard to conclude that LULUs caused the neighborhood deterioration.

12.4.3 PUSH FORCES: OTHER ENVIRONMENTAL RISKS

Solid waste facilities are among many environmental liabilities that people might take into account in their residential location choices. These may include urban air pollution, water pollution, hazardous waste facilities, major facilities discharging toxins, major wastewater dischargers, and uncontrolled hazardous waste sites (Superfund sites). Some environmental externalities may not impose health risks to humans but are nuisances to human life. Two examples are noise and odor. These environmental problems may affect changes in the socioeconomic characteristics of the communities in Houston over time. The distribution of these environmental externalities and how people respond to them may determine the dynamics of the relationship between population distribution and environmental risk distribution. Further investigations are needed in this direction.

The major industrial polluters are located in the southeastern Houston quadrant, particularly along the Houston Ship Channel, where there is a petrochemical industrial complex. The prevailing wind direction in Houston is from the south southeast (Ruffner and Bair 1985). Clearly, the neighborhoods in the north (particularly the northeast quadrant) of Houston are downwind of the industrial polluters.

The communities in the northeast quadrant report the highest degree of concern about environmental problems among the four quadrants. In an environmental survey conducted in 1983 (Bullard 1987), civic club leaders from the northeast quadrant had the highest percentages of reporting "severe problem" with air pollution, water pollution, noise, and litter, trash, and solid waste. Air pollution was their biggest concern, with 82.4% of the civic leaders from the northeast reporting a "severe problem." In contrast, only approximately 40% indicated a severe air pollution problem in the city (Bullard 1987). The distribution of these risks and people's perception of them might contribute to the dynamics of environmental equity.

12.5 CONCLUSIONS

Why are some environmental risks distributed disproportionately in the neighborhoods of minorities and the poor? A hypothesis was proposed in a recent study that market dynamics contributed to the current environmental inequity. That is, locally unwanted land uses (LULUs) make the host communities home to more poor people and people of color. This hypothesis was allegedly supported by a Houston case study, whereby its author analyzed the post-siting changes of the socioeconomic characteristics of the neighborhoods surrounding solid waste facilities. In this chapter, I have argued that an analysis of the post-siting changes in the LULU neighborhood characteristics alone is insufficient to test the causation hypothesis that market dynamics contributed to the current environmental inequity. Instead, I have presented an analytical framework for analyzing the issue of whether LULUs contributed to the current environmental inequity over time. I suggest that the pre-siting neighborhood dynamics and the characteristics of control neighborhoods be analyzed to pinpoint deviation of the host neighborhoods from the past and from other similar neighborhoods without LULUs. Furthermore, I propose that the analyst examines alternative hypotheses that theories of neighborhood change offer for explaining neighborhood changes and for the roles of LULUs in neighborhood changes. These alternative hypotheses should be examined when analyzing the relationship between LULUs and neighborhood changes in a metropolitan area.

Using this framework of analysis, I re-visited the Houston case. First, I found no evidence that provided support for the hypothesis that the presence of a LULU made the neighborhoods home to more blacks and poor people, contrary to the conclusion made by the previous study. Rather, there is no significant difference in the changes in the percentage of blacks and relative median family income between the LULU and "control" neighborhoods. This means that the forces driving neighborhood changes might be similar for the neighborhoods with and without LULUs. Second, examination of alternative hypotheses indicates that invasion–succession and other push forces might offer a better explanation.

As noted in the beginning of this chapter, understanding the post-siting effects of LULUs on the host neighborhoods is crucial for designing effective public policies that can be sustained in the long run. To evaluate post-siting effects as illustrated in theories of neighborhood changes and the Houston case study, we need a comprehensive examination of social, economic, institutional, and environmental factors that may drive neighborhood changes. We need to know how these factors interact temporally and spatially. A fair-share siting policy, for example, needs to consider not only the current socioeconomic and environmental characteristics of neighborhoods but also how these come into being and what will happen in the future. The Houston case study indicates that a single type of LULU may not be a determinant for changes of neighborhood characteristics. Rather, a mix of social, economic, institutional, and environmental dynamics might have come into play. Public policies should be developed so that these variables are addressed in a coordinated, integrated way.

The Houston case study demonstrates that it can be misleading to make an inference that the LULUs make host communities home to more black and poor people by analyzing the post-siting changes alone. The case study, however, does not reject the market dynamics hypothesis in general; that is, in the case of solid waste facilities in Houston, the effects of LULUs might be overshadowed by other forces. In other cases, the market dynamics hypothesis may hold true. Furthermore, the conclusions in the Houston case do not mean that the solid waste facilities in Houston did not have adverse impacts on the surrounding neighborhoods. Rather, they are not significant factors in making the host neighborhoods home for more black and poor people.

13 Equity Analysis of Transportation Systems, Projects, Plans, and Policies

> Urban transit systems in most American cities… have become a genuine civil rights issue — and a valid one — because the layout of rapid-transit systems determines the accessibility of jobs to the black community. If transportation systems in American cities could be laid out so as to provide an opportunity for poor people to get meaningful employment, then they could begin to move into the mainstream of American life.

Martin Luther King, Jr.

The modern civil rights movement has its root in fighting the inequity in transportation systems. A landmark event happened in the early evening of December 1, 1955 (Oedel 1997). After work, Rosa Parks boarded a crowded bus in Montgomery, Alabama. She sat down in the rows marked for whites, and later refused to relinquish her seat when sought by a white rider. Her arrest led to a bus boycott organized by her pastor, Rev. Martin Luther King, Jr. In 1954, the Supreme Court declared unconstitutional the infamous doctrine of "separate but equal," which the same court upheld when examining the Jim Crow segregated seating law of Louisiana in 1896 (Bullard and Johnson 1997). Later, the Supreme Court declared discrimination in interstate travel unconstitutional. To test this decision, John Lewis and a group of young people began a historic journey by bus from Washington through the Deep South in early 1960s. These Freedom Riders challenged racial segregation and discrimination along their campaign roads and highways. These events preceded the enactment of the Civil Rights Act of 1964. On June 15, 1999, Rosa Parks received the prestigious Congressional Gold Medal of Honor, the highest award bestowed by the U.S. government.

Although early struggles have focused on inequity in the operations and services of transit systems, little attention was given to the distributional impacts of transportation planning and policies until recently. On April 15, 1997, the U.S. Department of Transportation (U.S. DOT 1997a) issued the final order to address environmental justice in minority populations and low-income populations. The order states two principles to be observed:

1. Planning and programming activities that have the potential to have a disproportionately high and adverse effect on human health or the environment

shall include explicit consideration of the effects on minority populations and low-income populations.

2. Steps shall be taken to provide public access to public information concerning the human health or environmental impacts of programs, policies, and activities.

Specifically, U.S. DOT will take actions to prevent disproportionately high and adverse effects by

1. Identifying and evaluating environmental, public health, and interrelated social and economic effects,
2. Proposing measures to avoid, minimize, and/or mitigate disproportionately high and adverse environmental and public health effects and interrelated social and economic effects,
3. Considering alternatives, where such alternatives would result in avoiding and/or minimizing disproportionately high and adverse human health or environmental impacts, and
4. Eliciting public involvement opportunities and considering the results thereof, including soliciting input from affected minority and low-income populations in considering alternatives.

Any programs, policies, or activities that will have a disproportionately high and adverse effect on populations protected by Title VI ("protected populations") will be carried out only if

1. A substantial need for the program, policy, or activity exists, based on the overall public interest; and
2. Alternatives that would have less adverse effects would either (a) have other adverse social, economic, environmental, or human health impacts that are more severe, or (b) involve increased costs of extraordinary magnitude.

On December 2, 1998, the Federal Highway Administration (FHWA) issued order 6640.23 "FHWA Actions to Address Environmental Justice in Minority Populations and Low-Income Populations," establishing FHWA's policies and procedures for compliance with EO12898. In a memorandum dated October 7, 1999, FTA and FHWA administrators further clarified their implementation of Title VI requirements in metropolitan and statewide planning (Linton and Wykle 1999). Specific strategies were identified in a national conference entitled "Environmental Justice and Transportation: Building Model Partnerships," held in Atlanta on May 11–13, 1995. Some of the recommendations are specifically targeted at transportation and land-use systems. The recommendations also include research and analysis strategies that emphasize the importance of GIS, qualitative and quantitative data, computer models, and multiple and cumulative impacts.

In this chapter, we begin with a look at environmental impacts of transportation systems. Next, we discuss the regulatory environment for incorporating equity analysis in the transportation planning process and outline major methods and tools that can be used for conducting equity analysis in transportation planning. The first of these methods is based on refinement of transportation system performance measures, which have different strengths and weaknesses for equity analysis. We devote a lot of attention to equity analysis of mobility and accessibility, including basic concepts, measurement methods, their uses and limitations, empirical evidence about mobility and accessibility disparity, and spatial mismatch. In Section 13.5, we discuss use of a hedonic price method for evaluating distributional impacts on property values. In Section 13.6, we outline an integrated GIS and modeling approach to assess differential environmental impacts and examine a few analytical issues. In Section 13.7, we review the equity implications of some transportation policies such as congestion pricing. Finally, we discuss the Los Angeles MTA case and some analytical issues involved.

13.1 ENVIRONMENTAL IMPACTS OF TRANSPORTATION SYSTEMS

Highways and transit are a mixed blessing. Accessibility is important in the real estate market, where there is a well-known motto of "Location, Location, Location." Good accessibility is valued, but this does not mean living right next to an interstate highway. Proximity to highways or transit is "a double-edged sword" (Landis and Cervero 1999). On one hand, people value good accessibility to jobs and services because of highways and transit. In order to obtain good accessibility, households and employers look for their residential or non-residential locations a certain distance from highways or transit. On the other hand, households typically do not like to live right near highways because of negative externalities generated such as noise and pollution. Here is a tradeoff between accessibility and environmental quality. Those homebuyers who are aware of and sensitive to the negative externality of highways and transit and who believe that the benefits associated with locating near highways and transit are less than the associated cost will locate away from highways and transit if they can afford it. Those homebuyers who are unaware of or insensitive to the negative externality of highways and transit will include houses along highways and transit in their house-hunting domain. They will locate right near highways and transit if they perceive benefits such as good accessibility and good home value outweigh costs such as noise and air pollution.

The U.S. DOT's order defines adverse effects from transportation. "Adverse effects mean the totality of significant individual or cumulative human health or environmental effects, including interrelated social and economic effects, which may include, but are not limited to: bodily impairment, infirmity, illness or death; air, noise, and water pollution and soil contamination; destruction or disruption of man-made or natural resources; destruction or diminution of aesthetic values; destruction or disruption of community cohesion or a community's economic vitality; destruction or disruption of the availability of public and private facilities and services; vibration; adverse employment effects; displacement of persons, businesses, farms, or nonprofit organizations; increased traffic congestion, isolation, exclusion or separation of minority or low-income individuals within a given community or from the broader community; and the denial of, reduction in, or significant delay in the receipt of, benefits of DOT programs, policies, or activities" (U.S. DOT 1997a).

Transportation emissions continue to be a significant cause of air pollution, although today's cars are 70 to 90% cleaner than their 1970 counterparts. This happens partly because of the rapid increase in travel activity since 1970. Vehicle miles traveled have almost doubled in the U.S. from 1970 to 1990, tripled from 1960, and increased even faster in many metropolitan areas (U.S. EPA 1998d). In 1970, vehicle miles traveled totaled 1,114 billion, compared with 2,144 billion in 1990 and 2,405 billion in 1995. As shown in Table 10.1, in 1997, the transportation sector accounts for 76.6% (67,014,000 short tons) of all CO; 49.2% (11,595,000 short tons) of NOx; 39.9% (7,660,000 short tons) of VOCs; and 13% (522 tons) of lead (U.S. EPA 1998c). In 1993, mobile sources contributed 21% (1.7 million tons) of 166 air toxics, and on-road gasoline vehicles emitted 76% of all mobile source Hazardous Air Pollutants (HAPs) nationwide (U.S. EPA 1998d).

Transportation impacts decay with distance from roadways. Transportation-related air pollution is the most serious near roadways, and tends to reduce to the background level between 500 and 1000 meters (1640 to 3280 feet) from the roadway (FIIWA 1978). Traffic noise depends on the volume, speed, and composition of traffic, and often lessens to the background level at about 1000 feet (300 m) from a highway (Hokanson et al. 1981). Acting as barriers, the first row of houses along a transportation corridor absorbs most noise impacts (Stutz 1986). In old city centers, a street network is often in grids. Houses are usually attached and face a street. More or less, most houses in city centers are impacted by noise associated with vehicles. However, a street network consists of links to various functions with various traffic volumes and speed. Some streets serve a local purpose, others serve as minor arteries, and still others serve as major arteries. This means that environmental impacts from the transportation network are different for different neighborhoods. This has been commonly recognized in the real estate market.

Besides environmental impacts, transportation projects and plans also have a wide range of social and economic impacts on the communities. The U.S. Department of Transportation's Community Impact Assessment reference guide (FHWA 1996) identifies a variety of potential community impacts that need to be addressed (Table 13.1). Clearly, transportation projects, programs, and plans have much broader impacts.

13.2 INCORPORATING EQUITY ANALYSIS IN THE TRANSPORTATION PLANNING PROCESS

The Clean Air Act Amendments (CAAA) of 1990 mandate integration between transportation and air quality processes at all levels of governments. The planning process is the focal point for consideration of existing transportation system performance, transportation management strategies, CAAA requirements, transportation improvement program and implementation, funding, and other factors such as social, economic, and environmental factors. TEA-21 consolidates the previous sixteen planning factors into seven broad areas to be considered in the planning process (same as for statewide planning).

As mentioned earlier, FTA and FHWA administrators jointly issued a memorandum to further clarify their implementation of Title VI requirements in metropolitan

TABLE 13.1
Community Impacts of Transportation

Impact Category	Examples of Impacts
Social and psychological impacts	Redistribution of population, community cohesion and interaction, social relationships and patterns, isolation, social values, and quality of life
Physical and visual impacts	Barrier effect, sounds (noise and vibration), other physical intrusions (dust and odor), aesthetic character, compatibility with community characteristics
Land-use impacts	Farmland, induced development, consistency and compatibility with land-use plans and zoning
Economic impacts	Location decision of firms (move-in, move-out, close, stay-put), direct impacts of construction on local economy, tax base, property value, business visibility
Mobility and access impacts	Pedestrian and bicycle access, traffic-shifting, public transportation, vehicular access, parking availability
Public services impacts	Relocation or displacement of public facilities or community centers, inducing or reducing use of public facilities
Safety impacts	Pedestrian and bicycle safety, accidents, crime, emergency response
Displacement	Effect on neighborhood, residential displacements (characteristic of displaced population, types and number of dwellings displaced, residents with special needs), business and farm displacements (types and number of businesses and farms displaced), and relocation sites

Source: Federal Highway Administration. 1996. Community Impact Assessment: A Quick Reference for Transportation. Publication No. FHWA-PD-96-036. Washington, D.C.

and statewide planning (Linton and Wykle 1999). The memo emphasizes that the law applies equally to the processes and products of planning as well as to project development. They request actions to ensure compliance with Title VI during the planning certification reviews conducted for Transportation Management Areas (TMAs) and through the statewide planning finding rendered at approval of the Statewide Transportation Improvement Program (STIP). The reviews assess Title VI capability in terms of overall strategies and goals, service equity, and public involvement. Review questions address specific procedural and analytical capabilities for complying with Title VI. "Does the planning process have an analytical process in place for assessing the regional benefits and burdens of transportation system investments for different socio-economic groups? Does this analytical process seek to assess the benefit and impact distributions of the investments included in the plan and TIP (or STIP)?" In 2000, FHWA and FTA revised planning regulations (23 CFR 450 and 49 CFR 619) and environmental regulations (23 CFR 771 and 49 CFR 62) to propose appropriate procedural and analytical approaches for Title VI compliance.

These laws and regulations prompt a paradigm shift in transportation planning. It moves away from the old paradigm, which emphasizes how fast vehicles move. The new paradigm emphasizes how well people's travel needs are met economically efficiently, environmentally friendly, justly, and socially.

Equity analysis of transportation systems is more complicated than most other environmental justice issues. While most environmental justice controversies originate from specific site-based projects in a particular time, transportation systems involve much broader spatial and temporal dimensions. In the spatial dimension, transportation systems (either highway or transit) are linear and penetrate every community, compared with scattered points with limited impact areas of some LULUs in a few communities. In the temporal dimension, transportation systems have definite planning horizons well into 20 years in the future, as well as their legacy from the past. While transportation projects are carried out one by one in different communities, they generally have to go through transportation planning processes at the city, county, regional, or state levels. Those federally funded projects with regional significance require evaluation in the metropolitan planning process, which involves both project and policy levels. Clearly, equity analysis has to deal with these two levels.

Table 13.2 lists major methods and tools for conducting environmental justice analysis in transportation planning. They are built upon existing analytical capabilities of metropolitan planning organizations. Technical staff for MPOs need to adapt these methods and tools to deal with environmental justice issues in their regions. For example, many MPOs use a traditional four-step transportation modeling chain (see Figure 13.1) to forecast travel demand and evaluate air quality conformity. These models can be refined and adapted to provide measures for equity analysis,

---

TABLE 13.2
Methods for Conducting Environmental Justice Analysis in Transportation Planning

Categories	Methods/Tools
Transportation systems	Performance measures by population groups
Accessibility and mobility	Vehicle availability and vehicle availability modeling
	Accessibility measures
	• Distance/time-based measures
	• Cumulative-opportunity measures
	• Gravity-type accessibility index
	GIS
Property value impacts	Hedonic price methods/GIS
Environmental impacts	Land-use/transportation modeling
	Environmental modeling
	GIS
User benefits	Consumer welfare measures by population groups
	• Travel demand models/urban models
	Direct valuation of travel time saving, operating costs, and safety improvement
	• Travel demand models
	• Benefit-cost analysis
Fiscal impacts of transportation funding and pricing	Welfare economics and incidence of tax analysis

---

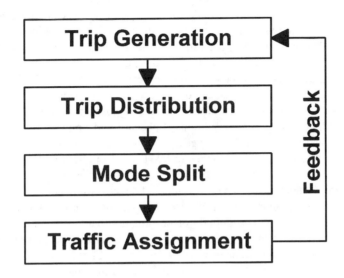

FIGURE 13.1 Traditional four-step transportation modeling.

such as accessibility. In particular, mode choice and trip generation submodels are very useful for equity analysis. A mode choice submodel can provide accessibility measures and consumer welfare measures (see Chapter 9 and Section 13.4 below). These measures can be stratified by income and race for equity assessment. Trip generation and vehicle availability submodels can provide measures for evaluating the impacts of transportation projects and plans on mobility by various population groups. Most data needed to carry out these analyses are available. In the following, we will discuss some of these methods in detail.

13.3 TRANSPORTATION SYSTEM PERFORMANCE MEASURES

With a shift in the planning paradigm, new performance measures are adopted to evaluate transportation systems. Vehicle Miles of Travel (VMT), vehicle trips, and average travel speed are three performance measures that are very important in linking transportation planning and air quality planning.

Planners and engineers have used a lot of variables to measure transportation system performance. For example, people often feel and complain about congestion during commuting, but it is not easy to define and measure congestion. Congestion is vaguely defined as "the level at which transportation performance is no longer acceptable due to traffic interference. The level of acceptable system performance may vary by type of transportation facility, geographic location and/or time of day" (Interim Final Rule on Management and Monitoring Systems). With this definition, the federal government gives state and local governments substantial flexibility to measure congestion.

Each performance measure shows different strengths and weaknesses which have various equity implications (Table 13.3). These measures can be segmented by population group for evaluating distributional impacts of transportation projects. For example, a VMT measure was used to assess the disproportionate impacts of the Barney Circle Freeway Modification Project in Washington, D.C. (Novak and Joseph 1996). The goals of the project were to complete a vital freeway link as a major commuter route in the District and to divert through traffic from residential streets to freeways. The project was controversial and opposed by a coalition of civic, neighborhood, and environmental groups (Sphepard and Sonn 1997). To measure the distributional impacts of the changes in freeways and residential streets, VMT-persons were estimated as the changes in VMT on each residential street link and each freeway segment multiplied by the population that is affected (Novak and Joseph 1996). Reduction in VMT in residential streets is a benefit that accrues to local residents, while an increase in VMT on freeways is a burden to local residents. VMT-persons measure the magnitude of benefits and burdens for local residents. Holding population constant, a census tract with a large VMT reduction in residential streets benefits more than a census tract with a small VMT reduction. VMT changes in each census tract were derived from traffic modeling. To assess distributional impacts, VMT-persons were estimated separately for minority and low-income populations and compared to the proportion of the total population that each group comprises in the area. Any disproportionate impacts were then spatially identified.

13.4 EQUITY ANALYSIS OF MOBILITY AND ACCESSIBILITY

13.4.1 CONCEPTS AND METHODS

Mobility, literally the ability to move around, "relates to the day-to-day movement of people and materials" (U.S. DOT 1997b). Is mobility a merit good or a right? The majority of 1,600 people randomly surveyed in New Mexico considered mobility as a right (Hamburg, Blair, and Albright 1995). They were asked the question: Do you believe that the ability to get where you want to go in a reasonable time and for a reasonable cost is or should be a basic right in the same sense as freedom of speech or the pursuit of happiness? Of the approximately 1,600 people surveyed, 58.9% responded yes, and 20.8% responded no. The survey also shows that females and low-income people are more likely to consider mobility as a right.

Mobility is an easy to define concept but is not easy to measure. The measures often used include vehicle availability and number of miles traveled or trips taken during a given time, the latter referred to as revealed mobility. Factors affecting mobility trends include income growth, household size, labor force participation (particularly women), aging, changing levels of immigration, residential and job location, changes in the nature of work and workplaces, and advances in information technologies (U.S. DOT 1997b).

Accessibility is a concept that is not easy to define and measure. When people talk about how accessible a place is, they relate that place to other places. Accessibility involves relative locations of activities or opportunities such as working places,

TABLE 13.3
Transportation Performance Measures and Their Equity Implications

Measure	Definition	Strengths	Weaknesses	Equity Implications
Person-miles of travel (PMT) or passenger-miles traveled	Vehicle volume × link length × vehicle occupancy	Can be aggregated to any spatial level (link, facility, subarea, region) and to any temporal level (peak-hour, peak-period, daily, weekly, monthly, annually) Can be used by mode, including trucks Reflects persons' real demand for travel	Does not directly address air quality impacts from vehicle trips	Favors long-distance travelers
Vehicle miles of travel (VMT)	Vehicle volume × link length	Can be aggregated to any spatial level and to any temporal level Can be used and aggregated by mode, including trucks Related to air quality, accessibility, and sustainability	Does not address non-motorized trips Generally used for car or truck, and would give misleading and difficult-to- interpret information if including transit vehicles	Favors users of motorized modes such as car, truck, or transit Favors particularly drivers with more mileage than average
Vehicle hours of travel (VHT)	Vehicle volume × travel time	Accounts for congestion More directly related to air quality Better proxy for mobility, accessibility and sustainability	Does not address non-motorized trips	Favors users of motorized modes
Average speed	Speed of person trips between origin and destination averaged for a period of time	Intuitively appealing Can be aggregated by trip purpose, mode, spatially, and temporally Related to air quality	Overemphasis on vehicle trips	Favors users of motorized modes
Average trip length, time	Distance and time of person trips between origin and destination averaged for a period of time	Intuitively appealing Can be aggregated by trip purpose, mode, spatially, and temporally Related to accessibility.	Requires Origin-Destination Survey which is expensive	Can show accessibility disparity

continued

TABLE 13.3 (CONTINUED)
Transportation Performance Measures and Their Equity Implications

Measure	Definition	Advantages	Disadvantages	Equity Implications
Congestion (hours of delay)	Free flow operating speed	Delay per vehicle at intersections is standard for measuring intersection level of service; Directly reflects traveler's perspective; Data are readily available; Easy and inexpensive to estimate	Difficult to define free flow speed; Free flow speed may change and affect the public's understanding	Favors travelers using congested roads
Congestion (volume/capacity ratio; level of service)	Volume/capacity; Level of Service is a set of descriptors (such as A to F) to measure transportation system performance and is defined based on travel time, cost, number of transfer, volume/capacity ratio, etc.		Difficult for the public to understand; Requires a good estimate of capacity, which is difficult; Cannot measure a volume that is greater than capacity in reality; Deficient in over-saturated conditions; Difficult to aggregate to a higher level of geography such as a region	Favors travelers using congested roads
Congestion (speed or travel time)	Defined by speed/travel time threshold range by facility. For freeways, severe congestion occurs if speeds less than 30 MPH	Directly addresses traveler perspective of congestion	Expensive to collect data	Favors travelers using congested roads
Vehicle Trips	Number of trips taken using vehicles.	Related directly to air quality	Does not address non-motorized trips	Favors users of motorized modes
Person Trips	Averaged number of trips per person.	Can be aggregated by trip purpose, mode, spatially, and temporally	Does not address accessibility	Can show mobility disparity

shopping centers, recreational facilities, residential locations, and so on. Accessibility is the ease in reaching them from an origin. Proximity does not necessarily mean accessibility. The fact that A is near B does not mean that one can get to B from A quickly because there may be some barriers between them, such as a river. The concept of accessibility includes not only relative spatial locations of activities or opportunities but also how fast one can travel from A to B by what mode at what time. It depends on the structure and performance of the transportation network. In a broader sense, the determinants of accessibility may encompass real and perceived travel cost, and modal, personal, and location attributes. For example, modal characteristics that affect perception include comfort, speed, directness, consistency, degree of physical effort, and extent of waiting.

Measuring accessibility is the subject of a great deal of empirical investigations, but there is no consensus on a universally accepted measure for accessibility. A variety of accessibility measures appear in the literature (Ingram 1971; Guy 1983; Song 1996). Song (1996) identifies and compares nine accessibility measures. In a concise manner, accessibility measures can be classified into three categories: distance- or time-based measures, cumulative opportunity measures, and gravity-type measures.

We can measure the relative accessibility of one location based on distance, travel time or cost, or composite or generalized cost of travel from other places. A common measure used in the early literature was distance from the center of the central business district (CBD) to a location in a city or metropolitan area. This measure assumes the predominant role of the CBD employment base in a metropolitan area's population distribution. It corresponds to a monocentric model of urban structure. Clearly, it fails to account for the decentralized nature of urban forms in the U.S. and other countries, particularly the polycentric structure that penetrates metropolitan America. Not surprisingly, it is one of the poorest accessibility measures (Song 1996). The measure of average commuting distance or time to work, popularly used as a transportation system performance measure, corrects the inadequacy of the monocentric assumption. However, it does not contain any information about the magnitude and distribution of opportunities such as employment at different destinations. Use of this measure is perplexed by the "commuting paradox" (Gordon, Richardson and Jun 1991). National and local surveys have repeatedly reported slightly increasing, or constant and stable commuting time over the years, although commuting distance increases significantly as does congestion in most metropolitan areas. Evidently, accessibility measures based on commuting time and distance will show different results. Further, the difference in commuting time between central city and suburban residents in urbanized areas is small, 18.2 min. vs. 20.8 min. from the 1990 NPTS data. Does accessibility really remain constant over time? Not necessarily. Is accessibility a little different between central city and suburban areas? Not necessarily. Of course, the distance measure can be refined to incorporate the importance and attractiveness of different destinations by varying weights. However, distance-based accessibility measures do not fare well overall in representing population distribution in an urban area (Song 1996).

A cumulative-opportunity measure is to estimate the number of opportunities within a specified amount of time. One example is the number of shopping facilities located within 10-min. travel by car or transit. Another example is the number of jobs reachable within an average commuting distance or time in a region, or within a proportion of such averages (Song 1996). For accessibility of a residential location to employment, we can estimate the number of employees within a 30-min. or 45-min. commute by car or transit. This type of measure has gained increasing popularity in analysis of the distributional effects of long-range regional transportation plans. In its equity evaluation of the 1998 Regional Transportation Plan, the Southern California Area Governments (SCAG) measures accessibility improvement by income and race/ethnicity in two modes: job opportunities accessible through trips less than 30 min. by transit and job opportunities accessible through trips less than 30 min. by auto. One strength of this measure is its easy, commonsense interpretation. It has gained endorsement from some environmental groups, which facilitates its increasing adoption in other metropolitan areas such as Washington, D.C. However, its major limitation is the assumption that each and every opportunity is equally important within the specified time. For example, a job reachable in 5 min. is treated the same as a job reachable in 30 min. In addition to the lack of spatial differentiation, the specified time or distance is often arbitrary and arbitrarily cuts off opportunities just outside the boundary. If this cut-off value is large enough, we will not easily detect any changes in accessibility caused by transportation improvement projects in the future. As a result of using this measure, distributional impacts could be distorted. In comparing with eight other accessibility measures, Song (1996) finds cumulative job opportunity within the average commuting distance to be the poorest accessibility measure in explaining population distribution in the Reno–Sparks metropolitan area in Nevada.

Gravity-type accessibility measures, as commonly used by transportation planners and engineers, take the form of opportunities weighted by a distance-decay function. The common assumption underlying this type of measure is that the importance of opportunities at the destination declines as the distance from an origin increases. They can be derived from a gravity trip distribution model or a logit destination choice model. For transportation modeling, a study area is divided into relatively homogeneous zones. The accessibility for zone i is

$$A_i = \sum_j W_j f(C_{ij})$$

where W_j = the amount of activity in zone j or the total trips attracted to zone j; $f(C_{ij})$ = impedance function or friction factor that represents the nonutility of travel or spatial separation between zones i and j. We obtain a normalized measure by dividing A_i by the total amount of activity in the study area.

For a common, simple example of evaluating access to jobs, W_j is the number of jobs in zone j, and $f(C_{ij})$ is typically a negative linear or nonlinear function of distance between zones i and j. Another example is exponent function ($d_{ij}^{-\beta}$). The most commonly used β-value is 1. β-value can also be calibrated in the trip distribution stage of the four-step travel demand modeling chain. This measure is sometimes referred to as the Hansen accessibility index, credited to Hansen (1959). Two other functional forms are an exponential distance decay function [$\exp(-\beta d_{ij})$] and a Gaussian function {$\exp[-\beta(d_{ij})^2]$}. For any β-value less than 1, the Gaussian function decays more gradually at the beginning and then more quickly, with increasing distance than an exponential or exponent function does. Song (1996) shows that the exponent accessibility measure with the most commonly used β value of 1 is not statistically inferior to any other measures in explaining population distribution. Overall, gravity-type accessibility measures perform better than other measures.

A measure of accessibility can also be derived for the multinomial logit model in the following form (Ben-Akiva and Lerman 1985):

$$V_n = 1/\beta \ \ln \ \Sigma_k \exp(\beta V_{kn})$$

where V_{kn} is the utility function for individual n for option k, and β is the parameter of the exponential function. This measure is, in fact, "the systematic component of the maximum utility of all alternatives in a choice set" and "a measure of the individual's expected utility associated with a choice situation" (Ben-Akiva and Lerman 1985:301). As shown in Chapter 9, it is also the average benefit perceived by the individual. It is often referred to as the log-sum index.

Gravity-type accessibility measures can be estimated by modes of transportation such as private automobiles and public transit. This is done simply by using different impedance measures by modes. To aggregate accessibility by modes, Shen (1998) proposed a general accessibility index, which takes into account the level of automobile ownership:

$$A_i^G = \alpha_i \ A_i^{auto} + (1 - \alpha_i) \ A_i^{trans}$$

where A_i^G = the general employment accessibility for residential location i;
        α_i = the percentage of workers in location i whose household has
               at least one automobile;
    A_i^{auto}, A_i^{trans} = employment accessibility for auto drivers and transit riders,
                respectively.

Gravity-type accessibility measures account for three major factors affecting accessibility: transportation system performance, quantity of opportunities, and proximity to opportunities. It could potentially include quality of opportunities, another factor affecting accessibility. Accessibility is also influenced by the socioeconomic and personal characteristics of travelers, which this measure does not take into account.

A cautionary note for using gravity-type accessibility is that the zonal system affects the results (Brunton and Richardson 1998). For a large zone structure, it is certain that some short trips, particularly intrazonal trips, are omitted. Thus different zonal systems will result in different trip length distributions. The analyst should be cautious in comparing gravity-type accessibility measures for two regions with zonal systems at different scales.

13.4.2 Using Accessibility for Equity Analysis

Accessibility measures described above can be used to conduct equity analysis for Transportation Improvement Programs (TIPs) and Long-Range Transportation Plans

(LRTPs) in the metropolitan planning process. The spatial distribution of accessibility changes associated with transportation improvement projects, programs, and plans can be related to the spatial distribution of population in the metropolitan areas. This helps answer the question of who gains and loses in a regional transportation plan. This analysis consists of three processes:

- Measure changes in the accessibility due to transportation improvement projects or plans;
- Correlate these changes with distributions of population by income and race/ethnicity;
- Compare protected populations with other population groups.

GIS-based highway accessibility can be measured in the following steps:

1. Obtain or build a road GIS layer, which includes, at least, the attributes of segment length and design (post) speed. Additional useful attributes include free-flow speed, congested speed, and turn penalty.
2. Calculate design travel time for each segment: time = segment length/design (post) speed, or calculate congested travel time = segment length/congested speed, or calculate free-flow travel time = segment length/free-flow speed.
3. Create a road network in GIS such as TransCAD, ArcView/Network Analyst, or ArcInfo. This network should include, at least, the calculated travel time data.
4. Identify origins and destinations in a GIS layer. Origins and destinations could both be TAZ centroids. Alternatively, origins could be block-group centroids, and destinations could be TAZs.
5. Run the shortest path procedure in a GIS to obtain the shortest travel time between origins and destinations.
6. Obtain data on the amount of activity in each destination zone such as total employment, retail employment, and households.
7. Calculate the accessibility index based on the aforementioned equation. This index could represent employment, retail, or residential accessibility, depending on the type of activity used in the calculation.

Use this procedure to calculate the accessibility measurements for a base-year road network. Then repeat for a year with transportation improvements. The difference between the two represents the accessibility changes caused by transportation improvements.

The accessibility index based on free-flow speed does not account for congestion in some segments of the road network, thereby representing an ideal situation and a high end of true index values. On the other hand, the accessibility index based on congestion speed represents the real-world situation. Congestion speed data can be obtained through traffic monitoring, which, however, has limited geographic coverage. Usually, we can obtain congestion speed through the four-step transportation modeling for a base year and any forecast year. Without access to a four-step transportation model and real-world speed-monitoring data, analysts could use post speed or design speed, which is readily available or easy to classify in the road network.

To estimate the accessibility index for a future year, the analyst needs not only data about transportation improvements in the plan but also socioeconomic forecast data such as total and retail employment, and households. It should be emphasized that the accessibility index takes into account not only the performance of the transportation infrastructure but also the amount of activities in the region. Therefore, any change in the accessibility index for a future year cannot be attributed to transportation improvements alone. To detect the contribution of transportation improvements, we can keep the amount of activities constant. Figure 13.2 shows the distribution of accessibility to jobs in the Baltimore metropolitan region.

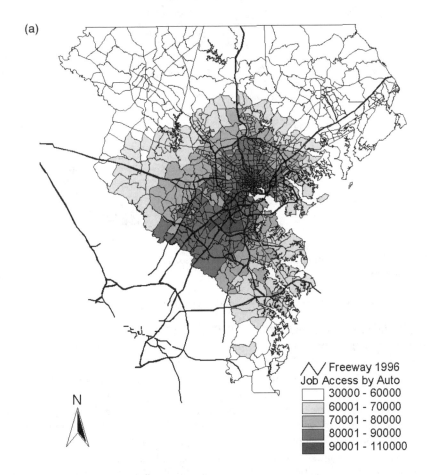

FIGURE 13.2a Distribution of accessibility to jobs in 1996 in the Baltimore metropolitan modeling domain.

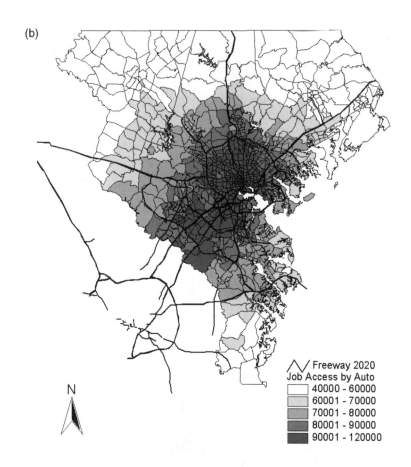

(b)

Freeway 2020
Job Access by Auto
40000 - 60000
60001 - 70000
70001 - 80000
80001 - 90000
90001 - 120000

N

FIGURE 13.2b Distribution of accessibility to jobs in 2020 in the Baltimore metropolitan modeling domain.

13.4.3 EMPIRICAL EVIDENCE ABOUT MOBILITY DISPARITY

Mobility increases with income. In 1990, people with household income under $10,000 traveled 16 mi a day (2.6 trips), less than half the distance by people with household income of $40,000 and over (38 mi and 3.6 trips) (U.S. DOT 1997b). The disparities are evident in the percentage of adults holding drivers' licenses and vehicle availability: in 1990, households with income under $10,000 had only 73% of adults holding drivers' licenses and had only 1 vehicle available, compared with 95% of adults and 2.3 vehicles for households with income $40,000 and over. In 1995, low-income households with income below $25,000 comprised 29% of all households but accounted for 65% of households without a vehicle.

Mobility disparities by race/ethnicity are significant. Regardless of income, whites travel more than any other races and Hispanics (of all races) (see Figure 13.3).

After controlling geographic locations (central city, suburban, and nonmetropolitan areas), the disparity remains (U.S. DOT 1997b). In 1990, white men in urban areas traveled 39 mi and made 3.4 trips a day on average, compared with 29.5 mi and 2.8 trips for Hispanic men and 24.1 mi and 3.0 trips for black men (Figure 13.4). In 1995, African Americans took 3.9 trips a day, compared to 4.4 daily trips per person for whites. Vehicle availability and licensing vary with race/ethnicity. In 1990, whites had the highest vehicle availability and percentage of adults holding drivers' licenses. In 1995, African Americans, while making up 11.8% of all households, accounted for 35.1% of households without a vehicle (FHWA 1997). This disparity affects the choice of transportation modes by race/ethnicity. Whites are more likely to make trips by private vehicle than minorities. In 1990, whites aged 16 to 64 in urban areas took 92% of their trips by car, compared with no more than 78% for blacks, 81% for other races, and no more than 82% for Hispanics (Rosenbloom 1995). Blacks were the most likely to take transit and walking trips, 8% and 12%, respectively, while whites were the least likely to take transit and walking trips (less than 2% and 5%, respectively). According to the 1995 NPTS data, African Americans made 76% of their trips by private vehicle, compared with 88% for whites (FHWA 1997). Conversely, African Americans were more likely to use transit and walking, accounting for 7 and 9% of their trips, respectively. Transit share of person trips was only 1% for whites and 3% for Hispanics.

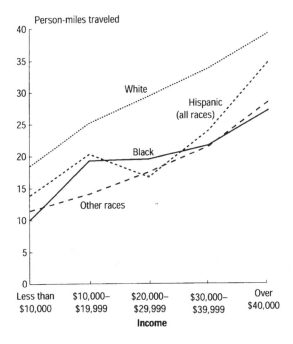

FIGURE 13.3 Average daily travel by race and income level:1990. *Source:* U.S. DOT (Department of Transportation, Bureau of Transportation Statistics). Transportation Statistics Annual Report 1997. BTS97-S-01. Washington, D.C.: U.S. Government Printing Office. 1997b.

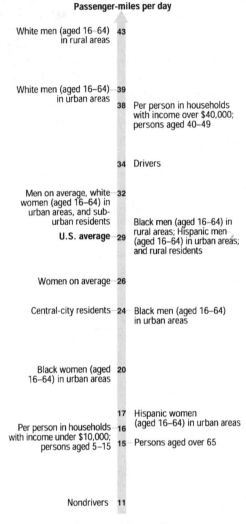

Passenger-miles per day

White men (aged 16–64) in rural areas — 43

White men (aged 16–64) in urban areas — 39

38 — Per person in households with income over $40,000; persons aged 40–49

34 — Drivers

Men on average, white women (aged 16–64) in urban areas, and sub-urban residents — 32

U.S. average — 29 — Black men (aged 16–64) in rural areas; Hispanic men (aged 16–64) in urban areas; and rural residents

Women on average — 26

Central-city residents — 24 — Black men (aged 16–64) in urban areas

Black women (aged 16–64) in urban areas — 20

17 — Hispanic women (aged 16–64) in urban areas

Per person in households with income under $10,000; persons aged 5–15 — 16

15 — Persons aged over 65

Nondrivers — 11

Passenger-miles per day

FIGURE 13.4 Miles of daily travel: 1990. *Source:* U.S. DOT (Department of Transportation, Bureau of Transportation Statistics). 1997b. Transportation Statistics Annual Report 1997. BTS97-S-01. Washington, DC: U.S. Government Printing Office. 1997b.

13.4.4 ACCESSIBILITY DISPARITY AND SPATIAL MISMATCH

While evidence about mobility disparity is abundant, accessibility disparity studies mostly focus on access to jobs and the spatial mismatch hypothesis. For many years, researchers have observed a clear contrast between the large number of jobs created in the suburbs and the high unemployment rate in inner city neighborhoods, partic-ularly among African Americans. This lack of connection between job and residence locations of potential job seekers has been a topic of debate about the so-called spatial mismatch hypothesis for three decades. More than 30 years ago, Kain (1968) found a positive relationship between the fraction of employment held by blacks and the proportion of black residents in various neighborhoods in Detroit and Chicago. The fraction of black employment declined with distance from the major black neighborhoods. From these findings, he believed that residential segregation contributed to high unemployment rates of blacks in inner cities.

After reviewing 20 years' empirical evidence on the "spatial mismatch" hypothesis, Holzer (1991) concluded that spatial mismatch had a significant effect on black employment, but the magnitudes of these effects remained unclear. A rich body of literature on the spatial mismatch hypothesis has produced mixed results, and research methodology may contribute to the conflicting findings (Sanchez 1999).

A critical analytical issue in this debate is the measurement of access to jobs. Commuting time and distance are the most frequently used measures. Gordon, Kumar, and Richardson (1989) compared average commute time of automobile commuters by income, race, place of residence, and city-size class. Their results show that neither nonwhite nor low-income commuters have longer commutes than others. However, Kasarda (1995) found that blacks had longer commutes than whites and central city to suburb commute times were exceptionally long for public transit users, particularly blacks. O'Regan and Quigley (1997) also reported longer commute times for blacks than for other ethnic groups. After controlling for travel modes and residence and workplace locations, they still found slightly longer commutes for blacks. These results provide support for the spatial mismatch hypothesis. Although finding a more than 2-min. longer commute on average for black workers after controlling other factors, Taylor and Ong (1995) attributed the extra time to slower travel speed rather than spatial mismatch. Overall, the bulk of research indicates that most researchers agree that blacks have a longer commute time than other groups, but they do not agree on what causes the differences or what the policy and planning implications of the differences in terms of transportation barriers to job access are (Shen 2000).

Two methodological issues weaken previous research. As discussed above, commute time and distance are poor accessibility measures, and are especially unreliable measures of relative access to jobs (Sanchez 1999). This inadequacy is worsened by the use of aggregate data at large geographic levels. Most analyses are based on the dichotomy of the central city and suburb. Shen (2000) argues that this dichotomy blurs the spatial variations of accessibility among neighborhoods and particularly misrepresents that of disadvantaged population groups. Such a spatial bias is evident in the country's 20 largest metropolitan areas. In 13 metropolitan areas, workers residing in the largest poverty enclave in the central city had the longest average commute times, usually 10% more than the other areas.

An improvement in the recent literature is use of gravity-type accessibility measures and fine-grained units of analysis. Sanchez (1999) used gravity-type accessibility measures for access to job and a two-stage least-square regression for estimating the relationship between labor participation levels and access to public transit in Portland, Oregon and Atlanta. He found that access to public transit was a significant factor in explaining labor participation and provided support for use of public transportation for addressing the spatial mismatch phenomenon. Shen (2000) used the general accessibility index based on a gravity formulation and multiple regression models to explain the variation in commute times. In the Boston metropolitan area, general employment accessibility was very significant in explaining variation in commute time; increasing the accessibility index by one standard deviation (0.44) would result in a decrease of the average commute time by 2 min. (0.4 standard deviation). These studies did not show how accessibility measures themselves vary by different population groups. There is some evidence that suggests accessibility disparities vary by race/ethnicity. In Los Angeles, accessibility was found to be lower for middle-class people than either low- or high-income people (Wachs and Kumagai 1973).

As mentioned earlier, some metropolitan planning organizations have recently adopted accessibility, particularly the cumulative-opportunity type, as an equity measure. The 1998 Regional Transportation Plan (RTP) of the Southern California Area Governments (SCAG) established eight categories of objectives and performance indicators: mobility, accessibility, environment, reliability, safety, livable communities, equity, and cost-effectiveness (SCAG 1998). As an objective in the RTP, accessibility is defined as ease of reaching opportunities and measured by percent of commuters who can get to work within 25 min. For equity evaluation, the RTP measures accessibility improvement by income and race/ethnicity in two modes: increased accessibility (trips < 30 min.) by transit and increased accessibility (trips < 30 min.) by auto. Results show that all groups benefit from increased accessibility when compared with the baseline scenario. The low-income group particularly benefits from the improved accessibility due to transit restructuring.

Access is also a priority in the Vision of the National Capital Region Transportation Planning Board (TPB), an official Metropolitan Planning Organization for the Washington metropolitan region. To address federal requirements on environmental justice, TPB staff used accessibility measures to evaluate the 1999 Financially Constrained Long-Range Transportation Plan (MWCOG 1999). It is cumulative-opportunity type accessibility, defined as jobs that can be reached by mode (auto, transit, and the fastest travel mode) within 45 min. The 45-min. travel time captures 78% of the work trips in 2000 and 68% of the work trips in 2020. Results show that changes in accessibility between 2000 and 2020, as a result of the long-range plan, will not disproportionately affect low-income and minority populations.

13.5 MEASURING DISTRIBUTIONAL IMPACTS ON PROPERTY VALUES

Transportation improvements change accessibility, which affects the location choice of households and firms. These impacts influence distributional patterns of activities. When making their location decisions, households and firms are willing to pay a certain amount to satisfy their demand for accessibility. Therefore, property value reflects the value of accessibility, and transportation investment benefits are capitalized in the land market.

Hedonic pricing method, as discussed in Chapter 4, can be used to measure the implicit price for accessibility. Including an accessibility index in the hedonic price function would yield an implicit price that households or firms are willing to pay for a certain degree of accessibility. This analysis involves

- Model specification,
- Variable estimation,
- Model estimation,
- Application of implicit price to analysis zones,
- Segmentation of analysis zones by population groups,
- Tabulation of the distribution of benefits by different groups.

As shown in Chapter 4, the hedonic price function is specified so that the dependent variable is housing price and the independent variables include structural, neighborhood, locational, and environmental characteristics. The accessibility index, as part of location characteristics, can be estimated for highway, transit, or their combination, using the method discussed previously. If the hedonic function is linear, the regression coefficients for accessibility variables represent the value of capitalized benefits per unit accessibility index. Since various locations have different degrees of accessibility, applying the regression coefficients to an analysis zone would give us the total value of capitalized benefits from transportation improvements for that particular zone. These capitalized benefits accrue to the residents in that particular zone. By analyzing population composition in all analysis zones, we are able to see how benefits are distributed among population groups.

Sanchez (1998) uses the hedonic price method to examine the distribution of capitalized transportation benefits within the Atlanta urbanized area. The geographic unit of analysis is block groups. A gravity-type highway accessibility index is used to measure the change in highway accessibility associated with the addition of major highway facilities. The dependent variable is owner-occupied house value per square foot (median house value for each block group divided by the average block-group house size from tax assessor's records). The independent variables include average house age, highway access, bus access, rail access, percent homes on public sewer, commercial and industrial land, property tax millage rate, sales taxes, per capita transit fares, standardized test scores, and student-to-teacher ratio. Results show that the average per square foot capitalized benefit is about \$4.33, which equivalently represents 6.83% (\$7,365) of the total value of the average-priced house (\$107,863). Net transportation benefits have an intrametropolitan variation; the central city receives the largest benefit while the suburbs obtain the smallest. The results do not indicate any significant bias toward particular income or racial groups. Unlike almost all other hedonic studies in the current literature, this study uses census data rather than armslength sales (i.e., sales not between people related by kin or business) data.

This type of analysis is useful for evaluating long-range transportation plans and transportation improvement programs at the metropolitan and regional level. However, it misses some microlevel impacts associated with transportation improvements. While transportation improvements benefit houses at large, major transportation facilities actually depress the property values of those houses in close proximity. Early studies indicate up to a 10% decline in property value for houses immediately abutting a freeway, especially an elevated freeway (FHWA 1976). Homes near freeway interchanges are also sold at a discount. In Alameda and Contra Costa counties of the San Francisco Bay Area, the 1990 sales price of a home declined $2.80 and $3.41, respectively, for every meter closer to a freeway interchange (Landis and Cervero 1999). On the other hand, new 4-lane roads generally raise land values by 3% to 5% for houses within a 0.5-mi radius, and the impact can extend further, as much as 15 mi (National Transit Institute 1998).

Residential property values and rail transit are also positively related. New transit systems or transit improvements also increase property values along the transit corridor. Boyce et al. (1972) found that the PATCO rapid transit line between Philadelphia and Lindenwold had a positive effect on residential property values. In Portland, Oregon, the average house price within a 500-m walking distance of selected light-rail stations was 10.6% ($4,300) higher than that of houses within the study area but beyond 500 m (Al-Mosaind, Dueker, and Strathman 1993). A hedonic price model reports a statistically weak negative price gradient with respect to distance for houses within 500 m ($21.75 per m from a station). A much lower price gradient was reported for houses near BART stations in San Francisco (Landis and Cervero 1999). Homes near BART stations command a premium of $2.29 per m from the nearest BART station (measured along the street network) in Alameda County and $1.96 in Contra Costa County. In Washington, D.C., Metro generated considerable land rent premium; these premiums were most dramatic in older, more deteriorated sections of downtown (Rybeck 1981). In Atlanta, elevated heavy-rail stations increased housing prices in lower-income neighborhoods nearby but decreased house prices in a higher-income neighborhood (Nelson 1992). The findings indicate that the balance of benefits and costs associated with proximity to an elevated transit station might manifest itself differently depending on the neighborhoods.

Evidence also shows positive impacts of transit on commercial property values adjacent to transit stations, although it is more sketchy than the residential literature. In the Midtown MARTA station area of Atlanta, the sales price per square meter of commercial buildings declined by $75 for each meter away from the center of transit stations and increased by $443 for location within special public interest districts (Nelson 1998). In Washington, D.C., Metro generated commercial rent premium of $2 per square foot for prime blocks that had a Metro station entrance or were across the street from one and $1 per square foot for blocks adjoining prime blocks (Rybeck 1981). In Santa Clara County, the lease premium was 3.3 cents per square foot for properties within 0.25 mi of a light-rail station and 6.4 cents per square foot for those between 0.25 and 0.5 mi (Weinberger 2000). The sales premiums were $8.73 and $4.87 per square foot, respectively.

Clearly, these capitalized benefits or costs can happen at the microscale, station-specific level. Therefore, equity impact assessment of these types requires a more fine-grained unit of analysis than the regional level. It appears that it is necessary to go to the block level, even the parcel level to detect these microscale impacts. A GIS-based methodology is essential. In particular, GIS can be used to geocode the exact locations of sales transactions, to measure the distance to the nearest transit stations and to the nearest freeway interchanges, to identify property parcels along the roadway, to estimate socioeconomic data at the fine scale using data at the larger geographic units, and to estimate accessibility measures.

13.6 MEASURING ENVIRONMENTAL IMPACTS

In Chapters 8 and 9, we discussed the integration of urban and environmental models in a GIS environment. This integration serves as a framework for evaluating the environmental impacts of transportation systems, projects, plans, and programs on different population groups. This methodology follows a sequential process (Table 13.4). Land and transportation models generate data about traffic speed, traffic volume, and vehicle trips, which are input to emission modeling, air quality modeling, and noise modeling. Ambient pollutant concentration and noise level data, which are produced from air quality and noise models, are then processed in a GIS environment and displayed as contour lines. By overlaying these contour lines with the distribution of protected populations, we are able to see the distributional impacts of transportation projects, plans, and programs.

A few analytical issues need the analyst's attention. First, it is essential to use a fine geographic unit of analysis to uncover some environmental impacts of transportation on different segments of population (Forkenbrock and Schweitzer 1999). As discussed earlier, air pollution and noise from transportation are generally constrained by the proximity of transportation facilities. Line-source dispersion models usually predict pollutant concentrations in the vicinity of highway; for example, CAL3QHCR predicts air pollutant concentrations in receptors located within 150 m of the roadway (Benson 1994). Just like measuring property value impacts, measuring differential environmental impacts requires demographic data at a fine unit of geography, such as the census-block level.

Second, sociodemographic data may not be available at the fine geographic unit, and assumptions and alternative data sources have to be used. At the census-block level, we still have to assume that population is homogeneous within a census block. As we know, the first row of houses abutting the roadway bears the brunt of adverse environmental impacts. This fact makes some researchers wonder whether it is necessary to use parcel level data. With the help of GIS and property databases, those houses adjacent to the roadways can be identified. However, demographic data are not available at the parcel level. If the study area is small, these data can be obtained through a survey. A large study area may preclude data collection through a survey due to budget constraints. Instead, some statistical methods may be used to derive estimates for a smaller geographic unit from available data at a larger geographic unit.

Third, while some transportation-related environmental impacts are localized such as carbon monoxide, others have a regional scope such as ozone. Reactive pollutants such as ozone and its precursors nitrogen oxides and VOCs are modeled differently from inert pollutants. Nitrogen oxides and VOCs themselves have health impacts. The modeling domain for ozone often covers a few states, while land-use transportation modeling often operates at the metropolitan area level. The analyst needs to develop an interface among different modeling domains and units of analysis for integrating urban and environmental models.

TABLE 13.4
Integrating Urban Models, Environmental Models, and GIS
for Transportation Equity Analysis

Modeling	Input	Output	Examples of Modeling Packages
Land-use/Transportation Modeling	Socioeconomic data	VMT	ITLUP
• Land-use model	Land-use data	Vehicle trips	(DRAM/EMPAL)
• Vehicle availability	Transportation network	Travel speed	MEPLAN
• Trip generation	and database	Traffic volume	TRANUS
• Trip distribution			TP+ (MINUTP)
• Modal split			TRANPLAN
• Trip assignment			TransCAD
Emission Modeling	Traffic mix	Vehicle emission	MOBILE
	Traffic speed	factors	EMFAC (California)
	Temperature		
	Precipitation		
Air Quality Modeling	Emission factors	Ambient	ISC3
	Traffic speed	pollutant	PART5
	Traffic volume	concentration	AP-42
	Queue lengths		CALINE, such as
	Lane widths		CAL3QHCR
	Wind speed		
	Wind direction		
Noise Modeling	Traffic volume	Noise level	STAMINA
	Traffic mix		
	Traffic speed		
Human Exposure Modeling	Human activity patterns	Exposure	Human Exposure Model
	Ambient pollutant concentration		SHAPE REHEX
GIS	Ambient pollutant concentration	Spatial relationship	ArcView/ArcInfo MapInfo
	Noise level	between	TransCAD/Maptitude
	Exposure	environmental	
	Transportation network	quality/exposure	
	Census geography	and population	
	Census data	groups	

13.7   EQUITY ANALYSIS OF TRANSPORTATION POLICIES

A large body of literature on welfare economics and public finance deals with public policies, or primarily the distribution of costs and benefits incurred by taxes and public infrastructure and services.

Transportation infrastructure and services are funded by a variety of taxes such as federal and state fuel gasoline tax, state use fees, state sales tax, local sales tax, federal and state income tax, and property tax. Giuliano (1994) reviews the literature on the incidence of taxes that are used to fund highway infrastructure and services. She concludes that these taxes are regressive overall.

The clash between some mainstream environmental and environmental justice groups and between environmental and equity goals is demonstrated in various road pricing or value pricing proposals. Economists and some environmental groups argue that auto driving has been underpriced and subsidized and does not account for external costs caused by driving. To combat air pollution problems that have eluded any solution in many metropolitan areas, they have proposed a variety of pricing mechanisms such as congestion pricing, tolls, Pay-As-You-Drive insurance, and VMT charges. They believe that these mechanisms will force drivers to face the true cost of driving and reduce the bias toward overconsumption of driving. Environmental justice groups fear that these mechanisms would price the poor and often people of color out of roads. Another argument against road pricing, like State Route 91 in southern California, is that it will encourage sprawl because any relief in congestion as a result of pricing mechanism will facilitate the rich moving further away.

Many believe that congestion and other pricing policies are not necessarily regressive. One reason is that peak period drivers tend to be overwhelmingly middle- and upper-income people (Hattum 1996; Lee 1989). In San Francisco and Minneapolis, low-income drivers account for just 2% and 3%, respectively, of commuter trips (Komanoff 1996). Another argument for congestion and other pricing mechanism is that they will reduce traffic congestion, air pollution, and automobile dependency and, thereby, benefit the low-income population. In particular, as discussed in Chapter 10, studies have shown that a low-income population tends to be exposed to a higher degree of air pollution nationwide.

Cameron (1994) evaluates efficiency and equity of existing transportation systems and a 5-cent-per-mile VMT in Southern California. The study measures transportation benefits by five income groups using the willingness-to-pay methodology. Transportation costs by income groups include automobile expenses (ownership, maintenance, fuel, and insurance), transit fares, transportation taxes and fees, health costs of air pollution, and traffic congestion costs. The distribution of net transportation benefits (benefits minus costs) for the current transportation systems is regressive: the lowest income group receives $650 (6% of the regional total), the lower-middle income group receives $1,430, the middle-income group receives $1,980 (19% of regional total), the upper-middle-income group receives $2,880, and the upper-income group receives $3,750 (35% of the regional total). Comparing it with personal income distribution, the current transportation system redistributes transportation benefits toward those in the middle. A 5-cent-per-mile VMT fee would make each income group better off: an increase in net transportation benefits from $90, $150, $180, $250, to $420, respectively, for the lowest-income to the highest-income group. This result is subject to a wide band of uncertainty because of highly debatable assumptions about the value of travel, time, and health. The author concludes that the result is plausible for the upper three, perhaps four, income groups, but highly questionable for the lowest income group (Cameron 1994). From the utilitarian perspective, this policy proposal is desirable since everyone seems to be better off. However, it aggregates the existing inequity, and is certainly unacceptable to egalitarians. It does not please Rawlians because the lowest income group receives little.

Litman (1997) presents a total cost perspective for transportation equity analysis. He divides transportation system users into four classes: non-drivers, low-income drivers, middle-income drivers, and upper-income drivers. Further, he divides transportation costs into six categories: user market (e.g., direct costs), user non-market (e.g., travel time and comfort), external market (e.g., subsidies), external environmental (e.g., air and water pollution), automobile dependency, and economic (e.g., changes in consumer prices and employment). Using this classification scheme, he summarizes equity impacts of an automobile-user price increase, transit subsidy, and traffic management strategies. Is a pricing mechanism such as congestion pricing or road user fees equitable? Litman (1997) concludes that how revenues are distributed determines the overall equity impacts of an automobile-user price increase. It can be progressive if revenues are targeted at low-income households. Transit subsidies are generally progressive. In particular, urban bus service is highly progressive, while commuter bus and suburban rail services are regressive. Traffic management strategies such as traffic calming and neotraditional street design are progressive.

In its 1998 long-range transportation plan, SCAG (1998) measures equity in terms of transportation investment benefits by various income groups. Equity performance indicators for various income groups consider time savings, the value of time saved, use of expenditures by mode and access to opportunities. Specifically, transportation benefits by various income groups over a planned period include hours saved (reduced delay) due to transportation investment (percent of total hours saved), monetary value of hours saved (percent of total savings), and percent of total expenditures on programs and projects that are used by various income groups.

SCAG analysis shows that 13% of households live in poverty (with household income less than $12,000) but receive only 2.3% of transportation investment benefits (as measured by value of time saved) in the 1996 to 2003 Regional Transportation Improvement Program. They are the dominant users of bus and urban rail. Under the baseline scenario, the poverty population will share 9% of the total hours saved and 2% of the total monetary value of time saved in 2020. Not surprisingly, high-income households with annual household income of $70,000 and above receive the largest benefit: 48% of total value of time saved and 25% of the hours saved. By either measure, the plan shows substantial improvements for the poverty population; shares increase to 16% and 4.5%, respectively, for hours saved and value of time saved. By contrast, the plan shows decreased shares in benefits for high-income households. Under both the baseline and plan, expenditures on programs and projects that are used by poverty population exceed expenditures on programs and projects that are used by high-income households. The poverty population's share of total expenditures increase by 2 percentage points to 28% under the plan, while high-income households' share remains at 15%. Poverty populations contribute almost 7% of all gasoline and sales taxes collected to fund transportation.

13.8 ENVIRONMENTAL JUSTICE OF TRANSPORTATION IN COURT

On August 31, 1994, The NAACP Legal Defense and Educational Fund, Inc. (LDF) filed a class action civil rights lawsuit against the Los Angeles County Metropolitan Transportation Authority (MTA), challenging MTA's proposal to increase the one-way cash fare for a bus ride from $1.10 to $1.35, to eliminate the $42 monthly bus pass, and to raise the fares on the Blue Line rail system. This lawsuit came at a time when a huge amount of money was earmarked for building the rapid rail system although the bus system was badly deteriorated in Los Angeles County. LDF filed the class action suit on behalf of the Labor/Community Strategy Center, the Bus Riders Union, the Southern Christian Leadership Conference, the Korean Immigrant Workers Advocates, and 350,000 poor minority bus riders. On October 28, 1996, both parties entered into a Consent Decree settling the civil rights class action (Garcia et al. 1996; Garcia and Graff 1999).

The lawsuit (Labor/Community Strategy Center et al. vs. Los Angeles County Metropolitan Transportation Authority et al.) challenged the MTA on two grounds:

- The MTA intentionally discriminated against poor minority bus riders,
- The MTA's actions had a discriminatory impact on poor people of color.

The complaint alleged that the MTA violated the Equal Protection Clause of the Fourteenth Amendment of the United States Constitution, 42 U.S.C. §§ 1981 and 1983, Title VI of the Civil Rights Act of 1964, and the regulations under that statute. Plaintiffs established the evidence of a discriminatory impact by showing

- Racial ridership disparities
- Adverse subsidy and service disparities

The case raised several analytical issues including the standard of disparity and comparison population. For a Title VI challenge, plaintiffs have the initial burden of demonstrating that the complained conduct, whether facially neutral or not, has adverse disparate impact on the protected class. Once a plaintiff has established a *prima facie* case, the burden shifts to the defendants to prove that the alleged conduct was justified by business necessity. Then, plaintiffs may rebut the claim of necessity by showing the existence of other nondiscriminatory alternatives.

How can we define a disparity? The Supreme Court established the *Castaneda* standard in *Castaneda v. Partida*, an equal protection jury discrimination case. The case challenged grand jury selection procedures of Hidalgo County, Texas, which resulted in grand juries that ranged between 39 and 50% Hispanic, although the pool of eligible grand jurors was 79% Hispanic. The Court held that a difference between an actual selection rate and an expected selection rate in a binomial distribution greater than two or three "standard deviations" was "suspect." The Court defined a standard deviation, which measures the fluctuation from the expected value, as the square root of the product of the total number in the sample times the probability of selecting a minority group member times the probability of selecting a majority group member. This *Castaneda* standard essentially says that any difference larger than two or three standard deviations between actual and expected values is statistically significant enough to be a disparity.

The MTA raised the following claims:

1. While the population of Los Angeles County as a whole is 60% minority, the ridership of MTA rail lines is more than 60% minority.
2. DOT guidelines classify a transportation line as a "minority line" based on the racial composition of the census tracts that the line traverses.
3. The ridership of MTA-operated rail lines consists of a majority of minorities.

Plaintiffs argued that courts had consistently emphasized the use of most likely affected target population as a comparison group and rejected the use of general population statistics as a proxy for the relevant target population to analyze disparate impact claims in Title VII employment discrimination cases. In Latimore v. Contra Costa County, the court held that defendants "inappropriately relied on a sample population comprised of all County residents, as opposed to the target population of County residents who are eligible to use [the County hospital for the poor]. In conducting a disparate impact analysis, the appropriate sample population should consist of those most likely affected by the action at issue."

In Labor/Community Strategy Center et al. vs. Los Angeles County Metropolitan Transportation Authority et al., the class consists of minority MTA bus riders, not minority residents generally. Therefore, the comparison group was minority MTA bus riders, which made up 80% MTA ridership. Although the minority ridership of MTA-operated rail lines apparently exceeds the minority population of the County as a whole, it is significantly smaller than the expected 80% minority MTA ridership, the legally relevant comparison.

MTA claimed that U.S. DOT guidelines classify a transportation line as a "minority line" based on the racial composition of the census tracts that the line traverses. Plaintiffs argued that the racial composition of a transit corridor does not necessarily represent the racial composition of actual ridership. For example, a transit line may have mostly white riders but may be classified as a "minority line" because it passes through minority census tracts. In this case, demographics of actual ridership is more relevant than that of the transit corridor. This focus on actual ridership is different from attention to the corridor communities when evaluating environmental impacts of a transit line.

MTA's claim that the ridership of MTA-operated rail lines consists of a majority of minorities raises another intricate question. That is, does a situation where the racial minority is the majority have any disparity? Courts focus on proportionality rather than the dichotomy of minority and majority. That is, it is disproportionate representation in the affected group that establishes adverse disparate impact, not the presence of a minority group in the majority in a particular situation.

13.9 SUMMARY

In this chapter, we have dealt with a wide range of issues in evaluating transportation equity. Equity concerns can arise at various stages of the transportation planning process, at the project and plan level, and during transportation system operations. Transit operation and service have long been a major concern in the civil rights movement, but equity implications of transportation planning have not received attention until recently. To incorporate equity issues into the transportation planning process, planners need to reach out and engage the disadvantaged groups of society for public participation on one hand and enhance their analytical capabilities to evaluate equity impacts of proposed projects and plans on the other hand.

As shown in this chapter, planners can adapt a variety of existing tools and methods for equity analysis. Transportation performance measures, which have been used to evaluate air quality impacts and long-range transportation plans, can be segmented into transportation users by race and income. Accessibility measures, particularly gravity-type measures, are very useful to evaluate the impacts of proposed transportation projects and plans on different population groups. These measures can be estimated with the help of the commonly used four-step travel demand forecasting models and GIS. Four-step transportation or land-use models may also generate consumer welfare measures by different populations. Hedonic price methods can be used to assess capitalized benefits of accessibility improvements in the land market. Integration of land use/transportation models, environmental models, and GIS has great potential for conducting a multi-objective analysis of efficiency, equity, and environmental sustainability.

14 Trends and Conclusions

14.1 INTERNET-BASED AND COMMUNITY-BASED TOOLS

We are in an era of information and high tech. The Internet and high technology tools help democratize the society. Environmental justice analysis has benefited a lot from high technology innovations, and we have talked about the roles of GIS in environmental justice analysis. We have recently seen a rapid growth of web-based databases and GIS servers, community-based GIS and mapping, and more user-friendly tools. In the following, we present an overview of three Internet-based or GIS-based tools that can help the public perform their own environmental justice analysis.

14.1.1 EPA's ENVIRONFACTS

Envirofacts is EPA's database warehouse that consists of individual databases from various EPA programs:

- Aerometric Information Retrieval System (AIRS) — air pollution information such as criteria air pollutant emissions data for major point sources and air quality data at monitoring sites;
- RCRIS and Biennial Reporting System (BRS) — hazardous waste information such as location of TSDFs and Large Quantity Generators (LQGs) and amount of waste generated or managed;
- CERCLIS — inactive hazardous waste site information such as Superfund sites and contaminated waste sites;
- TRI — toxic release information for TRI facilities;
- Risk Management Plans — for about 64,000 facilities nationwide;
- PCS (Permit Compliance System) — information about wastewater discharge facilities;
- SDWIS (Safe Drinking Water Information System) — information about drinking water;
- National Drinking Water Contaminant Occurrence Database;
- Drinking Water Microbial and Disinfection Byproduct Information.

Maps on Demand (MOD) is a set of web-based mapping applications (Enviro-Mapper, Query Mapper, SiteInfo, BasinInfo, CountyInfo, and ZipInfo) that allow users to generate environmental maps through access to the Envirofacts Warehouse. EnviroMapper provides some basic GIS functionality; for example, you can turn on and off layers and specify a layer for query. Users can visualize environmental data in Envirofacts, view detailed reports for EPA-regulated facilities, and generate maps dynamically. Three spatial levels are currently available: national, state, and county. EnviroMapper accesses EPA's spatial databases such as the National Shape File Repository. Mapping and GIS functionality include displaying multiple spatial layers, zooming, panning, identifying features, and querying single Envirofacts points. Query Mapper displays the results of Envirofacts queries and can be used to map facility locations and view the surrounding demographics, Geographic Retrieval and Analysis System (GIRAS) land use and land cover, and other features. This application is particularly useful for conducting site-based environmental justice analysis. SiteInfo and ZipInfo provide maps and reports about EPA-regulated facilities and demographic information at the site and ZIP code level and are also useful for community-based equity analysis.

EPA is enhancing its web-based GIS functionality, including development of an Internet Address Matching System for environmental data (Zhang and Dai 1999). This system was incorporated into EnviroMapper, and its initial application was the Region 5 Intranet Environmental Justice MapObjects Tool. Users can use this tool to view environmental justice data by using facility name, Superfund ID, longitude/latitude, or an address. GIS functionality includes data layer overlay, buffering according to the user-provided address and radius, and database query. Demographic data such as minority and low-income population are at the block-group level.

Envirofacts, particularly Maps on Demand, uses a variety of EPA's spatial data. The National Shape File Repository contains spatial data (in the shape file format) from the U.S. Geological Survey, the U.S. DOT, and the EPA Spatial Data Library System, Wessex, and Geographic Data Technologies. The EPA Spatial Data Library System is a repository for EPA's new and legacy geospatial data holdings (in ArcInfo format). These spatial data are at the county, state, and national levels and at the scale of 1:100,000 (county), 1:250,000 (state), and 1:2,000,000 (state and national). As discussed in Chapter 13, TRI has been notorious for its inaccuracy in facility locations (longitudes and latitudes). EPA's Location Data Improvement Project (LDIP) is intended to improve the quality of location data for EPA-regulated facilities and sites, operable units, and environmental monitoring and observation locations. The project's goal is to obtain and store these data by the end of calendar year 2000, and the Location Data Policy sets the goal for measurement accuracy as ±25 m. Through this project, EPA has established the Location Reference Tables (LRT) as a repository for location data. Currently, the LRT contains location data from AIRS/AFS, CERCLIS, PCS, RCRIS, and TRIS. Users can obtain latitude/longitude coordinates in the detailed facility report through Environfacts Query. Alternatively, users can use the EZ Query to build a tabular report or a Comma Separated Value (CSV) file for downloading. The LDIP is particularly important for conducting rigorous environmental justice analysis.

Envirofacts Warehouse provides a huge amount of data available to the general public; no doubt the data are becoming more and more accessible and more accurate. Now it is possible for an academic researcher to obtain much needed databases by downloading directly from the web. This certainly facilitates further research in the environmental justice area. For the general public who are interested in environmental justice issues, some rudimentary analysis can be done using the web. However, the EPA's web data are mostly in the form of proximity and emission measures and do not represent actual risks.

14.1.2 LANDVIEW™ III

LandView™ III consists of a database query and search engine and a mapping engine (MARPLOT for windows). MARPLOT stands for Mapping Application for Response, Planning, and Local Operational Tasks. LandView III databases include

- Demographic and socioeconomic data for the 1990 Census;
- EPA-regulated site locations and information;
- TIGER/Line® map data; and
- Miscellaneous public structures and facilities.

Geographic units for the census data are as follows:

- Legal entities such as States, Counties, MCDs, Incorporated Places, congressional districts;
- Statistical entities such as Metropolitan Areas (MAs), CDPs, Census Tracts/Block Numbering Areas, Block Groups, and Alaska Native Village.

EPA-regulated site locations and information include criteria air pollutant emissions data for major point sources and air quality data at monitoring sites; TSDF and Large Quantity Generator (LQG) locations and amounts of waste generated or managed; Superfund sites; TRI facilities; wastewater discharge facilities; watershed boundaries and watershed indices [data source — EPA's Index of Watershed Indicators (IWI)]; ozone non-attainment areas.

Other map layers include dams, airports, nuclear sites; highways and waters; schools, hospitals, religious institutions, and cemeteries; ZIP Codes; and brownfields pilots.

LandView III provides rudimentary functions such as mapping capabilities for displaying, searching, and identifying map objects, thematic mapping, and printing maps and reports. In LandView III, you are able to

- Identify the census tract and block group based on a street address or point location on a map;
- Identify the census tract and block group based on latitude and longitude data in a user file;
- Summarize the demographic and socioeconomic characteristics of the population within a radius from a given point;
- Query databases and map objects and export the search results to a file;
- Create a user-defined map layer.

For environmental justice analysis, you can use the proximity analysis tools in LandView III. You can select census-block groups, for example, within a mile from a facility and summarize population characteristics of those block groups.

14.1.3 ENVIRONMENTAL DEFENSE'S SCORECARD
(HTTP://WWW.SCORECARD.ORG/)

Scorecard is a web-based community right-to-know tool that allows the public to identify environmental risks in their communities. Scorecard provides detailed reports on the health risks of selected pollutants and environmental priorities in different areas of the country. Reports can be obtained at the national, state, county, ZIP code, and facility level. Scorecard covers the following major sources of pollution or exposures to toxic chemicals:

1. Six most common air pollutants (based on the National Emissions Trend database and the AIRS);
2. Almost 150 air toxic chemicals (based on the U.S. EPA Cumulative Exposure Project and health effects information);
3. Toxic chemical releases into the environment from manufacturing plants (based on TRI)
4. Animal waste generated by factory farms (based on livestock population data of the U.S. Department of Agriculture and waste factors).

Scorecard conducts a screening-level risk assessment, which incorporates potential exposure (ambient concentrations) and toxicity information. A chemical's toxicity information is based on EPA's risk assessment values or nationally applicable media quality standards. Risk assessment values are summary measures of the toxic potency of a chemical and have separate numbers for carcinogens (potencies) and non-carcinogens (reference doses or concentrations). These values are included in EPA's four databases: the Integrated Risk Information System, the Health Effects Assessment Summary Tables, the Office of Pesticide Programs Reference Dose and Cancer Potency tracking systems, and the Superfund Chemical Data Matrix. Scorecard also uses risk assessment values derived by California regulatory agencies. Media quality standards are legal limits on the chemical concentrations in air, water, or soil such as the NAAQS (see Chapter 10).

Exposure data come from monitoring in relevant environmental media (air, water, food) and model-based estimates. As discussed in Chapter 4, the EPA Cumulative Exposure Project estimates ambient concentrations of 144 hazardous air pollutants at the census-tract level for the entire U.S.

Scorecard uses risk assessment methodology (discussed in Chapter 4) to estimate the potential health risk associated with outdoor exposures to hazardous air pollutants. For cancer risks, Scorecard estimates an upper bound of added cancer risk by multiplying the estimated dose of a chemical an average individual would receive from its predicted concentration by its cancer potency. For noncancer risks, Scorecard derives a hazard index by dividing the estimated dose of a chemical an average individual would receive from its predicted concentration by its reference concentration. Additivity was assumed for multiple chemicals, and population-weighted averages were used for aggregation from census tracts to county, state, and national levels.

Scorecard can provide users with the top 20% of facilities (or zip codes, counties, or states) that have the largest pollution releases or waste generation. Scorecard ranks facilities or geographic areas using only TRI data. Scorecard ranks can be based on pounds of reported TRI chemicals, benzene-equivalents for cancer hazards, and toluene equivalents for noncancer hazards. Users can conduct ranking from 39 different categories, such as cancer and noncancer hazards, air and water releases of chemicals associated with recognized or suspected health effects, different types of environmental releases and transfers, or total production-related waste.

Scorecard represents the most sophisticated web-based methodology that presents environmental risk information to the public. Based on the risk assessment methodology, it has more accurate environmental risk measures than emission data. These measures can be compared against the same benchmarks, essentially compressing a huge amount of information into a few numbers. This represents a better communication tool. However, the public can easily get lost in technical jargon as the methodology becomes increasingly complicated. Other important caveats remain. Scorecard was not designed for the purpose of environmental justice. It does not contain socioeconomic and demographic data and does not have fine-grained spatial resolution, which is necessary for community-based environmental justice analysis. Although it uses ambient air toxic concentration data at the census-tract level, Scorecard warns users of uncertainties in the accuracy of exposure data that increase with increasingly smaller geographic units such as census tracts.

14.2 TRENDS AND CONCLUSIONS

Three decades ago, the distributional impact issue of air pollution received researchers' attention. Nothing seemed to happen in the following decade. Suddenly, toxic and hazardous wastes became buzzwords, and siting of hazardous waste management facilities put environmental justice issues in the national spotlight. Local communities were motivated and organized to confront environmental risks and, in particular, minority communities wrestled with the issue of the relationship between race and environmental hazards. However, the research community was not motivated at all. Most were busy, talking about efficiency and rationality. They did not seem to care about the equity issue until one day they found that environmental justice was on the national environmental policy agenda.

We have seen an intensive debate since early 1990. This debate is not simply whether there is a correlation between the distribution of environmental risks and exposures and the distribution of minority and low-income populations. As noted in Chapter 1 and throughout this book, this debate goes deeper, into some fundamental questions about inquiry and the best ways of knowing and acting. Should we rely on calculation or communication? What is really environmentally just? (See Chapter 2.) The century-dominating paradigm in epistemology — positivism — has been challenged. The phenomenological perspective or participatory research has been called upon to help deal with environmental justice issues. However, they are not a panacea. This does not mean that we should forget about the question of what constitutes justice or equity. Should we follow the utilitarian notion of equity, Rawls' theories of contractarian justice, the egalitarian notion of equality, or the libertarian notion of freedom? (See Chapter 2.) We all love justice, but we have different notions of justice or equity. We still need to know whether there is inequity, what it is, and why it exists.

We still have to wrestle with a wide range of methodological issues (see Chapter 3). The whole positivist process is subject to debate in environmental justice analysis. Contested issues include, among others, scientific reasoning, validity, causality, ecological fallacy vs. individualistic fallacy, comparison (control) population, units of analysis, independent variables, and statistical analysis. What is the appropriate unit of analysis to define an affected neighborhood? (See Chapter 6.) What is the appropriate control population as a comparison benchmark? How can we effectively measure environmental impacts? (See Chapter 4.) Who are the disadvantaged groups of the society? How can we quantify their distribution? (See Chapter 5.) Which statistics and statistical methods should we use? (See Chapter 7.) Should we care about who came first — residents or the LULU? (See Chapter 12.) What has happened to the LULU-host neighborhoods since the LULU's operation? What causes an inequity — market dynamics, discriminatory siting practice, unequal enforcement of environmental laws and land-use regulations, neighborhood invasion–succession and life cycle, uneven provision of municipal services, or discriminatory practices in the housing market? Is the inequity simply a product of urbanization and industrialization?

Some of the issues have been resolved, but a lot more remain. The debate on geographic units of analysis is more than census tracts vs. ZIP codes. Neither of them could serve environmental justice analysis adequately. In fact, none of the census geographic units fit well in the real world, where multiple and cumulative environmental impacts occur and individuals perceive these impacts differently. What we need is to consider the multiple dimensions of environmental impacts and the zone structure techniques that could effectively deal with the modifiable area/unit problem (MAUP). The debate on what constitutes an appropriate comparison (control) group is more complicated for a national level study than a local analysis. GIS and siting models are two promising tools that can make a contribution to the debate.

We have seen mixed evidence. This is not surprising at all. We live in a heterogeneous world. Case studies are useful, but you always can find cases with opposite results. That is the way the world works. That is why we should treat environmental justice issues locally.

We have seen an explosion of published papers on environmental justice issues over the past few years. We have also seen a lot of progress in the quality of these studies, although there are still methodological flaws in these peer-reviewed publications. In fact, many studies can be faulted on methodological grounds. Environmental justice analysis as a field of inquiry is still in its infancy and is in the pre-paradigm stage of the normal scientific development process according to Khun's notion.

We have seen several trends for shifting the environmental justice analysis:

- from positivism-dominated approach to combined positivism–participatory research,
- from the single discrimination/racism model to a multitheoretical, multi-equity criteria, and multidisciplinary perspective,
- from the proximity-based paradigm to the exposure/risk-based paradigm,
- from large geographic units to a fine-grained analysis,

- from statics analysis to both statics and dynamics analysis,
- from problem identification/remedy to pollution prevention,
- from evaluating existing associations due to past and current practice to assessing potential impacts that might occur because of the proposed future projects and plans,
- from reactive to proactive policies.

It is the time to break new ground for rigorous environmental justice analyses. This is an exciting time because the field has a lot of competing hypotheses, methods, and evidence. It is exciting because a lot of interesting work remains to be done, some of which have been presented within idealized frameworks in this book. It is exciting because high technology that has evolved over the past decade has provided many powerful tools so that researchers are equipped to reach higher and more sophisticated levels of analysis. It is exciting because we have a lot of challenges ahead.

We know more about inequity or lack thereof at the current time than at the time of facility siting. We know more about what spatial association is than why it comes into being. We have done almost nothing about the future. Until we do a much better job evaluating and preventing the impacts of our present and proposed actions, we will most likely find ourselves in the future in the same situation as we are today. We need a lot more data, more accurate data, more powerful and user-friendly modeling and GIS tools. We need these tools to be more accessible and user-friendly so the public can do their own analysis. We need to integrate these tools into a holistic analytical framework. We are not talking about a utopian world. It is becoming a reality.

References

Adler, K.J., et al. 1982. The Benefits of Regulating Hazardous Disposal: Land Values as an Estimator. U.S. Environmental Protection Agency. Washington, D.C. GPO.

Ahlbrandt, R.S., et al., 1977. Citizen Perceptions of Their Neighborhoods. *Journal of Housing* 7:338.

Al-Mosaind, M.A., K.J. Dueker, and J.G. Strathman. 1993. Light-rail Transit Stations and Property Values: A Hedonic Price Approach. Transportation Research Record 1400: 90–94.

Alonso, W. 1964. *Location and Land Use*. Cambridge, MA: Harvard University Press.

Alonso, W. 1967. A Reformulation of Classical Location Theory and Its Relation to Rent Theory. Papers, Regional Science Association 19:23–44.

Alonso, W. 1978. A Theory of Movements. pp. 195–212 in *Human Settlement Systems: International Perspectives on Structure, Change and Public Policy*, edited by N.M. Hansen. Cambridge, MA: Ballinger.

Anas, A. 1975. Empirical Calibration and Testing of A Simulation Model of Residential Location. *Environment and Planning* A 7:899–920.

Anas, A. 1981. The Estimation of Multinomial Logit Models of Joint Location and Mode Choice from Aggregated Data. *Journal of Regional Science* 21(2):223–242.

Anas, A. 1982. *Regional Location Markets and Urban Transportation: Economic Theory, Econometrics, and Policy Analysis with Discrete Choice Models*. New York: Academic Press.

Anas, A. 1983. Discrete Choice Theory, Information Theory and the Multinomial Logit and Gravity Models. *Transportation Research* 17B(1):13–23.

Anas, A. 1992. *NYSIM (The New York Area Simulation Model): A Model for Cost-Benefit Analysis of Transportation Projects*. New York: Regional Plan Association.

Anderson, A. B., D.L. Anderton, and J.M. Oakes. 1994. Environmental Equity: Evaluating TSDF Siting over the Past Two Decades. *Waste Age* 25(7):83–100.

Anderson, C.W. 1979. The Place of Principles in Policy Analysis. American Political Science Review 73(3):711–723. Reprinted in *Ethics in Planning*, M. Wachs (ed.), 1985, New Brunswick, N.J.: Center for Urban Policy and Research.

Anderson, T.R. 1955. Intermetropolitan Migration: a Comparison of the Hypotheses of Zipf and Stouffer. *American Sociological Review* 20:287–291.

Anderstig, C. and Mattson, L.-G. 1991. An integrated Model of Residential and Employment Location in a Metropolitan Region. *Papers in Regional Science* 70: 167–84.

Anderstig, C. and Mattson, L.-G. 1998. Modelling Land-use and Transport Interaction: Policy Analyses Using the IMREL Model. pp. 308–28 in *Network Infrastructure and the Urban Environment: Advances in Spatial Systems Modelling*, edited by L. Lundqvist, L.-G., Mattsson, and T.J. Kim. Berlin/Heidelberg: Springer Verlag.

Anderton, D.L. 1996. Methodological Issues in the Spatiotemporal Analysis of Environmental Equity. *Social Science Quarterly* 77(3):508–515.

Anderton, D.L. et al. 1994. Environmental Equity: The Demographics of Dumping. *Demography* 31(2):229–248.

Anderton, D.L., J.M. Oakes, and K.L. Egan. 1997. Environmental Equity in Superfund: Demographics of the Discovery and Prioritization of Abandoned Toxic Sites. *Evaluation Review* 21(1):3–26.

Anselin, L. 1999. Interactive Techniques and Exploratory Spatial Analysis. Chapter 17, pp. 253–266, in *Geographical Information Systems, Volume 1: Principles and Technical Issues*, edited by P.A. Longley et al., New York: John Wiley & Sons, Inc.

Andrews, F.M. 1981. A Guide for Selecting Statistical Techniques for Analyzing Social Science Data. Second edition. Ann Arbor, MI: Survey Research Center Institute for Social Research, the University of Michigan.

Anselin, L. 1993. Discrete Space Autoregressive Models. Chapter 46 in *Environmental Modeling with GIS*, edited by M.F. Goodchild, B.O. Parks, and L.T. Steyaert. Oxford, England: Oxford University Press.

Asch P. and J. J. Seneca. 1978. Some Evidence on the Distribution of Air Quality. *Land Economics* 54 (3):278–297.

Atlas, M.A. 1998. Mad about You: Community Activism and the Closing of Hazardous Waste Management Facilities. Presented at the 1998 Annual Meeting of the American Political Science Association, Boston, September 3–6, 1998.

ATSDR (Agency for Toxic Substances and Disease Registry). 1988. The Nature and Extent of Lead Poisoning in Children in the United States: Report to Congress. Centers for Disease Control: Atlanta, GA.

Bacow, L.S. and J. R. Milkey. 1982. Overcoming Local Opposition to Hazardous Waste Facilities: The Massachusetts Approach. *Harvard Environmental Law Review* 6:270–305.

Babbie, E. 1992. *The Practice of Social Research*. Sixth Edition. Belmont, CA: Wadsworth, Publ.

Bae, C.C. 1997. The Equity Impacts of Los Angeles' Air Quality Policies. *Environment and Planning* A 29(9):1563–1584.

Balagopalan, M. 1999. Communication of Health Risk Assessment by Integrating Geographic Information System (GIS) with Computer Dispersion Models. Proceedings from the 1999 Environmental Systems Research Institute Users Conference. http://www.esri.com/library/userc...99/proceed/pap599/p599.htm.

Balentine, et al. 1988. Analysis of Ozone Concentration and VOC Emission Trends in Harris County, Texas, 1981–1986, pp. 429–440 in *The Scientific and Technical Issues Facing Post-1987 Ozone Control Strategies*, edited by G.T. Wolff, J. L. Hanisch, and K. Schere. Pittsburgh, PA: Air & Waste Management Association.

Barker, M.L. 1976. Planning for Environmental Indices: Observer Appriasals of Air Quality. In *Perceiving Environmental Quality: Research and Applications*, edited by K.H. Craik and E.H. Zube, pp. 175–203, New York: Plenum Press.

Bartell, S.M., R.H. Gardner, and R.V. O'Neill. 1992. *Ecological Risk Estimation*. Boca Raton: Lewis Publishers.

Batten, D.F. and D.E. Boyce. 1986. Spatial Interaction, Transportation, and Interregional Commodity Flow Models. In *Handbook of Regional and Urban Economics*, Volume1, edited by P. Nijkamp, p. 357–406. Amsterdam: Elsevier Science.

Batty, M. 1976. *Urban Modelling*. Cambridge, UK: Cambridge University Press.

Batty, M. and Y. Xie. 1994. Modelling Inside GIS: Part 1. Model Structures, Exploratory Spatial Data Analysis and Aggregation. *International Journal of Geographical Information Systems* 8(3):291–307.

Baucus, M. 1993. Environmental Justice Act of 1993. S. 1161. Introduced in the 103rd Congress, June 24.

Baumol, W.J. and W.E. Oates. 1988. *The Theory of Environmental Policy*. 2nd Edition. Cambridge, UK: Cambridge University Press.

Beatley, T. 1984. Applying Moral Principles to Growth Management. *Journal of the American Planning Association* 50 (4):459–469.

Beauchamp, T. 1981. The Moral Adequacy of Cost/Benefit Analysis as the Basis for Government Regulation of Research. In *Ethical Issues in Government*, N. Bowie (ed.), Philadelphia: Temple University Press.

Bedau, H.A. 1967. Egalitarianism and the Idea of Equality, pp. 3–27 in *Equality*, J.R. Pennock and J.W. Chapman (eds.), Nomos IX, Yearbook of the American Society for Political and Legal Philosophy. New York: Atherton Press.

Been, V. with Gupta, F. 1997. Coming to the Nuisance or Going to the Barrios? A Longitudinal Analysis of Environmental Justice Claims. *Ecology Law Quarterly.* 24(1):1.

Been, V. 1993. What's Fairness Got to Do With It? Environmental Justice and Siting of Locally Undesirable Land Uses. *Cornell Law Review* 78 (Sept.): 1001–1085.

Been, V. 1994. Locally Undesirable Land Uses in Minority Neighborhoods: Disproportionate Siting or Market Dynamics. *The Yale Law Journal* 10:1383.

Been, V. 1995. Analyzing Evidence of Environmental Justice. *Journal of Land Use & Environmental Law*, 11(1):1–36.

Belcher G.D. and H.A. Hattemer-Frey. 1990. A Program for Calculating Health Risks from Hazardous Waste Incineration. *Risk Analysis* 10(1):185–188.

Bell et al. 1991. Methylene Chloride Exposure and Birth weight in Monroe County, New York. *Environmental Research* 55(1):31–39.

Ben-Akiva, M. and A. De Palm. 1986. Analysis of a Dynamic Residential Location Choice Model with Transaction Costs. *Journal of Regional Science* 26:321–341.

Ben-Akiva, M. and S.R. Lerman. 1985. *Discrete Choice Analysis: Theory and Application to Travel Demand.* Cambridge, MA: The MIT Press.

Benson, P. 1994. CALINE4—A Dispersion Model for Predicting Air Pollution Concentrations Near Roadways. Sacramento, CA: California Department of Transportation.

Bentham, Jeremy. [1789] 1948. *An Introduction to the Principles of Morals and Legislation.* Oxford: Basil Blackwell.

Berechman, J. and K.A. 1988. Small. Research Policy and Review 25. Modeling Lnad Use and Transportation: an Interpretive Review for Growth Areas. *Environment and Planning* A 20: 1285–1309.

Berry, B.J.L. (ed.). 1977. *The Social Burdens of Environmental Pollution: A Comparative Metropolitan Data Source.* Cambridge, MA: Ballinger.

Berry, M. and P. Bove. 1997. Birth Weight Reduction Associated with Residence Near a Hazardous Waste Landfill. *Environmental Health Perspectives* 105(8):856–861.

Bertuglia, C.S. et al. 1989. *Performance Indicators and Model-Based Urban Planning.* Croom Helm, London.

Bertuglia, C.S., G. Leonardi and A.G. Wilson (eds). 1990. *Urban Dynamics: Designing an Integrated Model.* London: Routledge.

Beveridge, W.I.B. 1950. *The Art of Scientific Investigation.* New York: Vintage Books.

Bingham, T., Anderson, D. W., and Cooley, P. C. 1987. Distribution of the Generation of Air Pollution. *Journal of Environmental Economics and Management* 14 (March): 30–40.

Birch, D.L. 1971. Toward a Stage Theory of Urban Growth. *Journal of the American Institute of Planners* 37:78–87.

Birkin, M. and A.G. Wilson. 1986. Industrial Location Models 1: review and in integrating framework. *Environment and Planning* A 18:175–205.

Boer, J.T. et al. 1997. Is there Environmental Racism? The Demographics of Hazardous Waste in Los Angeles County. *Social Science Quarterly* 78(4): 793–829.

Bogue, D. 1985. *Population in the United States: Historical Trends and Future Projections.* New York: Free Press.

Boots, B. 1999. Spatial Tessellations. Chapter 36 (pp. 503–526) in *Geographical Information Systems, Volume 1: Principles and Technical Issues*, edited by P.A. Longley, M.F. Goodchild. D.J. Maguire, and D.W. Rhine. New York: John Wiley & Son, Inc.

Bowen, W.M. 1999. Comments on " 'Every Breath You Take…': The Demographics of Toxic Air Releases in Southern California." *Economic Development Quarterly.* 13(2):124–134.

Bowen, W.M., et al. 1995. Toward Environmental Justice: Spatial Equity in Ohio and Cleveland. *Annals of the Associations of American Geographers* 85(4): 641–663.

Boyce, D. et al. 1972. *The Impact of Rapid Rail on Suburban Residential Property Values and Land Development Analysis of the Philadelphia High Speed Line.* Philadelphia: Regional Science Department, Wharton School, University of Pennsylvania.

Boyce, D.E. 1986. Integration of Supply and Demand Models in Transportation and Location: Problem Formulation and Research Questions. *Environment and Planning* A 18:485–89.

Boyce, et al. 1983. Implementation and Computational Issues for Combined Models of Location, Destination, Mode, and Route Choice. *Environment and Planning* A 15 (9):1219–30.

Brajer, V. and J.V. Hall. 1992. Recent Evidence on the Distribution of Air Pollution Effects. *Contemporary Policy Issues* 10:63–71.

Bratt, R. G. 1983. People and Their Neighborhoods: Attitudes and Policy Implications. pp133–150 in *Neighborhood Policy and Planning*, edited by P.L. Clay and R. M. Hollister. Lexington, MA: LexingtonBooks.

Braybrooke, D. and C. Schotch. 1981. Cost-benefit Analysis under the Constraint of Meeting Needs. The Moral Adequacy of Cost/Benefit Analysis as the Basis for Government Regulation of Research. In *Ethical Issues in Government*, N. Bowie (ed.), Philadelphia: Temple University Press.

Brody, D. J., et al. 1994. Blood Lead Levels in the U.S. Population: Phase 1 of the Third National Health and Nutrition Examination Survey (NHANES IIIm 1988 to 1991). *Journal of the American Medical Association* 272(4):277–283.

Brooks, N. and R. Sethi. 1997. The Distribution of Pollution: Community Characteristics and Exposure to Air Toxics. *Journal of Environmental Economics and Management* 32:233–250.

Brotchie, J. and R. Sharpe. 1974. A General Land Use Allocation Model: Application to Australian Cities. In *Urban Development Models*, Proceedings of the Third Land Use and Built Form Studies Conference, pp. 217–236. Harlow, Essex: Construction Press.

Brown, H.S. 1988. Management of Carcinogenic Air Emissions: A Case Study of a Power Plant. *Journal Air Pollution Control Association* 38(1):15–21.

Brunton, P.J. and A.J. Richardson. 1998. A Cautionary Note on Zonal Aggregation and Accessibility. Presented at the 77th Annual Meeting of the Transportation Research Board, Washington, D.C.

Bryant B. and P. Mohai (eds.). 1992. *Race and the Incidence of Environmental Hazards: A Time for Discourse.* Boulder, CO: Westview Press.

Bryant, B. 1995. Pollution Prevention and Participatory Research as a Methodology for Environmental Justice. *Virginia Environmental Law Journal* 14:589–611.

Bryant, B. 1998. Key Research and Policy Issues Facing Environmental Justice. Available at http://www.snre.umich.edu/~bbryant/Interview.html, accessed on February 24, 1998.

Bryant, P. 1989. Toxics and Racial Justice. *Social Policy.* Summer:48–52.

Bullard R.D. 1996. Environmental Justice: It's More Than Waste Facility Siting. *Social Science Quarterly* 77(3):493–499.

References

327

Bullard, R.D. 1993. Anatomy of Environmental Racism, pp. 25–35 in *Toxic Struggles: The Theory and Practice of Environmental Justice*, edited by R. Hofrichter. Philadelphia PA: New Society Publishers.

Bullard, R.D. 1983. Solid Waste Sites and the Houston Black Community. *Sociological Inquiry* 53 (Spring):273–288.

Bullard, R.D. 1987. *Invisible Houston: The Black Experience in Boom and Dust*. College Station: Texas A&M University Press.

Bullard, R.D. 1994. *Dumping in Dixie: Race, Class, and Environmental Quality*. Second Edition. Boulder: Westview Press.

Bullard, R.D. 1994. The Legacy of American Apartheid and Environmental Racism. *St. John's Journal of Legal Comment*, 445(9):467–469.

Bullard, R.D. and G.S. Johnson. 1997. Just Transportation. pp. 7–21 in *Just Transportation: Dismantling Race and Class Barriers to Mobility*, edited by R.D. Bullard and G.S. Johnson. Gabriola Island, BC: New Society Publishers.

Bullard, R.D. and B.H. Wright. 1993. Environmental Justice for All: Community Perspectives on Health and Research Needs. *Toxicology and Industrial Health* 9(5):821–842.

Bullock, H.A. 1957. *Pathways to the Houston Negro Market*. Ann Arbor: J.N. Edwards.

Bureau of the Census. 1909. A Century of Population Growth in the United States:1790–1900. By W.S. Rossiter. Washington, D.C.: Government Printing Office.

Bureau of the Census. 1960. Historical Statistics of the United States: Colonial Times to 1957. Washington, D.C.: Government Printing Office.

Bureau of the Census. 1983. 1980 Census of Population, volume 1 Characteristics of the Population, Chapter C, General Social and Economic Characteristics, Part 1, United States Summary, PC80-1-C1; Connecticut, PC80-1-C8; New Jersey, PC80-1-C32; New York, PC80-1-C34; Pennsylvania, PC80-1-C40.

Bureau of the Census. 1984. Neighborhood Statistics from the 1980 Census. Washington, D.C.

Bureau of the Census. 1992a. Census of Population and Housing, 1990: Summary Tape File 3 on CD-ROM Technical Documentation. Washington, DC: The Bureau of Census.

Bureau of the Census. 1992b. 1990 Census of Population, Volume 1, Characteristics of Population, General Social and Economic Characteristics, Houston, TX. Washington, D.C.: U.S. Government Printing Office.

Bureau of the Census. 1994. Geographic Areas Reference Manual. Washington, D.C.

Bureau of the Census. 1997. Participant Statistical Areas Program Guidelines: Census Tracts, Block Groups (BGs), Census Designated Places (CDPs), Census County Divisions (CCDs). U.S. Census 2000, Form D 1500. Washington, D.C.

Bureau of the Census. 1998. Poverty in the United States: 1997. Current Population Reports, Series P60–201. Washington, D.C.

Bureau of the Census. 1999. Current Population Survey. Washington, D.C. Poverty Thresholds are available at http://www.census.gov/hhes/poverty/threshold.html

Burby, R. J. and D.E. Strong. 1997. Coping with Chemicals. Blacks, Whites, Planners, and Industrial Pollution. *Journal of the American Planning Association*. 63(4):469–480.

Burgess, E.W. 1925. The Growth of the City, pp. 47–62 in R.E. Park et al. (eds.), *The City*. Chicago: University of Chicago Press.

Burnell, J.D. 1985. Industrial Land Use, Externalities, and Residential Location. *Urban Studies* 22: 399–408.

Caliper. 1996. TransCAD: Transportation GIS Software, User's Guide Version 3.0. Newton, MA: Caliper Corp.

Cameron, M. 1994. *Efficiency and Equity on the Road*. Environmental Defense Fund. Oakland, CA.

Campbell, D. T. and J.C. Stanley. 1966. *Experimental and Quasi-Experimental Design for Research*. Chicago: Rand McNally.

Can, A. 1992. Residential Quality Assessment: Alternative Approaches Using GIS. *The Annals of Regional Science* 26(1):97.

Carmines, E.G. and R.A. Zeller. 1979. *Reliability and Validity Assessment*. Beverly Hills, CA: Sage.

Chakraborty, J. and M. P. Armstrong. 1997. Exploring the Use of Buffer Analysis for Identification of Impacted Areas in Environmental Equity Assessment. *Cartography and Geographic Information Systems*, 24(3):145–157.

Chameides, W.L., R.D. Saylor, and E.B. Cowling. 1997. Ozone Pollution in the Rural United States and the New NAAQS. *Science* 276(5314):916.

Clark, D.E. and L.A. Nieves. 1994. An Interregional Hedonic Analysis of Noxious Facility Impacts on Local Wages and Property Values. *Journal of Environmental Economics & Management* 27:235–253.

Clarke, J. N. and A. K. Gerlak. 1998. Environmental Racism in the Sunbelt? A Cross-Cultural Analysis. *Environmental Management* 22(6):857–867.

Cleveland, W.S. and T. E. Graedel. 1979. Photochemical Air Pollution in the Northeast United States. *Science* 204:1273–1278.

Cleveland, W.S. et al. 1977. Geographical Properties of Ozone Concentrations in the Northeast United States. *Journal of the Air Pollution Control Association* 27 (4):325–328.

Clinton, W.J. 1994. Memorandum to the Heads of Departments and Agencies. Comprehensive Presidential Documents No. 279. Feb. 11, 1994.

Coase, R. 1960. The Problem of Social Cost. *Journal of Law and Economics* 3(October):1–44.

Coelho, J.D. and H.C.W.L. Williams. 1978. On the Design of Land Use Plans Through Locaitonal Surplus Maximization. Papers Regional Science Association 40:71–85.

Cole, L.W. 1992. Enpowerment as the Key to Environmental Protection: The Need for Environmental Property Law. *Ecology Law Quarterly* 19(4):619–683.

Cook, K.S. and K.A. Hegtvedt. 1983. Distributive Justice, Equity, and Equality. *Ann. Rev. Sociol.* 9:217–241.

Council on Environmental Quality (CEQ). 1971. Environmental Quality. Washington, D.C.

Council on Environmental Quality (CEQ). 1996. Environmental Quality. The 25th Anniversary Report. Available at http://www.whitehouse.gov/CEQ.

Council on Environmental Quality (CEQ). 1997. Environmental Justice Guidance under the National Environmental Policy Act. Executive Office of the President. Washington, D.C.

Crecine J.P. et. al. 1967. Urban Property Markets: Some Empirical Results and Their Implication for Municipal Zoning. *Journal of Law and Economics* 10:79–87.

Crecine, J.P., O.A. Davis, and R.E. Jackson. 1967. Urban Property Markets: Some Empirical Results and Their Implications for Municipal Zoning. *Journal of Law and Economics*. 10:79–99.

Croen et al. 1997. Maternal Residential Proximity to Hazardous Waste Sites and Risk for selected Congenital Malformation. *Epidemiology* 8(4):347–354.

Crump, K.S. 1984. An Improved Procedure for Low-Dose Carcinogenic Risk Assessment from Animal Data. *Journal of Environmental Pathology, Toxicology, and Oncology* 5–4:339–349.

Cutter, S.L. 1993. *Living with Risk: The Geography of Technological Hazards*. London: Edward Arnold.

Cutter, S.L., D. Holm, and L. Clark. 1996. The Role of Geographic Scale in Monitoring Environmental Justice. *Risk Analysis* 16(4):517–526.

Dale, L. et al. 1999. Do Property Values Rebound from Environmental Stigmas? Evidence from Dallas. *Land Economics* 75(2):311–26.

Darney, A.J. 1994. *Economic Indicators Handbook*. 2nd edition. Detroit: Gale Research Inc.

Davy, B. 1996. Fairness as Compassion: Towards a Less Unfair Facility Siting Policy. *Risk: Issues in Health, Safety & Environment* 7(Spring): 99–119.

de la Barra, T. 1989. *Integrated Land Use and Transport Modelling*. Cambridge: Cambridge University Press.

de la Barra, T. 1998. Improved Logit Formulations for Integrated Land Use, Transport and Environmental Models, pp. 288–307 in *Network Infrastructure and the Urban Environment: Advances in Spatial Systems Modelling*, edited by L. Lundqvist, L.-G., Mattsson, and T.J. Kim. Berlin/Heidelberg: Springer Verlag.

de la Barra, T., Pérez, B. and Vera, N. 1984. TRANUS-J: Putting Large Models into Small Computers. *Environment and Planning B: Planning and Design*, 11: 87–101.

de Neufville, J.I. 1987. Knowledge and Action: Making the Link. *Journal of Planning Education and Research* 6(2):86–92.

Declercq, F. A.N. 1996. Interpolation Methods for Scattered Sample Data: Accuracy, Spatial Patterns, Processing Time. *Cartography and Geographic Information Systems* 23(3):128–144.

Denton, N.A. and D.S. Massey. 1991. Patterns of Neighborhood Transition in Multiethnic World: U.S. Metropolitan Areas, 1970–1980. *Demography* 28:41,

Dohrenwend, B.P. et al. 1981. Stress in the Community: A Report to the President's Commission on the Accident at Three Mile Island, pp. 159–74 in *The Three Mile Island Nuclear Accident: Lessons and Implications*. T.H. Moss and D. L. Sills (eds) New York: New York Academy of Sciences.

Doll, R. and R. Peto. 1981. Avoidable Risks of Cancer in the U.S. *Journal of National Cancer Institute* 66(6): 1193–1308.

Domencich, T.A. and McFadden, D. 1975. *Urban Travel Demand: a Behavioral Analysis*. Amsterdam/Oxford: North Holland Publ..

Douglas, M. 1990. Risk as a Forensic Resource. *DAEDALUS* 119(4):1–10.

Downey, L. 1998. Environmental Injustice: Is Race or Income a Better Predictor? *Social Science Quarterly* 79(4): 766–778.

Duan, N. 1982. Models for Human Exposure to Air Pollution. *Environ. Int.* 8:305–309.

Duan, N. 1981. Micro-Environment Types: A Model for Human Exposure to Air Pollution. SIMS Technical Report No. 47 Stanford, CA: Stanford University.

Dubin, R. and C. Sung. 1990. Specification of Hedonic Regressions: Non-nested Tests on Measures of Neighborhood Quality. *Journal of Urban Economics* 27:97–104

Duncan, O.D. and B. Duncan. 1957. *The Negro Population of Chicago: a Study of Residential Succession*. Chicago: The University of Chicago Press.

Earickson, R.J. and I.H. Billick. 1988. The Areal Association of Urban Air Pollutants and Residential Characteristics: Louisville and Detroit. *Applied Geography* 8:5–23.

Echenique, M.H., et al. 1990. The MEPLAN models of Bilbao, Leeds and Dortmund. *Transport Reviews* 10:309–22.

Environmental Justice Resource Center (EJRC). 1998. NRC Finds Environmental Racism Rejects Facility Permit. http://www.ejrc.cau.edu.

Environmental Justice Resource Center (EJRC), Clark Atlanta University. 1996. Environmental Justice and Transportation: Building Model Partnerships Conference Proceedings. Atlanta, GA: Clark Atlanta University.

ESRI. 1997. *Getting to Know ArcView GIS: The Geographic Information System for Everyone*. Cambridge, England: GeoInformation International.

Evans, S.P. 1973. A Relationship between the Gravity Model for Trip Distribution and the Transportation Problem in Linear Programming. *Transportation Research* 7:39–61.

Fahsbender, J.J. 1996. An Analytical Approach to Defining the Affected Neighborhood in the Environmental Justice Context. *New York University Environmental Law Journal* 5:120–180.

Farley, R., and W.H. Frey. 1994. Changes in the Segregation of Whites from Blacks during the 1980s: Small Steps toward a More Integrated Society. *Amer. Socio. Rev.* 59:23–45.

FHWA (Federal Highway Administration). 1976. Social and Economic Effects of Highways. Washington, DC: Government Printing Office.

FHWA (Federal Highway Administration). 1978. Highway Air Quality Impact Appraisals Vol. I: Introduction to Air Quality Analysis. Washington, DC: Government Printing Office.

FHWA (Federal Highway Administration). 1996. Community Impact Assessment: A Quick Reference for Transportation. Publication No. FHWA-PD 96–036. Washington, D.C.

FHWA (Federal Highway Administration). 1997. Our Nation's Travel: 1995 NPTS Early Results Report. FHWA-PL-97-028. Washington, D.C.

Finkel, A.M. and D. Golding. 1993. Alternative Paradigms: Comparative Risk is Not the Only Model. *EPA Journal* 19:50–52.

Finley, B. and D. Paustenbach. 1994. The Benefits of Probabilistic Exposure Assessment: Three Case Studies Involving Contaminated Air, Water and Soil. *Risk Analysis* 14(1):53–73.

Fisher, P.F. and M. Langford.1995. Modeling the Errors in Areal Interpolation between Zonal Systems Using Monte Carlo Simulation. *Environment and Planning* A 27:211–224.

Fisher, P.F. and M. Langford.1996. Modeling Sensitivity to Accuracy in Classified Imagery: A Study of Areal Interpolation by Dasymetric Mapping. *The Professional Geography* 48(3):299–309.

Flachsbart, P.G. and S. Phillips. 1980. An Index and Model of Human Response to Air Quality. *Journal of Air Pollution Control Association* 30(7):759–768.

Foreman, C.H. Jr. 1998. *The Promise and Peril of Environmental Justice*. Washington, D.C.: Brookings Institution Press.

Forkenbrock, D.J. and L. A. Schweitzer. 1997. Environmental Justice and Transportation Investment Policy. Report #Mn/RC-97/09. Iowa City, IA: Public Policy Center, University of Iowa.

Forkenbrock, D.J. and L. A. Schweitzer. 1999. Environmental Justice in Transportation Planning. *Journal of the American Planning Association* 65(1):96–111.

Fotheringham, A.S. and D.W.S. Wong. 1991. The Modifiable Areal Unit Problem in Multivariate Statistical Analysis. *Environment and Planning* 23:1025–44.

Fowlkes, M.R. and P.Y. Miller. 1983. Love Canal: The Social Construction of Disaster. Final report for the Federal Emergency Management Agency, Award No. EMW-1-4048.

Freedman, D.A. 1991. Adjusting the 1990 Census. *Science* 252:1233–1236.

Freeman III, A.M. 1972. The Distribution of Environmental Quality, pp. 243–278 in *Environmental Quality Analysis*, edited by A.V. Kneese and B.T. Bower. Resources for the Future.

Freeman III, A.M. 1993. *The Measurement of Environmental and Resource Values: Theory and Methods*. Baltimore, MD: Johns Hopkins University Press for Resources for the Future.

Fujita, M. 1989. *Urban Economic Theory: Land Use and City Size*. Cambridge, England: Cambridge University Press.

Furuseth, O.J. 1990. Impacts of a Sanitary Landfill: Spatial and Non-Spatial Effects on the Surrounding Community. *Journal of Environmental Management* 31:269–277.

Galloway, T.D. and R.G. Mahayni. 1977. Planning Theory in Retrospect: The Process of Paradigm Change. *Journal of the American Institute of Planners* 43(January):62–70.

Galper, J. 1998. How to Measure Local Incomes. *American Demographics.* March:12–17.

GAO (U.S. General Accounting Office). 1983. Siting of Hazardous Waste Landfills and Their Correlation with the Racial and Socio-Economic Status of Surrounding Communities. Washington, D.C.: General Accounting Office.

GAO. 1995. Hazardous and Nonhazardous Waste: Demographics of People Near Waste Facilities. GAO/RCED-95-84. Washington, D.C.

Garcia, R. and T. J. Graff. 1999. Memorandum for Amicus Curiae Environmental Defense Fund in Response to MTA's Motion for Clarification and Modification of the Special Master's March 6, 1999, Memorandum Decision and Order.

Garcia, R., et al. 1996. Plaintiff's Revised Statement of Contentions of Fact and Law, Labor/Community Strategy Center, et al. vs. Los Angeles County Metropolitan Transportation Authority, et al., CASE NO. CV 94–5936 TJH (Mcx).

Garson, G.D. and R. S. Biggs. 1992. Analytical Mapping and Geographic Databases. Sage University Paper Series on Quantitative Applications in the Social Sciences, series no. 07–087. Newbury Park, CA: Sage.

Gelobter, M. 1992. Toward a Model of Environmental Discrimination, pp. 64–81 in *Race and the Incidence of Environmental Hazards: A Time for Discourse*, edited by B. Bryant and P. Mohai, Boulder, CO: Westview Press.

Gerrard, M. B. 1994. Whose Backyard, Whose Risk: Fear and Fairness in Toxic and Nuclear Waste Siting.

Geschild, S.A. et al. 1992. Risk of Congenital Malformations Associated with Proximity to Hazardous Waste Sites. *American Journal of Epidemiology* 135(11):1197–1207.

Getis, A. 1999. Spatial Statistics. Chapter 16 (pp. 239–251) in *Geographical Information Systems, Volume 1: Principles and Technical Issues*, edited by P.A. Longley, M.F. Goodchild. D.J. Maguire, and D.W. Rhine. New York: John Wiley & Son, Inc.

Gianessi, L., H.M. Peskin and E. Wolff. 1979. The Distributional Effects of Uniform Air Pollution Policy in the U.S. *Quarterly Journal of Economics* 93(May):281–301.

Giuliano, G. 1994. Equity and Fairness Considerations of Congestion Pricing. pp. 250–279 in *Curbing Gridlock*. Washington, D.C.: TRB, National Academy Press.

Glickman, T.S., D. Golding, and R. Hersh. 1995. GIS-based Environmental Equity Analysis: A Case Study of TRI Facilities in the Pittsburgh Area, pp. 95–114 in *Computer Supported Risk Management*, edited by W.A. Wallace and E.G. Beroggi. Dordrecht, The Netherlands: Kluwer.

Goldman B. A. and L. Fitton. 1994. Toxic Wastes and Race Revisited: An Update of the 1987 Report on the Racial and Socioeconomic Characteristics of Communities with Hazardous Waste Sites. Washington, D.C.: Center for Policy Alternatives.

Goldman et al. 1985. Low Birthweight, Prematurity, and Birth Defects in Children Living Near the Hazardous Waste Site. *Hazardous Waste Materials* 2(2):209–223.

Goldman, B.A. 1991. *The Truth about Where You Live: An Atlas for Action on Toxins and Mortality.* New York: Times Books.

Goldman, B.A. 1996. What is the Future of Environmental Justice. *Antipode* 28 (2):122–141.

Goldner, W. 1968. Projective Land Use Model (PLUM): A Model for the Spatial Allocation of Activities and Land Uses in a Metropolitan Region. TR-219, Bay Area Transportation Study Commission.

Goldstein, B.D. 1986. Critical Review of Toxic Air Pollutants-Revisited. *Journal of the Air Pollution Control Association* 36(4):367–370.

Goodchild, M.F., L. Anselim, and U. Deichman. 1993. A Framework for the Areal Interpolation of Socioeconomic Data. *Environment and Planning* A 25:383–397.

Goodman, A.C. 1977. A Comparison of Block Group and Census Tract Data in a Hedonic Housing Price Model. *Land Economics* 53: 483.

Goodwin, P.B. 1990. Demographic Impacts, Social Consequences, and the Transport Policy Debate. *Oxford Review of Economic Policy* 6(2):76–90.

Gordon, P., A. Kumar, and H. Richardson. 1989. The Spatial Mismatch Hypothesis: Some New Evidence. *Urban Studies* 26:315–326.

Gordon, P., H. W. Richardson, and M. Jun. 1991. The Commuting Paradox: Evidence from the Top Twenty. *Journal of the American Planning Association* 57(4):416–420.

Gore, A. 1992. Environmental Justice Act of 1992. Introduced in the U.S. Senate. 102nd Congress. June 3.

Gough, M. 1989. Estimating Cancer Mortality: epidemiological and toxicological methods produce similar assessments. *Environmental Sciences and Technology* 23(8):925–930.

Gould, L.C., et al. 1988. *Perceptions of Technological Risks and Benefits*. New York: Russell Sage Foundation.

Greenberg, M.R. and J. Hughes. 1993. The Impact of Hazardous Waste Superfund Sites on the Value of Houses Sold in New Jersey. *Annals of Regional Sciences* 26(1):147–153.

Greenberg, M.R. and R. Anderson. 1984. *Hazardous Waste Sites: The Credibility Gap*. New Brunswick, NJ: The Center for Urban Policy Research at Rutgers.

Greenberg, M.R., D. Schneider, and J. Martell. 1994. Hazardous Waste Sites, Stress, and Neighborhood Quality in USA. *The Environmentalist* 14(2): 93–105.

Greenberg, M.R., D.A. Krueckeberg, and C.O. Michaelson. 1978. *Local Population and Employment Projection Techniques*. New Brunswick, NJ: Center for Urban Policy Research.

Greenberg, M.R. 1993. Proving Environmental Inequity in the Siting of Locally Unwanted Land Uses. *Risk: Issues in Health and Safety* 4(3):235–252.

Greenberg, M.R. et al. 1989. Network Evening News Coverage of Environmental Risk. *Risk Analysis* 9(1): 119–126.

Grether, D.M. and P. Mieszkowski. 1980. The Effects of Nonresidential Land Uses on the Prices of Adjacent Housing: Some Estimates of Proximity Effects. *Journal of Urban Economics* 8:1–15.

Griffith, D.A. and C.G. Amrhein. 1991. *Statistical Analysis for Geographers*. Englewood Cliffs, NJ: Prentice-Hall.

Grigsby, W. et al. 1987. *The Dynamics of Neighborhood Change and Decline*. Elmsford, N.Y.: Pergaman Journals.

Grisinger, J. and J. C. Marlia. 1994. Development and Application of Risk Analysis Methods to Stationary Sources of Carcinogenic Emissions for Regulatory Purposes by the South Coast Air Quality Management District. *Journal of the Air and Waste Management Association* 44(2):145–152.

Gross, M. 1979. The Impact of Transportation and Land Use Policies on Urban Air Quality. Ph.D. Dissertation in City and Regional Planning, University of Pennsylvania.

Grossman K. 1992. From Toxic Racism to Environmental Justice. *E: The Environmental Magazine* 3(3):28–35.

Guest, A.M. and B. A. Lee. 1984. How Urbanites Define Their Neighborhoods. *Population & Environment* 7:32–53.

Guinée, J. and R. Heijungs. 1993. A Proposal for the Classification of Toxic Substances within the Framework of Life Cycle Assessment of Products. *Chemosphere* 26(10):1925–1944.

Guthe, W.G. et al. 1992. Reasssessment of Lead Exposure in New Jersey Using GIS Technology. *Environmental Research* 59:318–325.

Guy, C.M. 1983. The Assessment of Access to Local Shopping Opportunities: A Comparison of Accessibility Measures. *Environment and Planning B: Planning and Design* 10:219–238.

Haeberle, S.H. 1988. People or Place: Variations in Community Leaders' Subjective Definitions of Neighborhood. *Urban Affair Quarterly* 23: 616–24.

Haemisegger E.R., A.D. Jones, and F. L. Reinhardt. 1985. EPA's Experience with Assessment of Site-Specific Environmental Problems: A Review of IEMD's Geographic Study of Philadelphia. *Journal of Air Pollution Control Association* 35(8):809–815.

Haining, R. P. 1990. *Spatial Data Analysis in Social and Environmental Sciences.* Cambridge, England: Cambridge University Press.

Hall, J. V. et al. 1992. Valuing the Health Benefits of Clean Air. *Science* 253(2):812–817.

Hamburg, J.R., L. Blair, and D. Albright. 1995. Mobility as a Right. Transportation Research Record, #1499, pp52–55.

Hamilton, B.W. 1982. Wasteful Commuting. *Journal of Political Economy* 90:1035–1053.

Hamilton, J.T. 1995. Testing for Environmental Racism: Prejudice, Profits, Political Power? *Journal of Policy Analysis Manage.* 14(1):107–132.

Handy, F. 1977. Income and Air Quality in Hamilton, Ontario. *Alternatives* 6(3): 18–24.

Hansen, W.G. 1959. How Accessibility Shapes Land Use. *Journal of the American Institute of Planners* 25:73–76.

Harris, B. and M. Batty. 1993. Locational Models, Geographic Information and Planning Support Systems. *Journal of Planning Education and Research* 12: 184–198.

Harris, J.S. 1983. Toxic Waste Uproar: A Community History. *Journal of Public Policy* 4:181–201.

Harrison, D. 1975. *Who Pays for Clean Air: the Cost and Benefit Distribution of Federal Automobile Emission Controls.* Cambridge, MA: Ballinger.

Hart, S. 1995. A Survey of Environmental Justice Legislation in the States. *Washington University Law Quarterly* 73:1459–1475.

Harvey, D. 1996. *Justice, Nature and the Geography of Difference.* Oxford: Blackwell Publishers.

Hattum, V. D. 1996. Land Use and Equity Impacts of Congestion Pricing. *Progress*, VI (9):7.

Haynes, M.A. and B.D. Smedley (ed). 1999. *The Unequal Burden of Cancer: An Assessment of NIH Research and Programs for Ethnic Minorities and the Medically Underserved.* Institute of Medicine. Washington, D.C.: National Academy Press.

HCPC (Houston City Planning Commission). 1978. Housing Analysis — Low-moderate Income Areas. Houston, Texas.

HCPC (Houston City Planning Commission). 1980. Houston — Year 2000 Report and Map. Houston, Texas.

Held, V. 1984. *Rights and Goods.* New York: Free Press.

Herbert, J.C. and B.H. Stevens. 1960. A Model for the Distribution of Residential Activity in Urban Areas. *Journal of Regional Science* 2:21–36.

Heritage, J. (ed.). 1992. Environmental Protection—Has It Been Fair. *EPA Journal* 18 (March/April).

Hertwich, E.G., W. S. Pease, and T. E. Mckone. 1998. Evaluating Toxic Impact Assessment Methods: Which Works Best? *Environmental Science and Technology* 32(3):138A-144A.

Hicks, J.R. 1939. The Foundation of Welfare Economics" *Economic Journal* 49 (Dec.): 696–712.

Hird, J.A. 1993. Environmental Policy and Equity: The Case of Superfund. *Journal of Policy Analysis and Management* 12 (2):323–343.

Hogan, H. 1990. The 1990 Post-Enumeration Survey: An Overview. Proceedings of the Section on Survey Research Methods of the American Statistical Association (August): 518–523.

Hokanson, B. et al. 1981. *Measures of Noise Damage Costs Attributable to Motor Vehicle Travel*. Iowa City, IA: Univeristy of Iowa, Institute of Urban and Regional Research.

Hollander, M. and D.A. Wolfe. 1973. *Nonparametric Statistical Methods*: New York: John Wiley & Sons, Inc.

Holzer, H.J. 1991. The Spatial Mismatch Hypothesis: What has the Evidence Shown? *Urban Studies* 28(1): 105–122.

Hoover, E. and R. Vernon. 1959. *Anatomy of a Metropolis*. Garden City, N.Y.: Doubleday.

Horvath, A., et al. 1995. Toxic Emissions Indices for Green Design and Inventory. *Environmental Science and Technology* 29(2):A86–90.

Hotelling, H. 1929. Stability in Competition. *Economic Journal* 39:41–57.

Howe, E. 1990. Normative Ethics in Planning. *Journal of Planning Literature* 5(2):123–150.

Howe, H.L. 1988. A Comparison of Actual and Perceived Residential Proximity to Toxic Waste Sites. *Archives of Environmental Health* 43 (6): 415–419.

Hunt, J.D. 1994. Calibrating the Naples Land-use and Transport Model. *Environment and Planning B: Planning and Design*, 21:569–90.

Hunt, J.D. and D.C. Simmonds. 1993. Theory and Application of an Integrated Land-Use and Transport Modelling Framework. *Environment and Planning B: Planning and Design* 20(11): 221–44.

Hunt, J.D., D.S. Krieger, and E.J. Miller. 1999. Current Operational Urban Land-Use Transport Modeling Frameworks. Paper Presented at the 78th Annual Meeting of the Transportation Research Board, Washington, D.C.

Hunter, A. 1983. The Urban Neighborhood: Its Analytical and Social Contexts, pp. 3–20 in *Neighborhood Policy and Planning*, edited by P.L. Clay and R. M. Hollister. Lexington, MA: LexingtonBooks.

Ingberman, D.E. 1995. Siting Noxious Facilities: Are Markets Efficient? *Journal of Environmental Economics and Management* 29: S-20—S-33.

Ingram, D.R. 1971. The Concept of Accessibility: A Search for an Operational Form. *Regional Studies* 5:15–26.

Ingram, G.K. and G.R. Fauth. 1974. TASSIM: A Transportation and Air Shed Simulation Model. Final Report for the U.S. Department of Transportation DOT-OS-30099–5. Cambridge, MA: Harvard University.

Ingram, G.K. and A. Pellechio. 1976. *Air Quality Impacts of Changes in Land Use Patterns: Some Simulation Results for Mobile Source Pollutants*. Cambridge, MA: Harvard University Dept. of City and Regional Planning.

Institute of Medicine.1999. Toward Environmental Justice: Research, Education, and Health Policy Needs. Committee on Environmental Justice. Washington, D.C.: National Academy Press.

Israel, B.D. 1995. An Environmental Justice Critique of Risk Assessment. *New York University Environmental Law Journal* 3:469.

Isserman, A.M. 1977. The Accuracy of Population Projections for Subcounty Areas. *Journal of the American Institute of Planners* 43:247–259.

Isserman, A.M. 1984. Projection, Forecast, and Plan. *Journal of the American Planning Association.* 50 (2):208–221.

Jacoby, L.R. 1972. Perception of Air, Noise and Water Pollution in Detroit. Michigan Geographical Pub. No. 7. Ann Arbor: University of Michigan.

Jargowsky, P.J. 1994. Ghetto Poverty among Blacks in the 1980s. *Journal of Policy Analysis and Management* 13(2):288.

Jerrett, M. et al. 1997. Environmental Equity in Canada: an Empirical Investigation into the Income Distribution of Pollution in Ontario. *Environment and Planning* A 29(10):1777–1800.

Jia, C.Q., A.D. Guardo, and D. Mackay. 1996. Toxics Release Inventories: Opportunities for Improved Presentation and Interpretation. *Environmental Science and Technology* 30(2):86A-91A.

Johnson, B.L., R.C. Williams, and C.M. Harris. 1992. *National Minority Health Conference: Focus on Environmental Contamination.* Princeton, NJ: Princeton Scientific Publishing.

Johnson, T. and R.A. Paul. 1982. The NAAQS Exposure Model (NEM) Applied to Carbon Monoxide. Office of Air Quality Planning and Standards, Research Triangle Park, NC.

Johnston, R.A., C.J. Rodier, and M. Choy. 1998. Transportation, Land Use, and Air Quality Modeling. Pages 306–315 in the Conference Proceedings of Transportation, Land Use, and Air Quality: Making the Connection, edited B.S. Easa and D. Samdahl, May 17– 20, 1998, Portland, Oregon. Reston, VA: ASCE.

Johnston, R.A., C. Rodier and D. Shabazian. 1998. Projecting Greenhouse Gas Emissions from Energy Use in Travel and Buildings Using an Integrated Urban Model. Paper presented at the 77th Annual Meeting of Transportation Research Board, January 11–15, 1998. Washington, D.C.

Jud, G.D. and D. G. Bennet. 1986. Public Schools and the Pattern of Interurban Residential Mobility. *Land Economics* 62(4):362–370.

Kain, J.F. 1968. Housing Segregation, Negro Employment, and Metropolitan Decentralization. *The Quarterly Journal of Economics* 82:175–197.

Kain, J.F. and W.C. Apgar, Jr. 1985. *Housing and Neighborhood Dynamics: A Simulation Study.* Cambridge, MA: Harvard University Press.

Kaldor, N. 1939. Welfare Propositions of Economics and Interpersonal Comparisons of Utility. *Economic Journal* 49(Sept.) 549–552.

Kanemoto, Y. Externalities in Space, pp. 43–103 in *Urban Dynamics and Urban Externalities*, edited by T. Miyao and Y. Kenemoto, Chur, Switzerland: Hardwood Academic.

Kant, O. [1785] 1964. *Groundwork of the Metaphysic of Morals.* Translated by H.J. Paton. New York: Harper Torchbooks.

Kasarda, J. 1995. Industrial Location and Changing Location of Jobs, pp. 215–266 in *State of the Union: America in the 1990s, Volume 1: Economic Trends*, edited by R. Farley. New York: Russell Sage Foundation.

Kennedy, P. 1992. *A Guide to Econometrics.* Third Edition. Cambridge, MA: The MIT Press.

Ketkar, K. 1992. Hazardous Waste Sites and Property Values in the State of New Jersey. *Applied Economics* 24:647.

Kiel, K.A. 1995. Measuring the Impact of the Discovery and Cleaning of Identified Hazardous Waste Sites on House Values. *Land Economics* 71(Nov.):428–435.

Kiel, K.A. and K.T. McClain. 1995. House Prices during Siting Decision Stages: The Case of an Incinerator from Rumor through Operation. *Journal of Environmental Economics and Management* 28(March):241–255.

Kim, T.J. 1989. *Integrated Urban Systems Modeling: Theory and Applications.* Dordrecht/Boston/London: Kluwer.

King, M.L. Jr. 1986. A Testament of Hope. pp. 313–328 in J.M. Washington (ed.), *A Testament of Hope: The Essential Writings and Speeches of Martin Luther King, Jr.* San Francisco: Harper Collins.

Kish, L. 1959. Some Statistical Problems in Research Design. *American Sociological Review* XXIV(6):328–338.

Kitchin, R.M. and A.S. Fotheringham. 1997. Aggregation Issues in Cognitive Mapping. *The Professional Geographer.* 49(3): 269–280.

Klosterman, R. E. 1990. *Community Analysis and Planning Techniques.* Savage, MD: Rowman & Littlefield.

Knox, E.G. and E.A. Gilman. 1997. Hazard Proximities of Childhood Cancers in Great Britain from 1953–80. *Journal of Epidemiology and Community Health* 51:151–159.

Kohlhase, J.E. 1991. The Impact of Toxic Waste Sites on Housing Values. *Journal of Urban Economics* 30(1):1–26.

Kolodny, R. 198 . Some Policy Implications of Theories of Neighborhood Change, pp. 93–110 in *Neighborhood Policy and Planning*, edited by P.L. Clay and R. M. Hollister. Lexington, MA: LexingtonBooks.

Komanoff, C. 1996. Road Pricing— A Key to Transportation Justice. *Progress*, VI (9):6.

Korc, M. E. 1996. A Socioeconomic Assessment of Human exposure to Ozone in the South Coast Air Basin of California. *Journal of the Air & Waste Management Association* 46 (6):547–557.

Krimsky, S. 1992. The Role of Theory in Risk Studies, pp. 3–22 in *Social Theories of Risk*, edited by S. Krimsky and D. Golding, Westport, CT: Praeger.

Kroch, E. 1975. TASSIM: An Application to Los Angeles. Harvard University Report DOT-OS-30099–7. Washington, D.C.:Department of Transportation.

Krotscheck, C. and M. Narodoslawsky. 1996. The Sustainable Process Index: A New Dimension in Ecological Evaluation. *Ecological Engineering* 6(4):241.

Krupnick, A. J. and P.R. Portney. 1991. Controlling Urban Air Pollution: A Benefit-Cost Assessment. *Science* 252(4):522–528.

Kuhn, T. 1970. *The Structure of Scientific Revolution.* 2nd edition. Chicago: University of Chicago Press.

Lafferty, R.N. and H.E. Frech. 1978. Community Environment and the Market Value of Single Family Homes: the Effect of the Dispersion of Land Uses. *Journal of Law and Economics* 21: 381–394.

Landis, J. and R. Cervero. 1999. Middle Age Sprawl: BART and Urban Development. *Access*. Spring, 1999. no. 14: 2–19.

Landis, J.D. 1994. The California Urban Futures Model: a new generation of metropolitan simulation models. *Environment and Planning B: Planning and Design* 21:399–422.

Lau, L.J. 1986. Functional Forms in Econometric Model Building. Chapter 25 in *Handbook of Econometrics* Vol. III, edited by Z. Griliches and M.D. Intriligator. Amsterdam: North Holland.

Lavelle M. 1992. Community Profile:Chicago — An Industrial Legacy. *National Law Journal: Unequal Protection — The Racial Divide in Environmental Law.* September 21, 1992. 15(3):S3

Lawrence, M.F. and M. H. Tegenfeldt. 1997. The Use of Economic and Demographic Forecasts by Metropolitan Planning Organizations. Presented at the 1997 Annual Meeting of Transportation Research Board, Washington, D.C.

Lazarsfeld, P. 1959. Problems in Methodology, in R.B. Merton (ed.) *Sociology Today.* New York: Basic Books.

Lee, B.A. and P.B. Wood. 1991. Is Neighborhood Racial Succession Place-Specific? *Demography*, 21, 24.

Lee, C. (ed.) 1992. Proceedings of the First National People of Color Environmental Leadership Summit. New York: United Church of Christ Commission for Racial Justice.

Lee, D. 1989. *Highway Pricing as a Tool for Congestion Management.* Cambridge, MA: Transportation Systems Center.

Lever, W.F. 1985. Theory and Methodology in Industrial Geography, pp. 10–39 in *Progress in Industrial Geography*, edited by M. Pacione. London: Croom Helm.

Lindell, M. and T. C. Earle. 1983. How Close is Close Enough: Public Perceptions of the Risks of Industrial Facilities. *Risk Analysis* 3:245–253.

Linton, G.J. and K.R. Wykle. 1999. Action: Implementing Title VI Requirements in Metropolitan and Statewide Planning. U.S. Department of Transportation FTA/FHWA Memorandum dated October 7, 1999.

Litman, T. A. 1997. *Evaluating Transportation Equity*. Victoria, British Columbia: Victoria Transport Policy Institute.

Liu, F. 1996. Urban Ozone Plumes and Population Distribution by Income and Race: a Case Study of New York and Philadelphia. *Journal of the Air & Waste Management Association* 46(3):207–215.

Liu, F. 1997. Dynamics and Causation of Environmental Equity, Locally Unwanted Land Uses, and Neighborhood Changes. *Environmental Management* 21(5):643–656.

Liu, F. 1998. Who Will be Protected by EPA's New Ozone and Particulate Matter Standards? *Environmental Science and Technology* 32(1)32A–39A.

Lober, D.J. 1995. Resolving the Siting Impasse: Modeling Social and Environmental Locational Criteria with a Geographic Information System. *Journal of the American Planning Association* 61(4):482–495.

Lober, D.J. and D. Green. 1994. NYMVY or NIABY: A Logit Model of Public Opposition. *Journal of Environmental Management* 40(33–50).

Los, M. 1979. Combined Residential-Location and Transportation Models. *Environment and Planning* A 11:1241–1265.

Louvar, J.F. and B.D. Louvar 1998. *Health and Environmental Risk Analysis: Fundamentals with Applications*. Upper Saddle River, NJ: Prentice Hall.

Lowry, I. 1964. *A Model of Metropolis*. RM-4035–RC, RAND Corporation, Santa Monica, Calif.

Lucy, W. 1981. Equity and Planning for Local Services. *Journal of the American Planning Association* 47(4):447–457.

Lurman, F.W. et al. 1989. Development and Application of a New Regional Human Exposure (REHEX) Model. Paper Presented at the 82nd Annual Meeting of the Air Pollution Control Association, Anaheim, CA, June 25–30, 1989.

Mackett, R.L. 1990. The Systematic Application of the LILT Model to Dortmund, Leeds and Tokyo. *Transport Reviews*, 10:323–38.

Mackett, R.L.1991a. A Model-Based Analysis of Transport and Land-Use Policies for Tokyo. *Transport Reviews*, 11:1–18.

Mackett, R.L. 1991b. LILT and MEPLAN: a comparative analysis of land-use and transport policies for Leeds. *Transport Reviews* 11:131–54.

Macmillan W.D. 1993. Urban and Regional Modelling: Getting It Done and Doing It Right. *Environment and Planning* A 25(Anniversary): 56–68.

Marans, R.W. 1976. Perceived Quality of Residential Environments: Some Methodological Issues, pp. 123–147 in *Perceiving Environmental Quality: Research and Applications*, edited by K.H. Craik and E.H. Zube, New York: Plenum Press.

Martin, R. L. 1974. On Spatial Dependence, Bias and the Use of First Spatial Differences in *Regression Analysis*. Area 6:185–194.

Martinez, F.J. 1992. The Bid-Choice Land Use Model: an Integrated Economic Framework. *Environment and Planning* A 24(6):871–885.

Maryland Department of Planning. 1998. Smart Growth Options for Maryland's Tributary Strategies. Managing Maryland's Growth Watershed Planning Series. Baltimore: Maryland.

Maser, S., H. Riker, and R. Rosett. 1977. The Effects of Zoning and Externalities on the Price of Land: An Empirical Analysis of Monroe County, New York. *Journal of Law and Economics* 20:111–132..

Mattsson, L.G. 1984a. Equivalence between Welfare and Entropy Approaches to Residential Location. *Regional and Urban Economics* 14: 147–173.

Mattsson, L.G. 1984b. Some Applications of Welfare Maximization Approaches to Residential Location. Paper of the Regional Science Association (European Congress) 55:103–150.

Mattsson, L.G. 1987. Urban Welfare Maximization and Housing Market Equilibrium in a Random Utility Setting. *Environment and Planning* A19:247–260.

McCarty, L. S. and D. Mackay. 1993. Enhancing Ecotoxicological Modeling and Assessment. *Environmental Science and Technology* 27(9):1719–1728.

McClelland, G.H., W.D. Schultz, and B. Hurd. 1990. The Effect of Risk Beliefs on Property Values: A Case Study of a Hazardous Waste Site. *Risk Analysis* 10(4):485–497.

McCoy and Associates, Inc. 1990. 1990 Outlook for Hazardous Waste Management Facilities: A Nationwide Perspective. *Hazardous Waste Consultant*, March/April, pp. 4-1–4-6.

McEntire, D. 1960. *Residence and Race*. Berkeley, CA: University of California Press.

McFadden, D. 1973. Conditional Logit Analysis of Qualitative Choice Behaviour, pp. 105–142 in *Frontiers in Econometrics*, edited by P. Zarembka, New York: Academic Press.

McFadden, D. 1978. Modeling the Choice of Residential Location. in A. Karlquist, et al. (eds). *Spatial Interaction Theory and Planning Models*. New York: North-Holland.

McGinnis, J.M. and W.H. Foege 1993. Actual Causes of death in the United States. *Journal of the American Medical Association* 270:2207–2212.

McHarg, I. 1969. *Design with Nature*. Garden City, NY:Doubleday/The Natural History Press.

McKone, T.E. 1992. A Review of RISKPRO Version 2.1. *Risk Analysis* 12(1):151–159.

McMaster, R.B., H. Leitner, and E. Sheppard. 1997. GIS-based Environmental Equity and Risk Assessment: Methodological Problems and Prospects. *Cartography and Geographic Information Systems*, 24(3):172–189.

Mendelsohn, R. et al. 1992. Measuring Hazardous Waste Damages with Panel Models. *Journal of Environmental Economics and Management* 22(3):259–271.

Michaels, R.G. and V. K. Smith. 1990. Market Segmentation and Valuing Amenities with Hedonic Models: The Case of Hazardous Waste Sites. *Journal of Urban Economics* 28 (2):223–242.

Mill, John S. [1863] 1985. *Utilitarianism*. New York: Macmillan.

Miller, E.J., DS. Kriger, and J.D. Hunt. 1999. A Research and Development Program for Integrated Urban Models. Paper Presented at the 78th Annual Meeting ot the Transportation Research Board, Washington, D.C.

Mitas, L. and H. Mitasova. 1999. Spatial Interpolation. Chapter 34 (pp. 481–492) in *Geographical Information Systems, Volume 1: Principles and Technical Issues*, edited by P.A. Longley, M.F. Goodchild. D.J. Maguire, and D.W. Rhine. New York: John Wiley & Son, Inc.

Mitchell, R. C. 1980. Patterns and Determinants of Aversion to the Local Siting of Industrial, Energy and Hazardous Waste Dump Facilities by the General Public. *Resources for the Future*.

Miyamoto, K. and Kitazume, K. 1989. A Land-Use Model Based on Random Utility/Rent-Bidding Analysis (RURBAN). Transport Policy, Management and Technology — Towards 2001. Selected Proceedings of the Fifth World Conference on Transport Research, Yokohama. Ventura: Western Periodicals, Vol. IV, 107–21.

Mohai, P. 1995. The Demographics of Dumping Revisited: Examining the Impact of Alternate Methodologies in Environmental Justice Research. *Virginia Environmental Law Journal* 14(615–653).

Mohai, P. and B. Bryant. 1992. Environmental Racism: Reviewing the Evidence, pp. 163–176 in *Race and the Incidence of Environmental Hazards: A Time for Discourse*, edited by B. Bryant and P. Mohai, Boulder, CO: Westview Press.

Montgomery, L.E. and O. Carter-Pokras. 1993. Health Status by Social Class and/or Minority Status: Implications for Environmental Equity Research. *Toxicology and Industrial Health* 9(5):729–773

Moore, M. 1981. Realms of Obligation and Virtue. In *Public Duties: The Moral Obligations of Government Officials*, J. Leishman, L. Liebman, and M. Moore (eds.). Cambridge, MA.: Harvard University Press. Reprinted in *Ethics in Planning*, M. Wachs (ed.), 1985, New Brunswick, N.J.: Center for Urban Policy and Research.

Moore R. and L. Head. 1993. Acknowledging the Past, Confronting the Present: Environmental Justice in the 1990s, pp. 118–127 in *Toxic Struggles: The Theory and Practice of Environmental Justice*, edited by R. Hofrichter. Philadelphia PA: New Society Publishers.

Motallebi, N., and P.D. Allen. 1991. Characterization of Population Exposure to Ozone During the Southern California Air Quality Study, pp. 527–537 in *Tropospheric Ozone and the Environment*, edited by R. L. Bergland, D.R. Lawson and D. J. McKee. Pittsburgh, PA: Air & Waste Management Association.

Munro, I.C. and D.R. Krewski. 1981. Risk Assessment and Regulatory Decision-making. *Food Cosmet. Toxicol.* 19:549–560.

MWCOG (Metropolitan Washington Council of Governments). 1999. An Accessibility Analysis of the 1999 Financially Constrained Long-Range Transportation Plan and Impacts on Low-Income and Minority Populations. Draft. Washington, D.C.

Myers, D. 1992. *Analysis with Local Census Data: Portraits of Change*. Boston, MA: Academic Press.

Napton, M.L. and F.A. Day. 1992. Polluted Neighborhoods in Texas: Who Live There? *Environment and Behavior* 24(4):508–526.

National Research Council. 1983. Risk Assessment in the Federal Government: Managing the Process. Washington, D.C.: National Academy Press.

National Research Council. 1991. Human Exposure Assessment for Air Borne Pollutants: Advances and Opportunities. Washington, D.C.

National Transit Institute. 1998. Coordinating Transportation and Land Use: Course Manaual. Rutgers University and U.S. DOT.

Nelson, A. C. 1992. Effects of Elevated Heavy-rail Transit Stations on House Prices With Respect to Neighborhood Income. *Transportation Research Record* 1359:127–32.

Nelson, A.C. 1998. Transit Stations and Commercial Property Values: Case Study with Policy and Land Use Implications. Paper presented at the 77th Annual Meeting of Transportation Research Board, Preprint CD-ROM, Washington, D.C.

Nelson, A.C., J. Genereux, and M. Genereux. 1992. Price Effects of Landfills on House Values. *Land Economics* 68(4):359–365.

Nelson, J.P. 1981. Three Mile Island and Residential Property Values: Empirical Analysis and Policy Implications. *Land Economics* 57(3): 363–372.

New York State Parks Management and Research Institute. 1993. Locational Verification and Correction of 1989 New York State TRI Facilities. Final Report, A Report Prepared for the U.S. Environmental Protection Agency, Office of Toxic Substances.

Nieves, L.A.. 1993. Economic Impacts of Noxious Facilities: Incorporating the Effects of Risk Aversion. *Risk: Issues in Health, Safety and Environment* 4:35.

Norman, G. 1993. Of Shoes and Ships and Shredded Wheat, Of Cabbages and Cara: The Contemporary Relevance of Location Theory, pp. 38–68 in *Does Economic Space Matter?* edited by H. Ohta and I. Thisse, New York: St. Martin's Press.

Novak, W. J. And R.E. Joseph. 1996. Environmental Justice: Methodologies to Assess Disproportionate Impacts. Paper Presented at the 75th Annual Meeting of Transportation Research Board. Washington, D.C.

Nozick, R. L. 1974. *Anarchy, State and Utopia*. New York: Basic Books.

NRC (National Research Council). 1991. Rethinking the Ozone Problem in Urban and Regional Air Pollution. Washington, D.C.: National Academy Press.

Nyerges, T., M. Robkin, and T.J. Moore. 1997. Geographic Information Systems for Risk Evaluation: Perspectives on Applications to Environmental Health. *Cartography and Geographic Information Systems*, 24(3):123–144.

O'Neill, W.D. 1981. Estimation of A Logistic Growth and Diffusion Model Describing Neighborhood Change. *Geographic Analysis* 13: 391–97.

Oakes, J.M., D.L. Anderton, and A. B. Anderson. 1996. A Longitudinal Analysis of Environmental Equity in Communities with Hazardous Waste Facilities. *Social Science Research* 25:125–148.

Oates, W.E. 1969. The Effects of Property Taxes and Local Public Spending on Property Values: An Empirical Study of Tax Capitalization and the Tiebout Hypothesis. *Journal of Political Economy* 71:957–970.

OECD (Organization for Economic Co-operation and Development). 1978. Urban Environmental Indicators. Paris, France.

Oedel, D. G. 1997. The Legacy of Jim Crow in Macon, Georgia, pp. 97–109 in *Just Transportation: Dismantling Race and Class Barriers to Mobility*, edited by R.D. Bullard and G.S. Johnson. Gabriola Island, BC: New Society Publishers.

Openshaw, S. and J. Taylor. 1979. A Million or So Correlation Coefficients: Three Experiments on the Modifiable Areal Unit Problem, pp. 27–44 in *Statistical Applications in the Spatial Sciences*, Edited by N. Wrigley. London: Pion.

O'Regan, K. M. and J.M. Quigley.1997. Accessibility and Economic Opportunity. Unpublished manuscript, Yale University.

Ott, W.R. 1978. *Environmental Indices: Theory and Practice*. Ann Arbor, MI: Ann Arbor Science.

Ott, W.R. 1990. Total Human Exposure: Basic Concepts, EPA Field Studies, and Future Research Needs. *Journal of the Air and Waste Management Association* 40(7):967–975.

Ott, W.R., J. Thomas, and D. Mage. 1988. Validation of the Simulation of Human Activity and Pollutant Exposure (SHAPE) Model Using Paired Days from the Denver, Colorado, Carbon Monoxide Field Study. *Atmospheric Environment* 22 (11): 2101.

Paustenbach, D. J. 1989. A Survey of Health Risk Assessment, pp. 27–124 In *The Risk Assessment of Environmental and Human Health Hazards: A Textbook of Case Studies*, edited by D.J. Paustenbach. New York: John Wiley & Sons.

Paustenbach, D. J., et al. 1990. The Current Practice of Health Risk Assessment: Potential Impact on Standards for Toxic Air Contaminants. *Journal of the Air and Waste Management Association* 40(12):1620–1630.

PBQD (Parsons Brinckerhoff Quade and Douglas). 1999. Land Use Impacts of Transportation: A Guidebook. National Cooperative Highway Research Program Report 423A. Washington, D.C.: National Academy Press.

Peirce, N. 1991. Edge City: the Best America Can Do? Nation's Cities Weekly.

Perhac, R. M., Jr. 1999. Environmental Justice: The Issue of Disproportionality. *Environmental Ethics* 21(Spring):81–92.

Perlin, S.A. et al. 1995. Distribution of Industrial Air Emissions by Income and Race in the United States: An Approach Using the Toxic Release Inventory. *Environmental Science and Technology* 29(1):69–80.

Pigou, A.C. 1926. *The Economics of Welfare*. London: Macmillan and Co.

Pilisuk, M. and C. Acredolo. 1988. Fear of Technological Hazards: One Concern or Many? *Social Behaviour* 3:17–24.

Pittenger, D. 1976. *Projecting State and Local Populations*. Cambridge, MA: Ballinger.

Polednak, A.P. and D.T. Janerich. 1989. Lung Cancer in Relation to Residence in Census Tracts with Toxic-Waste Disposal Sites: A Case-Control Study in Niagara County, New York. *Environmental Research* 48:29–41.

Pollock, III, P.H. and M.E. Vittas. 1995. Who Bears the Burdens of Environmental Pollution? Race, Ethnicity and Environmental Equity in Florida. *Social Science Quarterly* 76(2):294–310.

Prastacos, P. 1986a. An Integrated Land-Use-Transportation Model for the San Francisco Region: 1. Design and Mathematical Structure. *Environment and Planning* A 18:307–322.

Prastacos, P. 1986b. An Integrated Land-Use-Transportation Model for the San Francisco Region: 2. Empirical Estimation and Results. *Environment and Planning* A 18:511–528.

Pratt et al. 1993. An Indexing System for Comparing Toxic Air Pollutants Based on Their Potential Environmental Impacts. *Chemosphere* 27(8):1359–79.

President Clinton. 1994. Executive Order 12898, Federal Actions to Address Environmental Justice in Minority Populations and Low-Income Populations. Federal Register Vol. 59, No. 32 (February 16):7629–7633 .

Presidential/Congressional Commission on Risk Assessment and Risk Management. 1997a. Framework for Environmental Health Risk Management. Final Report Volume 1.

Presidential/Congressional Commission on Risk Assessment and Risk Management. 1997b. Risk Assessment and Risk Management in Regulatory Decision-Making. Final Report Volume 2.

Pulido, L., S. Sidawi, and R.O. Vos. 1996. An Archaeology of Environmental Racism in Los Angeles. *Urban Geography* 17(5):419–439.

Putman, S.H. 1983. *Integrated Urban Models: Policy Analysis of Transportation and Land Use*. London: Pion.

Putman, S.H. 1991. *Integrated Urban Models 2: New Research and Application of Optimization and Dynamics*. London: Pion.

Putman, S.H. 1994a. Improving DRAM Calibrations: The Effects of Data Quality, Spatial Variation and Lagged Variables. Philadelphia, PA: S.H. Putnam Associates.

Putman, S.H. 1994b. Integrated Transportation and Land Use Models: An Overview of Progress with DRAM and EMPAL, with Suggestions for Further Research. Presented at Transportation Research Board, Washington, D.C.

Rai, K. and J. van Ryzin. 1979. Risk Assessment of Toxic Environmental Substances Using a Generalized Multi-hit Dose Response Model, pp. 99–117 in N. Breslow and A. Whittemore (eds.), *Energy and Health*. Philadelphia: SIAM Press.

Rao, S.T., G., Sistla, and J.Y. Ku. 1988. Temporal and Spatial Variability of Ozone Concentrations in the New York Metropolitan Region, pp. 176–198 in *The Scientific and Technical Issues Facing Post-1987 Ozone Control Strategies*, edited by G.T. Wolff, J. L. Hanisch, and K. Schere. Pittsburgh, PA: Air & Waste Management Association.

Rao, S.T., G. Sistla, and R. Henry. 1992. Statistical Analysis of Trends in Urban Ozone Air Quality. *Journal of Air & Waste Manage. Association* 42 (9):1204–1211.

Rawls, J. 1971. *A Theory of Justice*. Boston, MA: Harvard University Press.

Rayner, S. 1992. Cultural Theory and Risk Analysis, pp. 83–116 in *Social Theories of Risk*, edited by S. Krimsky and D. Golding. Westport, CT: Praeger.

Reichert, A.K., M. Small, and S. Mohanty. 1992. The Impact of Landfills on Residential Property Values. *The Journal of Real Estate Research* 7(3): 297–314.

Reilly W.J. 1931. *The Law of Retail Gravitation*. New York: Pilsbury.

Renn, O. 1992. Concepts of Risk: A Classification, pp. 53–82 in *Social Theories of Risk*, edited by S. Krimsky and D. Golding, Westport, CT: Praeger.

Ringquist, E.J. 1997. Equity and the Distribution of Environmental Risk: The Case of TRI Facilities. *Social Science Quarterly* 78(4):811–829.

Rios, R., G.V. Poje, and R. Detels. 1993. Susceptibility to Environmental Pollutants among Minorities. *Toxicology and Industrial Health* 9(5):797–820.

Ripley, B.D. 1981. *Spatial Statistics*. New York: John Wiley & Sons, Inc.

Rose, J.B. 1993. A Critical Assessment of New York City's Fair Share Criteria. *Journal of the American Planning Association* 59(1):97–100.

Rosenbaum, A.S., D.A. Axelrad, and J.P. Cohen. 1999. National Estimates of Outdoor Air Toxics Concentrations. *Journal of the Air and Waste Management Association* 49(10):1138.

Rosenbloom, S. 1995. Travel by Women. Pages 2-1 to 2-57 in *1990 NPTS National Personal Transportation Survey: Demographic Special Reports*. Washington, D.C.: U.S. Department of Transportation Federal Highway Administration.

Ruffner, J.A. and F.E. Bair. 1985. *The Weather Almanac*, 5th edition. Detroit, Michigan: Gale Research Co.

Russell D. 1989. Environmental Racism: Minority Communities and Their Battle against Toxics. *The Amicus Journal* (Spring): 22–32.

Rybeck, W. 1981. Transit-reduced Land Values: Development and Revenue Implications. *Economic Development Commentary* 5(4):23–27.

Saarinen, T.F. and R.U. Cooke. 1971. Public Perception of Environmental Quality in Tucson, Arizona. *Journal of Arizona Academy Science*. 6:250–274.

Sadd, J.L. et al. 1999a. "Every Breath You Take…": The Demographics of Toxic Air Releases in Southern California. *Economic Development Quarterly* 13(2):107–123.

Sadd, J.L. et al. 1999b. Response to Comments by William M. Bowen, *Economic Development Quarterly* 13(2):135–140.

Sager T. 1990. *Communicate or Calculate: Planning Theory and Social Science Concepts in a Contingency Perspective*. Stockholm: NORDPLAN.

Sanchez, T.W. 1998. Equity Analysis of Personal Transportation System Benefits. *Journal of Urban Affairs* 20(1): 69–86.

Sanchez, T.W. 1999. The Connection between Public Transit and Employment: The Cases of Portland and Atlanta. *Journal of the American Planning Association* 65(3):284–296.

Sandman, P.M. 1985. Getting to Maybe: Some Communications Aspects of Siting Hazardous Waste Facilities. Reprinted in *Readings in Risks*, edited by T.S. Glickman and M. Gough, Washington, D.C.: Resources for the Future.

SAS Institute Inc. 1990. SAS Procedure Guide, Release 6.03 edition. Cary, NC.: SAS Institute.

Savage, I. 1993. Demographic Influences on Risk Perceptions. *Risk Analysis* 13(4): 413.

SCAG (Southern California Council of Governments). 1998. Community Link 21: Regional Transportation Plan. Los Angeles.

Scheffe, R.D. and R.E. Morris. 1993. A Review of the Development and Application of the Urban Airshed Model. *Atmospheric Environment* 27B(1):23–39.

Schwab, W. A. 1987. The Predictive Value of Three Ecological Models: A Test of the Life-cycle, Arbitrage, and Composition Models of Neighborhood Change. *Urban Affair Quarterly* 23: 295–308.

Schwirian, K.P. 1983. Models of Neighborhood Change. *Ann. Rev. Socio.* 9:83–102.

Seinfield, J.H. 1975. *Air Pollution: Physical and Chemical Fundamentals.* New York: McGraw-Hill.

Seinfield, J.H. 1986. *Atmospheric Chemistry and Physics of Air Pollution.* New York: John Wiley & Sons.

Senior, M.L. and A.G. Wilson. 1974. Exploration and Synthesis of Linear Programming and Spatial Interaction Models of Residential Location. *Geographical Analysis* 6:209–238.

Sexton, K. et al. 1992. Estimating Human Exposures to Environmental Pollutants: Availability and Utility of Exiting Databases. *Archive of Environmental Health* 47(6):398–407.

Sexton, K. et al. 1993. Air Pollution Health Risks: Do Class and Race Matter? *Toxicology and Industrial Health* 9(5):843–878.

Sexton, K., K. Olden, and B.L. Johnson. 1993. "Environmental Justice": The Central Role of Research in Establishing a Credible Scientific Foundation for Informed Decision Making. *Toxicology and Industrial Health* 9(5):685–727.

Shaw et al. 1992. Congenital Malformations and Birthweight in Areas with Potential Environmental Contamination. *Archives of Environmental Health* 47(2):147–154.

Shelton, et al. 1989. *Houston: Growth and Decline in a Sunbelt Boomtown.* Philadelphia, PA: Temple University Press.

Shen, Q. 1998. Location Characteristics of Inner-city neighborhoods and employment accessibility of low-wage Workers. *Environment and Planning* B 25:345–365.

Shen, Q. 2000. Spatial and Social Dimensions of Commuting. *Journal of the American Planning Association* 66(1):68–82.

Short, J.F. 1984. The Social Fabric at Risk: Toward the Social Transformation of Risk Analysis. *Amer. Socio. Rev.* 12(1):33–51.

Shprentz, D.S., Bryner, G.C., and Shprentz, J.S. 1996. Breath-Taking. Premature Mortality Due to Particulate Air Pollution in 239 American Cities. Natural Resources Defense Council. Washington, D.C.

Slater, C.M. and G.E. Hall (eds.). 1993. *1993 County and City Extra: Annual Metro, City and County Data Book.* Lanham, MD: Berman Press.

Slovic, P. 1987. Perception of Risk. *Science* 236:280–285.

Slovic, P. 1992. Perception of Risk: Reflections on the Psychometric Paradigm, pp. 117–152 in *Social Theories of Risk*, edited by S. Krimsky and D. Golding, Westport, CT: Praeger.

Slovic, P. B. Fischhoff, and S. Lichtenstein. 1979. Rating the Risks. Reprinted in *Readings in Risks*, edited by T.S. Glickman and M. Gough, Washington, D.C.: Resources for the Future.

Slovic, P. B. Fischhoff, and S. Lichtenstein. 1985. Characterizing Perceived Risk, pp. 91–125 in *Perilous Progress: Managing the Hazards of Technology*, edited by R.W. Kates, C. Hohenemser, and J.X. Kasperson, Boulder, CO: Westview.

Small, K.A. and H.S. Rosen. 1981. Applied Welfare Economics with Discrete Choice. *Econometrica* 49:105–130.

Smith, P.J. and L.D. McCann. 1981. Residential Land Use Change in Inner Edmonton. *Annals of the American Association of Geographers* 71:536–51.

Smith, S. and T. Sincich. 1992. Evaluating the Forecast Accuracy and Bias of Alternative Projections for States. *International Journal of Forecasting* 8(3): 495–508.

Smith, V.K. and W.H. Desvousges. 1986. The Value of Avoiding a LULU: Hazardous Waste Disposal Sites. *Review of Economics and Statistics* 68(2):293–299

Smith, V.K. and J. Huang. 1995. Can Markets Value Air Quality? A Meta-Analysis of Hedonic Property Value Models. *Journal of Political Economy* 103(1):209–227.

Smith, W.S., J.J. Schueneman, and L.D. Zeidberg. 1964. Public Perception to Air Pollution in Nashville, Tennessee. *Journal of Air Pollution Control Association* 14(10):418–423.

Smolen, G., G. Moore, and L.V. Conway. 1992. Economic Effects of Hazardous Chemical and Proposed Radioactive Waste Landfills on Surrounding Real Estate Values. *The Journal of Real Estate Research* 7(3): 283–296.

Song, S. 1996. Some Tests of Alternative Accessibility Measures: A Population Density Approach. *Land Economics* 72(4):474–482.

Sphepard, F.L. and P.K. Sonn. 1997. A Tale of Two Cities, pp. 42–52 in *Just Transportation: Dismantling Race and Class Barriers to Mobility*, edited by R.D. Bullard and G.S. Johnson. Gabriola Island, BC: New Society Publishers.

Spicer, C.W., D.W. Joseph, P.R. Sticksel, and G.F. Ward. 1979. Ozone Sources and Transport in the Northeastern United States. *Environmental Science and Technology* 13(8):975–985.

Star, J.L. and J.E. Estes. 1990. *Geographic Information Systems: An Introduction.* Englewood Cliffs, NJ: Prentice-Hall.

Stevens, J.B. and D.L. Swackhamer. 1989. Environmental Pollution: A Multimedia Approach to Modeling Human Exposure. *Environmental Science and Technology* 23(10):1180–1186.

Stewart, J.Q. 1948. Demographic Gravitation: Evidence and Applications. *Sociometry* 11:31–58.

Stiglitz, J. E. 1993. *Economics.* New York: W.W. Norton & Company.

Stockwell, J.R. et al. 1993. The U.S. EPA Geographic Information System for Mapping Environmental Releases of Toxic Chemical Release Inventory (TRI) Chemicals. *Risk Analysis* 13 (2):155–164.

Stouffer, S.A. 1940. Intervening Opportunities: A Theory Relating Mobility and Distance. *American Sociological Review* 5(December):845–867.

Strange, W. 1992. Overlapping Neighborhoods and Housing Externalities. *Journal of Urban Economics* 32

Stretesky, P. and M.J. Hogan. 1998. Environmental Justice: An Analysis of Superfund Sites in Florida. *Social Problems* 45(2):269–286.

Stutz, F. 1986. Environmental Impact. Pp329–343 in *The Geography of Urban Transportation*, edited by S. Hanson. New York: The Guilford Press.

Stull, W.J. 1975. Community Environment, Zoning, and the Market Value of Single Family Homes. *Journal of Law and Economics* 18:535–558.

Sui, D.Z. and J.R. Giardino. 1995. Applications of GIS in Environmental Equity Analysis: A Multi-scale and Multi-Zoning Scheme Study for the City of Houston, Texas, USA. pp. 950–959 in Proceedings GIS/LIS'95.

Summerhays, J. 1991. Evaluation of Risks from Urban Air Pollutants in the Southeast Chicago Area. *Journal of the Air and Waste Management Association* 41(6):844–850.

Suter, II, G.W. 1993. *Ecological Risk Assessment.* Boca Raton: Lewis Publishers.

Szasz, A. 1994. EcoPopulism: Toxic Waste and the Movement for Environmental Justice. Minneapolis, MN: Univeristy of Minnesota Press.

Taeuber, K. E. and A.F. Taeuber. 1965. *Negroes in Cities: Residential Segregation and Neighborhood Change.* Chicago, IL: Aldine.

Tal, A. 1997. Assessing the Environment's Attitudes Toward Risk Assessment. *Environmental Science and Technology* 31(10):470A–476A.

Talen, E. 1998. Visualizing Fairness: Equity Maps for Planners. *Journal of the American Planning Association* 64 (1):22–38.

Taub R.P., D.G. Talor, and J.D. Dunham. 1987. *Paths of Neighborhood Change: Race and Crime in Urban America*. Chicago: The University of Chicago Press.

Taylor, B. and P. Ong. 1995. Spatial Mismatch or Automobile Mismatch? An Examination of Race, Residence and Commuting in U.S. Metropolitan Areas. *Urban Studies* 32:1453–1473.

Tayman, J. 1996. The Accuracy of Small-Area Population Forecasts Based on a Spatial Interaction Land-Use Modeling System. *Journal of the American Planning Association* 62(1):85–98.

Thayer, M., H. Albers, and M. Rahmatian. 1992. The Benefits of Reducing Exposure to Waste Disposal Sites: A Hedonic Housing Value Approach. *The Journal of Real Estate Research* 7(3): 265–282.

The Economist. 1992a. Let Them Eat Pollution. Page 66. Februrary, 8, 1992.

The Economist. 1992b. Pollution and the Poor. Page 18. Februrary, 15, 1992.

Thomas, J. et al. 1984. A Sensitivity Analysis of the Enhanced Simulation of Human Air Pollution and Exposure (SHAPE) Model. EPA-600/4–85–036. Research Triangle Park, NC: Environmental Monitoring Systems Laboratory, EPA.

Thompson, K.M., D.E. Burmaster, and E.A.C. Crouch. 1992. Monte Carlo Techniques for Quantitative Uncertainty Analysis in Public Health Risk Assessments. *Risk Analysis* 12(1):53–60.

Tiebout, C. 1956. A Pure Theory of Local Expenditures. *Journal of Political Economy* 64: 412–424.

Tobler, W. 1979. Cellular Geography, pp. 379–386 in *Philosophy in Geography*, S. Gale and G. Olsson, (eds.) Dordrecht: Reidel.

Tomlin, C.D. 1990. *Geographic Information Systems and Cartographic Modelling*. Englewood Cliffs, NJ: Prentice-Hall.

U.S. Department of Health and Human Services. 1999. Annual Update of the HHS Poverty Guidelines. Federal Register. Vol. 64, No. 52 (March 18), pp. 13428–13430.

UCC (United Church of Christ Commission for Racial Justice). 1987. Toxic Wastes and Race in the United States: A National Report on the Racial and Socio-Economic Characteristics of Communities with Hazardous Waste Sites. New York: United Church of Christ.

UNESCO (United Nations Educational, Scientific, and Cultural Organization). 1975. *Race, Science, and Society*. L. Kuper (ed). Paris, France: UNESCO Press.

U.S. DOT (Department of Transportation). 1997a. Department of Transportation Order to Address Environmental Justice in Minority Populations and Low-Income Populations. OST Docket No. OST–95–141 (50125). Federal Register, Vol. 62, No. 72 (April 15), pp. 18377–18381.

U.S. DOT (Department of Transportation, Bureau of Transportation Statistics). 1997b. *Transportation Statistics Annual Report 1997*. BTS97–S-01. Washington, D.C.: U.S. Government Printing Office.

U.S. EPA. 1980. Hazardous Waste Facility Siting: A Critical Problem (SW-865. Washington, D.C.: Government Printing Office.

U.S. EPA. 1986. Guidelines for Carcinogen Risk Assessment. Federal Register (September 24) 51:33992–35003.

U.S. EPA. 1987. Unfinished Business: A Comparative Assessment of Environmental Problems, Overview Report. Washington, D.C.

U.S. EPA. 1989a. National Air Pollutant Emission Estimates 1940–1987. EPA-450/4–88–022. U.S.Environmental Protection Agency, Washington, D.C.

U.S. EPA. 1989b. The Toxics Release Inventory, A National Perspective, 1987. EPA 560/4–89–005. Washington, D.C.:U.S. Government Printing Office.

U.S. EPA. 1991a. HEM-II User's Guide. EPA/450/3–91–0010. Research Triangle Park, NC: EPA Office of Air Quality Planning and Standards.

U.S. EPA 1991b. Federal Register 56: 7134.

U.S. EPA. 1992a. Environmental Equity: Reducing Risk for All Communities. Environmental Protection Agency, Office of Policy, Planning and Evaluation. EPA230–R-92–008 and 008A. Washington, D.C.

U.S. EPA. 1992b. VOC/PM Speciation Data System, Version 1.5. Office of Air Quality Planning and Standards, Research Triangle Park, NC.

U.S. EPA. 1992c. The Hazard Ranking System Guidance Manual, Interim Final, Washington, D.C.

U.S. EPA. 1992d. National Air Quality and Emissions Trends Report. Research Triangle Park, NC: EPA Office of Air Quality Planning and Standards.

U.S. EPA. 1993a. A Guidebook to Comparing Risks and Setting Environmental Priorities. Washington, D.C.

U.S. EPA. 1993b. Guideline on Air Quality Models (Revised). EPA-450/2–78–027R. Research Triangle Park, NC: U.S. EPA Office of Air and Radiation.

U.S. EPA. 1994a. 1992 Toxic Release Inventory Public Data Release. EPA-745–R-94–001. Washington, D.C.: U.S. EPA.

U.S. EPA. 1994b. National Air Quality and Emissions Trends Report, 1993. EPA 454/R-94–026. Research Triangle Park, NC: EPA Office of Air Quality Planning and Standards.

U.S. EPA. 1994c. User's Guide to MOBILE5 (Mobile Source Emissions Factor Model). Office of Mobile Sources. EPA–AA–AQAB–94–01. Ann Arbor, MI.

U.S. EPA. 1995a. Environmental Justice Strategy: Executive Order 12898. Washington, D.C.

U.S. EPA. 1995b. User's Guide to PART5: A Program for Calculating Particulate Emissions from Motor Vehicles. Office of Mobile Sources. EPA–AA–AQAB–94–02. Ann Arbor, MI.

U.S. EPA. 1995c. Federal Register 60: 7824.

U.S. EPA. 1996a. Proposed Guidelines for Carcinogen Risk Assessment. EPA/600/P-92/003C. Washington, D.C.: U.S. Government Printing Office. Available through http://www.epa.gov/ORD/WebPubs/carcinogen/.

U.S. EPA. 1996b. An SAB Report: The Cumulative Exposure Project: Review of the Office of Planning, Policy, and Evaluation's Cumulative Exposure Project (Phase 1) by the Integrated Human Exposure Committee. EPA-SAB-IHEC-ADV-96–004. Washington, D.C.: Science Advisory Board.

U.S. EPA. 1996c. Regulatory Impact Analysis for Proposed Ozone National Ambient Air Quality Standard; U.S. Environmental Protection Agency: Research Triangle Park, NC.

U.S. EPA. 1996d. Regulatory Impact Analysis for Proposed Particulate Matter National Ambient Air Quality Standard; U.S. Environmental Protection Agency: Research Triangle Park, NC.

U.S. EPA. 1996e. The Hazardous Waste Permitting Process: A Citizen Guide. EPA530–F-96–007.

U.S. EPA. 1996f. National Ambient Air Quality Standards for Particulate Matter, Proposed Rule. Pages 65637–656714. National Ambient Air Quality Standards for Ozone, Proposed Rule. Pages 65715–65750. Federal Register. 61(241) December 18, 1996.

U.S. EPA. 1996g. Notification of Regulated Waste Activity. Washington, D.C.

U.S. EPA. 1997a. Guidance on Cumulative Risk Assessment. Part 1 Planning and Scoping. Science Policy Council. Washington, D.C.

U.S. EPA. 1997b. RCRA: Reducing Risk from Waste. EPA530–K-97–004.

U.S. EPA. 1998a. Interim Guidance for Investigating Title VI Administrative Complaints Challenging Permits. EPA Office of Civil Rights, Washington, D.C.

U.S. EPA. 1998b. Final Guidance for Incorporating Environmental Justice Concerns in EPA's NEPA Compliance Analyses. EPA Office of Federal Activities. Washington, DC. Available at http://es.epa.gov/oeca/ofa/ejepa.html, October 15, 1998.

U.S. EPA. 1998c. National Air Quality and Emissions Trends Report, 1990–1997. Office of Air Quality Planning and Standards. Available on the EPA website at http://www.epa.gov/oar/agtrn97/.

U.S. EPA. 1998d. Smart Travel. EPA-420-K-97-002. Office of Air and Radiation.

U.S. EPA. 1999a. Sociodemographic Data Used for Identifying Potentially Highly Exposed Populations. EPA/600/R-99/060. National Center for Environmental Assessment-W, Office of Research and Development, Washington, D.C.

U.S. EPA. 1999b. Toxics Release Inventory: Public Data Release Report. Washington, D.C. Available at http://www.epa.gov/tri/tri97/pdx/index.html

U.S. Nuclear Regulatory Commission's (NRC) Atomic Safety and Licensing Board. 1997. Final Initial Decision – Louisiana Energy Services Docket 5/1/97. http://www.nrc.gov/OPA/reports/ lesfnl.htm.

U.S. Postal Service. 1982. Publication 100: History of the U.S. Postal Service 1775–1981. Washington, D.C.

Vaccaro, V. 1981. Cost-benefit Analysis and Public Policy Formulation. In *Ethical Issues in Government*, N. Bowie (ed.), Philadelphia: Temple University Press.

von Thünen, J.H. 1826. *Der Isolierte Staat in Beziehung auf Landwirtschaft und Nationale-konomie*. Hamburg. English translation by C.M. Wartenberg (1966), P. Hall, (ed.) *von Thünen's Isolated State*. London: Pergamon.

Wachs, M. 1993. Learning from Los Angeles: Transport, Urban Form, and Air Quality. *Transportation* 20: 329–354.

Wachs, M. and T.G. Kumagia. 1973. Physical Accessibility as Social Indicator. *Socioeconomic Planning Science* 7:437–456.

Waddell, P. 1998. Development and Calibration of the Prototype Metropolitan Land Use Model. Report to the Oregon Department of Transportation. Seattle, WA: Urban Analytics.

Wall, G. 1973. Public Response to Air Pollution in South Yorkshire, England. *Environment and Behavior* 5:219–248.

Wallace, L. 1993. A Decade of Studies of Human Exposure: What Have We Learned? *Risk Analysis* 13 (2): 135–139.

Waller, L.A., T. A. Louis, and B.P. Carlin. 1997. Bayes Methods for Combining Disease and Exposure Data in Assessing Environmental Justice. *Environmental and Ecological Statistics* 4:267–281.

Webber, M.J. 1984. *Industrial Location*. Beverly Hill: Sage.

Weber, A. 1909. *Uber den Standort der Industrien*. Tubigen: J.C.B. Mohr. Translated by C.J. Friedrich (1929) as *Weber's Theory of Industrial Location*. Chicago: University of Chicago Press.

Webster, F. V., P.H. Bly and N.J. Paulley (eds.). 1988. *Urban Land-Use and Transportation Interaction Policies and Models*. Aldershot: Avebury.

Webster, F.P. and N. Paulley. 1990. An International Study on Land-Use and Transport Interaction. *Transport Reviews* 10:287–309.

Wegener, M. 1986. Transport Network Equilibrium and Regional Deconcentration. *Environment and Planning* A 18:437–56.

Wegener, M. 1994. Operational Urban Models: State of the Art. *Journal of the American Planning Association* 60(1):17–29.

Wegener, M. 1999. Land Use and Transport Model Integration: Progress and Future Directions, pp. 5–20 in Proceedings of the Oregon Symposium on Integrating Land Use and Transport Models. Portland, Oregon, September 30, 1998. Salem, OR: Oregon Department of Transportation.

Wegener, M., Mackett, R.L. and Simmonds, D.C. 1991. One City, Three Models: Comparison of Land-Use/Transport Policy Simulation Models for Dortmund. *Transport Reviews* 11:107–29.

Weinberger, R. 2000. Commercial Property Values and Proximity to Light Rail: Calculating Benefits with a Hedonic Price Model. Paper Presented at the 79th Annual Meeting of Transportation Research Board. Washington, D.C.

Weisberg, B. 1993. One City's Approach to NIMBY. *Journal of the American Planning Association* 59(1):93–97.

Weiss, N.A. 1999. *Introductory Statistics*. 5th edition. Reading, MA: Addison-Wesley.

Weisskopf, M. 1992. Even Cash for Trash Fails to Slow Landfill Backlash: Public Resistance Widens U.S. Garbage Gap. *Washington Post*, No. 115, pp. A1, A6–A7.

Wenz, P.S. 1988. *Environmental Justice*. Albany, NY: State University of New York Press.

Werner, R.J. 1997. Toxics Releases and Demography in Minneapolis/St. Paul: A GIS Exploration. Proceedings from the 1998 Environmental Systems Research Institute Users Conference.
http://www.esri.com/library/userc...c97/proc97/to150/pap122/pap122.htm.

Wernette, D.R., and L.A. Nieves. 1992. Breathing Polluted Air: Minorities are Disproportionately Exposed. *EPA Journal* 18 (2): 16–17.

White, M. C., P.F. Infante, and K.C. Chu. 1982. A Quantitative Estimate of Leukemia Mortality Associated with Occupational Exposure to Benzene. Risk Analysis 2(3):195–204. Reprinted in *Readings in Risk*, edited by T.S. Glickman and M. Gough. Washington, D.C.: Resource for the Future.

White, M. J. 1987. *American Neighborhoods and Residential Differentiation*. New York: Russell Sage Foundation for the National Committee for Research on the 1980 Census.

Wildavsky, A. and K. Dake. 1990. Theories of Risk Perception: Who Fear What and Why? *Deadalus* 119(4):41–60.

Wildgen, J.K. 1998. Environmental Justice in Louisiana's Industrial Corridor. Proceedings from the 1998 Environmental Systems Research Institute Users Conference.
http://www.esri.com/library/userc...c98/PROCEED/TO200/PAP158/P158.HTM.

Williams H.C. W.L. and M.L. Senior 1978. Accessibility, Spatial Interaction and the Spatial Benefit Analysis of Land-Use Transportation Plans, pp. 253–288 in A. Karlqvist, L. Lundqvist, F. Snickers and J.W. Weibull (eds.) *Spatial Interaction and Planning Models*, North Holland, Amsterdam.

Williams, H.C.W.L. 1977. On the Formation of Travel Demand Models and Economic Evaluation Measures of User Benefit. *Environment and Planning* A 9:285–344.

Wilson, A.G. 1970. *Entropy in Urban and Regional Modelling*. London: Pion.

Wilson, A.G. 1976. *Urban and Regional Models in Geography and Planning*. London: John Wiley.

Wilson, A. G. et al. 1981. *Optimization in Locational and Transport Analysis*. Chichester: Wiley.

Wolff, G.T. and P.L. Lioy. 1980. Development of an Ozone River Associated with Synoptic Scale Episodes in the Eastern United States. *Environmental Science and Technology* 14 (10):1257–1260.

Wolff, G.T., et al. 1977. Anatomy of Two Ozone Transport Episodes in the Washington, D.C., to Boston, Mass., Corridor. *Environmental Science and Technology* 11(5):506–510.

Wolter, K.M. 1991. Accounting for America's Uncounted and Miscounted. *Science* 253:12–15.

Wright, T. 1998. Sampling and Census 2000: The Concepts. *American Scientist* 86 (3): 245–253.

Wrigley, N. et al. 1996. Analysing, Modeling, and Resolving the Ecological Fallacy, pp. 23–40 in *Spatial Analysis: Modeling in a GIS Environment*, P. Longley and M. Batty, (eds.) Cambridge, England: GeoInformation International.

Young, H.P. 1994. *Equity in Theory and Practice*. Princeton, NJ: Princeton University Press.

Zankel, K.L., R.P. Brower, and P.M. Dunbar. 1990. Risks Due to Atmospheric Emissions of Toxic Pollutants from Power Plants. Air and Waste Management Association, 83rd Annual Meeting.

Zeiss, C. 1991. Municipal Solid Waste Incinerator Impacts on Residential Property Values and Sales in Host Communities. *Journal of Environmental Systems* 229:238.

Zhang, Y. and Q. Dai. 1999. Internet Address Matching for Environmental Data. Proceedings from the 1999 Environmental Systems Research Institute Users Conference. Available at http://www.esri.com/library/userc...c99/proceed/papers/pap549/p549.htm.

Zimmerman R. 1993. Social Equity and Environmental Risk. *Risk Analysis* 13(6):649–666.

Zimmerman R. 1994. Issues of Classification in Environmental Equity: How We Manage is How We Measure. *Fordham Urban Law Journal* XXI(3):633–669.

Zipf, G.K. 1946. The P1P2/D Hypothesis: On the Intercity Movement of Persons. *American Sociological Review* 11(December):677–686.

Zupan, J.M. 1973. *The Distribution of Air Quality in the New York Region*. Baltimore, MD: Johns Hopkins University Press for Resources for the Future.

Index

A

Acceptable daily intakes (ADI), 73
Accessibility, 287, 292, *See also* Transportation
    systems
    distributional impacts on property values,
        304–307
    as equity measure, 304
    litigation, 311–312
    measuring, 295–297
    planning and equity analysis, 297–299
    public facility siting, 174–175
    racial/ethnic disparities, 302–304
    spatial mismatch hypothesis, 302–304
ACE2588, 176
Additive exposure effects, 87
Adjacency analysis, 168
AERAM, 75
Aerometric Information Retrieval System (AIRS),
    207, 220, 315, 316
African Americans, *See also* Racial or ethnic
    minorities
    air quality disparities, 204, 221–223, 224, 228,
        230
    census category, 97
    distribution at time of siting, 272, *See also*
        Dynamics analysis
    hazardous waste site location disparities, 246
    health disparities, 105
    mobility disparities, 301
    spatial mismatch effect on employment,
        302–303
    spatial population distribution, 108
    Superfund or CERCLIS site distribution, 255,
        256
    toxics release facility distribution, 264
    uncontrolled waste site distribution, 252
Age
    air quality disparities, 223, 226, 227, 229, 230
    census measures, 107
    city vs. non-city, 230
    health risks and, 105–106
Aggregation, 49, 133, 146, 169, 243
Air pollution, 195–235, *See also* Air quality
    economic impacts, 65, 202
    health impacts, 64, 195, 197
    sensitive subpopulations, 195, 197
    transportation emissions, 288

Air quality, 195–198, *See also* Toxics Release
    Inventory
    ambient monitoring, 84–85
    ambient standards, 196, 198, 219–235, *See
        also* National Ambient Air Quality
        Standards
        city vs. non-city nonattainment areas,
            230–232
        major findings, 233
        methods, 220–221
        nonattainment areas overview, 221–223
        policy implications, 234–235
        spatial distribution of nonattainment areas,
            223–229
    income relationship, 215, 217
    LandView III tool, 317
    measuring transportation-associated
        environmental impacts, 307
    models, 69
    monitoring, 202
    population distribution, 199–204
        evidence, 203–204
        methods, 202
        neighborhood change theories, 201
        residential location theory, 199
        spatial interaction, 199
    population distribution, spatial interaction
        modeling, 205–219
        Houston results, 215–219
        index construction and data preparation,
            207–210
        Los Angeles results, 213–215, 217–219
        methods, 205–207
        model estimation, 210–213
    prevailing visibility index, 207
    risk perceptions, 196, 198, 200–201
    urban model integration, 191–192
    urban/rural disparities, 203
Air quality index, 191, 192, 207
AIRS database, 207, 220, 315, 316
Alaska census geography, 122
Alaska Native Village Statistical Areas, 120
Alaska Natives, 97
Allegheny, Pennsylvania, 117
Alonso, W., 34, 182, 183
Ambient monitoring, 84–85
American Indian Entities, 120

Methodological issues, 45–60, *See also* Dynamics
  analysis; Environmental justice analysis;
  Risk assessment; Statistical methods or
  models; Units of analysis
  cause-effect relationships, 10, 45, 51–52
  classical experimental design, 49
  dialectic approach, 46
  factors to consider, 59–60
  generalization and fallacies, 49–51
  integrated framework, 55–60
  interpretative or phenomenological
    epistemology, 46, 319
  macro-analysis framework, 55–57
  micro-analysis framework, 55
  participative framework, 11, 46, 52–53,
    55–60, 319
  positivism, 11, 45–47, 52–53, 55, 319–320
  post-siting dynamics, *See* Dynamics analysis
  reasoning approaches, 47
  relationship of method to results, 53–54
  scientific reasoning, 47
  stochasticity, 86–87
  underlying ideologies and motivations, 55
  validity, 47–51
Metropolitan Areas (MAs), 120, 126, 130, 133
Metropolitan districts, 132–133
Metropolitan planning organizations (MPOs)
  demographic forecasting, 110
  household income data, 104–105
  travel demand models, 193
Metropolitan Statistical Areas (MSAs), 120, 126,
  133, 221
Metropolus, 176
METROSIM, 189
Michigan Coalition, 3
Michigan statewide study, 265
Microenvironments, 65, 71, 90
Microsimulation, 187, 194
Mill, John Stuart, 20
Minimum cost theory, 33
Minneapolis, 267
Minor Civil Divisions (MCDs), 119, 121, 122,
  255
Miyamoto, K., 189
MOBILE, 68, 90, 193
Mobility, 218–219, 292, 300–301
Mode, 146
Modeling, *See* Exposure, modeling; Risk
  assessment, modeling; Spatial
  interaction modeling; Statistical
  methods or models; Urban modeling;
  specific models or simulations
Modifiable areal unit problem (MAUP), 51, 118,
  320
Mohai, P., 265

Monte Carlo simulation, 87
Moran's I statistic, 158, 176
Mortality, racial/ethnic and income disparities,
  105
Mothers of East Los Angeles (MELA), 2
Multicollinearity, 155–156
Multi-hit model, 73
Multinomial logit model, 183–184, 297
Multiple environmental exposure, 62, 87
Multiple Point Gaussian Dispersion Algorithy
  with Terrain Adjustment (MPTER), 70
Multi-stage model, 73, 74
MUS, 188
Myers, D., 114

N

NAACP Legal Defense and Educational Fund,
  Inc. (LDF), 311
NAAQS, *See* National Ambient Air Quality
  Standards
NAAQS Exposure Model (NEM), 72
Napton, M. L., 204, 217
National Academy of Sciences (NAS), 66
National Ambient Air Quality Standards
  (NAAQS), 196, 198, 203, 219–235
  city vs. non-city nonattainment areas,
    230–232
  major finding, 233
  methods, 220–221
  NAAQS Exposure Model (NEM), 72
  nonattainment areas overview, 221–223
  policy implications, 234–235
  spatial distribution of nonattainment areas,
    223–229
National Drinking Water Contaminant
  Occurrence Database, 315
National Environmental Justice Advisory
  Council, 4
National Environmental Policy Act (NEPA), 6–7
National Priority List (NPL) sites, 118, 251, *See*
  Superfund sites
National Research Council (NRC), 66
National Shape File Repository, 316
Native Americans, *See* American Indians
Natural neighborhood interpolation, 166
Negative binomial event count model, 262
Neighborhood change, 31, 37–43, 201
  geographical unit of analysis, 141–142
  institutional model, 41–43
  integrated analytical framework, 58
  invasion-succession model, 38–39, 279–280
  life-cycle model, 39–40, 106, 280–281

3 6 7 11 14